MEMOIRS
of the
American Mathematical Society

ISSN 0065-9266 (print)
ISSN 1947-6221 (online)

Number 1307

AF620968

The Irreducible Subgroups of Exceptional Algebraic Groups

Adam R. Thomas

November 2020 • Volume 268 • Number 1307 (fourth of 6 numbers)

AMS AMERICAN MATHEMATICAL SOCIETY

Library of Congress Cataloging-in-Publication Data

Cataloging-in-Publication Data has been applied for by the AMS.
See http://www.loc.gov/publish/cip/.
DOI: https://doi.org/10.1090/memo/1307

Memoirs of the American Mathematical Society

This journal is devoted entirely to research in pure and applied mathematics.

Subscription information. Beginning with the January 2010 issue, *Memoirs* is accessible from www.ams.org/journals. The 2020 subscription begins with volume 263 and consists of six mailings, each containing one or more numbers. Subscription prices for 2020 are as follows: for paper delivery, US\$1085 list, US\$868 institutional member; for electronic delivery, US\$955 list, US\$764 institutional member. Upon request, subscribers to paper delivery of this journal are also entitled to receive electronic delivery. If ordering the paper version, add US\$20 for delivery within the United States; US\$80 for outside the United States. Subscription renewals are subject to late fees. See www.ams.org/help-faq for more journal subscription information. Each number may be ordered separately; *please specify number* when ordering an individual number.

Back number information. For back issues see www.ams.org/backvols.

Subscriptions and orders should be addressed to the American Mathematical Society, P. O. Box 845904, Boston, MA 02284-5904 USA. *All orders must be accompanied by payment.* Other correspondence should be addressed to 201 Charles Street, Providence, RI 02904-2213 USA.

Copying and reprinting. Individual readers of this publication, and nonprofit libraries acting for them, are permitted to make fair use of the material, such as to copy select pages for use in teaching or research. Permission is granted to quote brief passages from this publication in reviews, provided the customary acknowledgment of the source is given.

Republication, systematic copying, or multiple reproduction of any material in this publication is permitted only under license from the American Mathematical Society. Requests for permission to reuse portions of AMS publication content are handled by the Copyright Clearance Center. For more information, please visit www.ams.org/publications/pubpermissions.

Send requests for translation rights and licensed reprints to reprint-permission@ams.org.

Excluded from these provisions is material for which the author holds copyright. In such cases, requests for permission to reuse or reprint material should be addressed directly to the author(s). Copyright ownership is indicated on the copyright page, or on the lower right-hand corner of the first page of each article within proceedings volumes.

Memoirs of the American Mathematical Society (ISSN 0065-9266 (print); 1947-6221 (online)) is published bimonthly (each volume consisting usually of more than one number) by the American Mathematical Society at 201 Charles Street, Providence, RI 02904-2213 USA. Periodicals postage paid at Providence, RI. Postmaster: Send address changes to Memoirs, American Mathematical Society, 201 Charles Street, Providence, RI 02904-2213 USA.

© 2021 by the American Mathematical Society. All rights reserved.

This publication is indexed in *Mathematical Reviews®*, *Zentralblatt MATH*, *Science Citation Index®*, *Science Citation Index™-Expanded*, *ISI Alerting Services℠*, *SciSearch®*, *Research Alert®*, *CompuMath Citation Index®*, *Current Contents®/Physical, Chemical & Earth Sciences.*
This publication is archived in *Portico* and *CLOCKSS*.
Printed in the United States of America.

∞ The paper used in this book is acid-free and falls within the guidelines established to ensure permanence and durability.
Visit the AMS home page at https://www.ams.org/

10 9 8 7 6 5 4 3 2 1 26 25 24 23 22 21

Contents

Abstract

This paper is a contribution to the study of the subgroup structure of exceptional algebraic groups over algebraically closed fields of arbitrary characteristic. Following Serre, a closed subgroup of a semisimple algebraic group G is called irreducible if it lies in no proper parabolic subgroup of G. In this paper we complete the classification of irreducible connected subgroups of exceptional algebraic groups, providing an explicit set of representatives for the conjugacy classes of such subgroups. Many consequences of this classification are also given. These include results concerning the representations of such subgroups on various G-modules: for example, the conjugacy classes of irreducible connected subgroups are determined by their composition factors on the adjoint module of G, with one exception.

A result of Liebeck and Testerman shows that each irreducible connected subgroup X of G has only finitely many overgroups and hence the overgroups of X form a lattice. We provide tables that give representatives of each conjugacy class of connected overgroups within this lattice structure. We use this to prove results concerning the subgroup structure of G: for example, when the characteristic is 2, there exists a maximal connected subgroup of G containing a conjugate of every irreducible subgroup A_1 of G.

Received by the editor August 23, 2016, and, in revised form, September 30, 2016 and February 20, 2017, and August 24, 2017.

Article electronically published on February 17, 2021.

DOI: https://doi.org/10.1090/memo/1307

2010 *Mathematics Subject Classification.* Primary 20G05, 20G15, 20G41.

The author is indebted to Prof. M. Liebeck for his help in producing this paper. He would also like to thank Dr A. Litterick and Dr T. Burness for their comments on previous versions of this paper. Finally, the author would like to thank the anonymous referee for their careful reading of this paper and many insightful comments and corrections.

The author is affiliated with the Mathematics Institute, Zeeman Building, University of Warwick, Coventry, CV4 7AL, UK. Email: adam.r.thomas@warwick.ac.uk.

©2021 American Mathematical Society

CHAPTER 1

Introduction

Let G be a reductive connected algebraic group over an algebraically closed field K of characteristic p. A closed subgroup X of G is said to be *G-completely reducible* (or *G-cr* for short) if, whenever it is contained in a parabolic subgroup P of G, it is contained in a Levi subgroup of P. This definition, due to Serre in [**Ser05**], generalises the standard notion of a completely reducible subgroup of $\mathrm{GL}(V)$. Indeed, if $G = \mathrm{GL}(V)$, a subgroup X is G-completely reducible if and only if X acts completely reducibly on V.

The concept of G-complete reducibility offers a bridge between many different branches of mathematics. These include the theory of buildings, Kac-Moody groups, geometric invariant theory, Lie algebras and representation theory. The definition for spherical buildings was introduced in the original paper of Serre [**Ser05**]. This was extended to subgroups of Kac-Moody groups by Caprace in [**Cap09**] and more generally for twin buildings in [**Daw**]. Complete reducibility has been studied from a geometric point of view in a series of works by Bate, Martin, Röhrle, et al. in which the authors apply ideas from geometric invariant theory (see for example [**BGM19**], [**BMR05**], [**BMRT13**]). There is also a natural generalisation of complete reducibility to subalgebras of Lie algebras of algebraic groups, as introduced by McNinch in [**McN07**], and studied further in [**BMRT11**] and [**ST18**].

The notion of complete reducibility for subgroups of algebraic groups is very familiar in characteristic 0. Indeed, in characteristic 0 a subgroup X of G is G-cr if and only if X is reductive [**Ser05**, Proposition 4.2]. In positive characteristic, a G-cr subgroup is still reductive [**Ser05**, Proposition 4.1] but the converse need not be true. However, non-G-cr reductive connected subgroups are inherently a low characteristic phenomenon: all reductive connected subgroups are G-cr when $p \geq a(G)$, where $a(G)$ is equal to $\mathrm{rank}(G) + 1$ if G is simple and equal to the supremum of $(1, a(G_1), \ldots, a(G_r))$ otherwise, where G_i are all the simple quotients of G, as proved by Jantzen, McNinch and Liebeck–Seitz (see [**Ser05**, Theorem 4.4]). There are also results concerning the complete reducibility of finite subgroups of G, see [**Gur99**, Theorem A], [**GHT17**, Theorem 1.9], and [**Lit18**, Corollary 5].

This paper is concerned with an important subset of G-cr subgroups, namely the G-irreducible subgroups. The definition, again introduced by Serre in [**Ser05**], is as follows. A closed subgroup X of G is called *G-irreducible* if it is not contained in any proper parabolic subgroup of G. We also say that the subgroup is *irreducible* if G is clear from the context. It is immediate from the definition that when $G = \mathrm{GL}(V)$, a subgroup X is G-irreducible if and only if V restricted to X is an irreducible X-module. In [**LT04**], Liebeck and Testerman studied G-irreducible connected subgroups when G is semisimple. They showed, amongst other things, that all irreducible connected subgroups are semisimple and have only a finite number of overgroups in G.

The G-irreducible connected subgroups play an important role in determining both the G-cr and non-G-cr connected subgroups of G. The G-cr subgroups of G are simply the L-irreducible subgroups of L for each Levi subgroup L of G (noting that G is a Levi subgroup of itself). To determine the non-G-cr subgroups of G, one strategy is as follows. Let P be a proper parabolic subgroup with unipotent radical Q and Levi complement L. Then for each L-irreducible subgroup X of L, determine the complements to Q in QX that are not Q-conjugate to X (if any exist). Any non-G-cr connected subgroup will be of this form for some L-irreducible connected subgroup X. This strategy has been used in [**Ste10**], [**Ste13**] and more recently in [**LT18**].

We now restrict our attention to the case where G is a simple algebraic group over an algebraically closed field K of characteristic p (setting $p = \infty$ for characteristic 0). If G is of classical type then determining the G-irreducible subgroups reduces to representation-theoretic considerations by [**LT04**, Lemma 2.2]. In particular, when (G, p) is not $(D_n, 2)$, the action of a subgroup X on the natural module for G determines whether or not X is G-irreducible. In any case, when G is of small rank (at most 8 suffices for the purpose of this paper) it is possible to determine the conjugacy classes of G-irreducible connected subgroups.

Now suppose that G is of exceptional type. The simple G-irreducible connected subgroups have already been classified through a series of works by various authors. Firstly, work of Liebeck–Seitz in [**LS96**, Theorem 1] shows that all reductive connected subgroups X are G-cr under the assumption that $p > N(X, G)$, where $N(X, G)$ is a prime depending on the type of X and G and always at most 7. They use this result to classify the simple connected subgroups of rank at least 2 and Lawther–Testerman used this in [**LT99**] to classify the subgroups of type A_1, both when $p > N(X, G)$. In these cases, the G-irreducible subgroups are those G-cr subgroups with trivial connected centraliser and so one can find all of the G-irreducible simple connected subgroups under their assumptions on p. In [**Ste13**], Stewart classified the F_4-irreducible simple connected subgroups of F_4 of rank at least 2 without any assumption on p. Amende in [**Ame05**] determined the G-irreducible subgroups of type A_1 when G is not of type E_8. Finally, work of the author in [**Tho15**], [**Tho16**] completed the classification of the simple G-irreducible connected subgroups of G.

This paper completes the classification of G-irreducible connected subgroups of G. Our main theorem is the following (here $\mathrm{Aut}(G)$ denotes the group of algebraic automorphisms of G).

THEOREM 1. *Let G be a simple exceptional algebraic group and X be a G-irreducible connected subgroup of G. Then X is $Aut(G)$-conjugate to exactly one subgroup in Tables* 1–11.5 *and each subgroup in the tables is G-irreducible.*

We also determine the composition factors of each irreducible connected subgroup in the action on the adjoint and minimal modules for G; these can be found in Tables 12.1–12.5. By "minimal module" we mean the smallest dimensional non-trivial module for G (which coincides with the adjoint module when G is of type E_8). The dimensions of such a module are 7 (6 if $p = 2$), 26 (25 if $p = 3$), 27 and 56 for $G = G_2$, F_4, E_6 and E_7, respectively.

We explain how to read Tables 1–11.5 in Section 11, where they are presented, but let us make some important remarks about them now. The irreducible connected subgroups X of G listed in the tables are given an identification number

$G(\#n)$ (ID number for short). In reference to Theorem 1, we only count each subgroup with ID number n once in Tables 1–11.5, even if it appears multiple times; the ID number n appears in italics each time the subgroup is repeated. These repeats are necessary to give the lattice structure of the connected overgroups of irreducible subgroups of G and we discuss this further in Section 11. We also provide Tables 11.1A–11.5A in Section 11 to help recover this lattice structure. They give the conjugacy classes of immediate connected overgroups for irreducible subgroups of G. In particular, they make it easier to find all the repetitions of a subgroup X in Tables 1–11.5. Another important remark to make is that we list large collections of diagonal irreducible connected subgroups in separate tables. We do this to improve the readability of Tables 1–11.5 and in order to condense the presentation of the large number of conjugacy classes of such diagonal subgroups.

A large part of this paper is devoted to proving Theorem 1. We do this by proving Theorems 5.1–9.1 which classify the irreducible subgroups of G_2–E_8, respectively. As mentioned, the author already completed the classification of the simple G-irreducible connected subgroups in [**Tho15**], [**Tho16**] and so the main focus is on the non-simple irreducible connected subgroups. We discuss the strategy used for the proofs in detail in Section 4. The strategy involved is different to that used in [Loc. cit.] with the main distinction being our methods for finding the irreducible subgroups of each maximal connected subgroup of G. This difference allows us to study the lattice structure of the connected overgroups of each irreducible connected subgroup; the overgroups can be read off from Tables 1–11.5 as aforementioned. This lattice structure allows us to prove Corollaries 5 and 7 below, as well as Corollary 10.3 in Section 10.1. Moreover, we believe that the presentation of the lattice structure of all G-irreducible connected subgroups will be beneficial to future readers.

In the remainder of this introduction we present many corollaries of Theorem 1. To do this we require notation used throughout the paper to describe representations of algebraic groups, diagonal subgroups, identification of irreducible subgroups etc. This is explained in Section 2.

For the first two of these corollaries we need the following definition. Let X and Y be semisimple subgroups of a semisimple algebraic group G and let V be a G-module. Then we say that X and Y *have the same composition factors on* V if there exists an isomorphism from X to Y sending the set of composition factors of $V \downarrow X$ to the set of composition factors $V \downarrow Y$ (counted with multiplicity).

The first of our corollaries shows that if G is a simple exceptional algebraic group then, with one exception, conjugacy between G-irreducible connected subgroups is determined by their composition factors on the adjoint module for G, which we denote by $L(G)$.

Corollary 1. *Let G be a simple exceptional algebraic group and X and Y be G-irreducible connected subgroups of G. If X and Y have the same composition factors on $L(G)$ then either:*

(1) *X is conjugate to Y in $Aut(G)$, or*
(2) *$G = E_8$, X, Y are of type A_2, $p \neq 3$, $X \hookrightarrow A_2^2 < \bar{D}_4^2$ via $(10, 10^{[r]})$ and $Y \hookrightarrow A_2^2 < \bar{D}_4^2$ via $(10, 01^{[r]})$ (or vice versa) where $r \neq 0$ and A_2^2 is irreducibly embedded in $\bar{D}_4^2$. In the notation of Table 11.5 the subgroup $X = E_8(\#53)$ and the subgroup $Y = E_8(\#54)$.*

We also deduce that for G not of type E_8, the Aut(G)-conjugacy class of a G-irreducible connected subgroup of G is determined by its composition factors on the minimal module for G.

COROLLARY 2. *Let G be a simple exceptional algebraic group not of type E_8, and let X and Y be G-irreducible connected subgroups of G. If X and Y have the same composition factors on a minimal module for G then X is conjugate to Y in Aut(G).*

From the lists of composition factors provided in Tables 12.1–12.5, one can determine the G-irreducible connected subgroups X for which $V \downarrow X$ is multiplicity-free when V is either the minimal or adjoint module for G. This is a specific case of a more general project of Liebeck, Seitz and Testerman; see [**LST15**] for further details.

The next corollary highlights interesting subgroups that are not G-irreducible but are M-irreducible for some reductive, maximal connected subgroup M. When we say a reductive, maximal connected subgroup we mean a reductive subgroup that is maximal among all closed connected subgroups; these have been classified by Liebeck–Seitz and are listed in Theorem 3.1. Here "interesting" means that the M-irreducible subgroup is not M_1-reducible for some other reductive, maximal connected subgroup M_1 nor contained in a proper Levi subgroup of G.

We explain some notation we only use in Table 1. A subgroup X is said to be "embedded via λ_i" in a simple classical group M if $V_M(\lambda_1) \downarrow X = V_X(\lambda_i)$. This determines X up to M-conjugacy unless M is of type D_n, in which case there may be two classes. Indeed, this happens for each of the subgroups of D_8 given in Table 1. However, we distinguish between the two classes in each case by their composition factors on $V_{D_8}(\lambda_7)$ which leads to the definition of $B_4(\ddagger)$ and $A_1C_4(\ddagger)$ in Section 9.1. Every subgroup X of D_8 listed in Table 1 is contained in $B_4(\ddagger)$ or $A_1C_4(\ddagger)$ and hence the conjugacy class of X is uniquely determined.

COROLLARY 3. *Let G be a simple exceptional algebraic group and X be a connected subgroup of G. Suppose that whenever X is contained in a reductive, maximal connected subgroup M of G it is M-irreducible and assume that such an overgroup M exists. Assume further that X is not contained in a proper Levi subgroup of G. Then either:*

(1) *X is G-irreducible, or*
(2) *X is Aut(G)-conjugate to a subgroup in Table 1. Such X are non-G-cr and satisfy the above hypothesis.*

Table 1: Non-G-cr subgroups that are irreducible in every (and at least one) maximal, reductive overgroup.

G	Max. M	p	M-irreducible subgroup X
G_2	$A_1\tilde{A}_1$	2	$A_1 \hookrightarrow M$ via $(1,1)$
F_4	$A_2\tilde{A}_2$	3	$A_2 \hookrightarrow M$ via $(10,01)$
E_6	A_2G_2	3	$A_2 \hookrightarrow A_2A_2 = E_6(\#48)$ via $(10,10)$

E_7	A_1G_2	7	$A_1 \hookrightarrow A_1A_1 = E_7(\#334)$ via $(1,1)$
	$\bar{A}_2A_5$	3	$A_2 \hookrightarrow \bar{A}_2A_2 = E_7(\#303)$ via $(10,10)$
	A_7	2	C_4 embedded via λ_1
			D_4 embedded via λ_1
			B_3 embedded via 001
			A_1B_2 embedded via $(1,01)$
			A_2 embedded via 11
			A_1^3 embedded via $(1,1,1)$
			$A_1A_1 \hookrightarrow A_1^3$ via $(1_a, 1_a^{[r]}, 1_b)$ $(r \neq 0)$
			$A_1 \hookrightarrow A_1^3$ via $(1, 1^{[r]}, 1^{[s]})$ $(0 < r < s)$
	G_2C_3	2	$G_2 \hookrightarrow G_2G_2 = E_7(\#320)$ via $(10,10)$
E_8	D_8	2	$B_4(\ddagger)$ embedded via λ_4
			$B_2^2 < B_4(\ddagger)$ embedded via $(01,01)$
			$B_2 \hookrightarrow B_2^2$ via $(10, 10^{[r]})$ $(r \neq 0)$, $(10,02)$ or $(10, 02^{[r]})$ $(r \neq 0)$
			$A_1^2B_2 < B_4(\ddagger)$ embedded via $(1,1,01)$
			$A_1B_2 \hookrightarrow A_1^2B_2$ via $(1, 1^{[r]}, 10)$ $(r \neq 0)$
			$A_1^4 < B_4(\ddagger)$ embedded via $(1,1,1,1)$
			$A_1A_1^2 \hookrightarrow A_1^4$ via $(1_a, 1_a^{[r]}, 1_b, 1_c)$ $(r \neq 0)$
			$A_1A_1 \hookrightarrow A_1^4$ via $(1_a, 1_a^{[r]}, 1_a^{[s]}, 1_b)$ $(0 < r < s)$
			$A_1A_1 \hookrightarrow A_1^4$ via $(1_a, 1_a^{[r]}, 1_b, 1_b^{[s]})$ $(0 < r \leq s)$
			$A_1 \hookrightarrow A_1^4$ via $(1, 1^{[r]}, 1^{[s]}, 1^{[t]})$ $(0 < r < s < t)$
			$A_1C_4(\ddagger)$ embedded via $(1, \lambda_1)$
			$A_1D_4 < A_1C_4(\ddagger)$ embedded via $(1, \lambda_1)$
			$A_1B_3 < A_1C_4(\ddagger)$ embedded via $(1,001)$
			$A_1A_2 < A_1C_4(\ddagger)$ embedded via $(1,11)$
	A_8	3	A_2^2 embedded via $(10,10)$
			$A_2 \hookrightarrow A_2^2$ via $(10, 10^{[r]})$ $(r \neq 0)$ or $(10, 01^{[r]})$ $(r \neq 0)$
	G_2F_4	7	$G_2 \hookrightarrow G_2G_2 = E_8(\#1035)$ via $(10,10)$

The corollary gives examples showing that one cannot generalise [**BMR05**, Theorem 3.26] in certain ways for exceptional algebraic groups. This theorem states that in good characteristic every H-cr subgroup is G-cr for a regular reductive subgroup H of G, where a subgroup is regular if it is normalised by a maximal torus of G. We say that a prime p is *good* for a simple exceptional algebraic group G if $p \geq 5$ for G of type G_2, F_4, E_6 or E_7, and $p \geq 7$ for G of type E_8. We say a prime is *bad* for G if it is not good.

Firstly, one cannot allow arbitrary bad characteristics. The subgroup $A_1 < G_2$ given in Table 1 is $\bar{A}_1\tilde{A}_1$-irreducible yet non-G_2-cr when $p = 2$, and $\bar{A}_1\tilde{A}_1$ is a regular subgroup of G_2.

Secondly, if one considers reductive, maximal connected subgroups of G, many of these are regular. However, we do not have such a result for arbitrary reductive, maximal connected subgroups. For example, the subgroup $A_1 \hookrightarrow A_1A_1 < A_1G_2 <$

E_7 via $(1,1)$ from Table 1 is A_1G_2-irreducible yet non-E_7-cr when $p=7$, which is even a good characteristic for E_7.

It is natural to ask whether G-irreducible subgroups of a certain type exist, especially in small characteristics. When G is a simple exceptional algebraic group it is shown in [**LT04**, Theorem 2] (corrected in [**Ame05**, Theorem 7.4]) that G-irreducible connected subgroups of type A_1 exist, except for $G=E_6$ when $p=2$. We extend this result to subgroups of type A_1^n.

COROLLARY 4. *Let G be a simple exceptional algebraic group of rank l. Then G contains a G-irreducible connected subgroup of type A_1^n for all $n \leq l$, unless $G=E_6$. For $G=E_6$, there exists a G-irreducible subgroup of type A_1^n if and only if $p \neq 2$ and $n \leq 3$.*

Given the existence of irreducible subgroups of type A_1^n, we study their overgroups. The next result shows the existence of a reductive, maximal connected subgroup that contains representatives of each conjugacy class of G-irreducible subgroups of type A_1^n in small characteristics, with one exception.

COROLLARY 5. *Let G be a simple exceptional algebraic group and $p=2$ or 3. Then there exists a reductive, maximal connected subgroup M containing representatives of every $Aut(G)$-conjugacy class of G-irreducible subgroups of type A_1^n, unless $G=F_4$ and $p=3$ (in which case two reductive, maximal connected subgroups are required). The following table provides examples of such overgroups M.*

Table 2: Maximal connected overgroups for G-irreducible subgroups of type A_1^n.

G	$p=3$	$p=2$
G_2	$\bar{A}_1\tilde{A}_1$	$\bar{A}_1\tilde{A}_1$
F_4	B_4 and A_1C_3	B_4
E_6	C_4	—
E_7	$\bar{A}_1D_6$	$\bar{A}_1D_6$
E_8	D_8	D_8

We also prove similar results for G-irreducible subgroups of type A_2^n.

COROLLARY 6. *Let G be a simple exceptional algebraic group of rank l. Then for G not of type E_7, there exists a G-irreducible connected subgroup of type A_2^n if and only if $n \leq \frac{l}{2}$. For G of type E_7, there exists a G-irreducible subgroup of type A_2^n if and only if $p \neq 2$ and $n \leq 2$.*

COROLLARY 7. *Let G be a simple exceptional algebraic group. Then there exists a reductive, maximal connected subgroup M containing representatives of every $Aut(G)$-conjugacy class of G-irreducible subgroups of type A_2^n, unless (G,p) is one of the following: $(G_2,3)$, $(E_6, p\neq 2)$, $(E_7, p \geq 5)$ or $(E_8, p \neq 3)$ (in all cases at most three reductive, maximal connected subgroups are required). The following table provides examples of such overgroups M.*

Table 3: Maximal connected overgroups for G-irreducible subgroups of type A_2^n.

G	$p \geq 5$	$p = 3$	$p = 2$
G_2	$\bar{A}_2$	$\bar{A}_2$ and $\tilde{A}_2$	$\bar{A}_2$
F_4	$\bar{A}_2\tilde{A}_2$	$\bar{A}_2\tilde{A}_2$	$\bar{A}_2\tilde{A}_2$
E_6	$\bar{A}_2^3$ and A_2	$\bar{A}_2^3$, A_2G_2 and G_2	$\bar{A}_2^3$
E_7	$\bar{A}_2A_5$ and A_2	$\bar{A}_2A_5$	—
E_8	$\bar{A}_2E_6$ and D_8	$\bar{A}_2E_6$	$\bar{A}_2E_6$ and D_8

CHAPTER 2

Notation

In this chapter we present the notation used throughout the paper. At many points in this paper we use the results of [**Tho15**] and [**Tho16**] and therefore have tried to be consistent with the notation in those papers, where possible. In particular, when we come to describe the identification number given to an irreducible connected subgroup X we have chosen to keep the same identification number given to X in [**Tho16**] when X is of type A_1.

Firstly we note that by a subgroup of an algebraic group we always mean a closed subgroup. Similarly, all representations of algebraic groups are assumed to be rational.

Let G be a simple algebraic group over an algebraically closed field K. Let Φ be the root system of G and Φ^+ be a fixed set of positive roots in Φ. Write $\Pi = \{\alpha_1, \dots, \alpha_l\}$ for the simple roots of G and $\lambda_1, \dots, \lambda_l$ for the fundamental dominant weights of G, both with respect to the ordering of the Dynkin diagram as given in [**Bou68**, p. 250]. We sometimes use $a_1 a_2 \dots a_l$ to denote a dominant weight $a_1\lambda_1 + a_2\lambda_2 + \cdots + a_l\lambda_l$. We denote by $V_G(\lambda)$ (or just λ) the irreducible G-module of dominant high weight λ. Similarly, the Weyl module of high weight λ is denoted by $W(\lambda) = W_G(\lambda)$ and the tilting module of high weight λ is denoted by $T(\lambda)$. Another module we refer to frequently is the adjoint module for G; we recall that we denote this by $L(G)$. We let

$$V_7 := W_{G_2}(\lambda_1),\ V_{26} := W_{F_4}(\lambda_4),\ V_{27} := V_{E_6}(\lambda_1),\ V_{56} := V_{E_7}(\lambda_7).$$

For G-modules V and W we write $V + W$ for the module $V \oplus W$ and let V^* denote the dual module of V. If $Y = Y_1 Y_2 \dots Y_k$, a commuting product of simple algebraic groups, then $(V_1, \dots, V_k)$ denotes the Y-module $V_1 \otimes \cdots \otimes V_k$, where each V_i is an irreducible Y_i-module. The notation $\bar{X}$ denotes a subgroup of G that is generated by long root subgroups of G. If the root system of G has short roots then $\tilde{X}$ denotes a subgroup generated by short root subgroups of G.

Suppose that $\mathrm{char}(K) = p < \infty$, recalling our convention that $p = \infty$ represents characteristic 0. Let $F : G \to G$ be the standard Frobenius endomorphism (acting on root groups $U_\alpha = \{u_\alpha(c) \mid c \in K\}$ by $u_\alpha(c) \mapsto u_\alpha(c^p)$) and V be a G-module afforded by a representation $\rho : G \to \mathrm{GL}(V)$. If r is a positive integer then $V^{[r]}$ denotes the module afforded by the representation $\rho^{[r]} := \rho \circ F^r$. Let $M_1, \dots, M_k$ be G-modules and $n_1, \dots, n_k$ be positive integers. Then $M_1^{n_1}/\dots/M_k^{n_k}$ denotes a G-module having the same composition factors as $M_1^{n_1} + \cdots + M_k^{n_k}$. Furthermore, $V = M_1|\dots|M_k$ denotes a G-module with a socle series as follows: $M_k \cong \mathrm{Soc}(V) = \mathrm{Soc}^1(V)$ and for $i > 0$, the module M_{k-i} is isomorphic to $\mathrm{Soc}^{i+1}(V) = \mathrm{Soc}(V/N_i)$ where N_i is the inverse image in V of $\mathrm{Soc}^i(V)$ under the quotient mapping $V \to V/N_{i-1}$ (so $N_0 = 0$ and $N_1 = M_k$). Sometimes, to make things clearer, we will use a tower of modules

$$\frac{\frac{M_1}{M_2}}{M_3}$$

to denote $V = M_1|M_2|M_3$.

We need a notation for diagonal subgroups of $Y = H_1H_2 \dots H_k$, a commuting product with all of the subgroups H_i simple and of the same type. Let H be a simply connected algebraic group of type H_1 and $\hat{Y} = H \times H \times \cdots \times H$, the direct product of k copies of H. Then we may regard Y as $\hat{Y}/Z$, where Z is a subgroup of the centre of $\hat{Y}$, and H_i is then regarded as the image of the ith projection map. A diagonal subgroup of $\hat{Y}$ is a subgroup $\hat{X} \cong H$ of the following form: $\hat{X} = \{(\phi_1(h), \dots, \phi_k(h)) \mid h \in H\}$ where each ϕ_i is a surjective endomorphism of H. A diagonal subgroup X of Y is the image of a diagonal subgroup of $\hat{Y}$ under the natural map $\hat{Y} \to Y$. To describe such a subgroup it therefore suffices to give a surjective endomorphism, ϕ_i, of H for each i. By [**GLS98**, Section 1.15], $\phi_i = \alpha_i\theta_iF^{r_i}$ where α_i is an inner automorphism, θ_i is a graph automorphism and F^{r_i} is a power of the standard Frobenius endomorphism. We only wish to distinguish these diagonal subgroups up to conjugacy. Therefore, we assume each α_i is trivial and give a (possibly trivial) graph automorphism θ_i of H and a non-negative integer r_i, for each $1 \leq i \leq k$.

Such a diagonal subgroup X is denoted by

$$X \hookrightarrow Y \text{ via } (\lambda_1^{[\theta_1 r_1]}, \lambda_1^{[\theta_2 r_2]}, \dots, \lambda_1^{[\theta_k r_k]}),$$

where λ_1 is the first dominant weight of X. We often abbreviate this to

$$X \text{ via } (\lambda_1^{[\theta_1 r_1]}, \dots, \lambda_1^{[\theta_k r_k]})$$

if the group Y is clear. Unless X is of type D_n ($n \geq 4$), a graph automorphism is uniquely determined (up to conjugacy) by the image of λ_1 (including the exceptional graph automorphisms of B_2, F_4 when $p = 2$ and G_2 when $p = 3$, which takes λ_1 to $2\lambda_2$, $2\lambda_4$ and $3\lambda_2$, respectively). In these cases, instead of writing $\lambda_1^{[\theta_i r_i]}$ we write $\mu^{[r_i]}$ where μ is the image of λ_1 under θ_i. The only time we need a diagonal subgroup of a product of subgroups of type D_n is when dealing with $Y = \bar{D}_4^2$. We give a notation for the standard graph automorphisms of $\bar{D}_4$: we let τ denote a standard triality automorphism induced by the permutation $(\alpha_1, \alpha_3, \alpha_4)$ and let ι denote a standard involutory automorphism induced by the permutation (α_3, α_4).

For clarity, note that field twists $r, s, t, \dots$ are not assumed to be distinct. This is consistent with [**Tho16**] but not with [**Tho15**].

We extend this notation to describe certain semisimple subgroups of the form $X = X_1X_2 \dots X_n$ of $Y = H_1H_2 \dots H_k$ ($n < k$) where each X_i is of type H and the projection of X to each H_i is surjective. Any such subgroup is a commuting product of diagonal subgroups of distinct subsets of the H_i. For this reason, we extend our use of the term "diagonal subgroup" to include such subgroups X. For example, consider diagonal subgroups isomorphic to A_1^2 contained in A_1^4. They are either a commuting product of one A_1 factor and a diagonal subgroup of A_1^3, or a commuting product of a diagonal subgroup of A_1^2 and another diagonal subgroup of the other A_1^2. Therefore, our notation needs to distinguish which of the subgroups H_i each of the simple factors of X project non-trivially to. We give the first factor of X the label a, the second factor of X the label b and so on. Then for each i such that X_1 has non-trivial projection to H_i we give a subscript a to $\lambda_1^{[\theta_i r_i]}$. For each j such that X_2 has non-trivial projection to H_j we give a subscript b to $\lambda_1^{[\theta_j r_j]}$ and so on. For example, consider $X = A_1A_1$ and $Y = A_1^4$ with the first A_1 factor of X embedded diagonally in the first two factors of Y (with field twists 0 and r) and the second A_1 factor embedded diagonally in the last two factors of Y (with field twists 0 and s). Then we write $X \hookrightarrow Y$ via $(1_a, 1_a^{[r]}, 1_b, 1_b^{[s]})$.

We make another natural extension of this notation. Let X and Y be as above and suppose that $S = S_1 \dots S_l$ is a semisimple group with no factors isomorphic to H. Then a subgroup XS of YS is denoted by

$$XS \hookrightarrow YS \text{ via } (\lambda_1^{[\theta_1 r_1]}, \lambda_1^{[\theta_2 r_2]}, \dots, \lambda_1^{[\theta_k r_k]}, \nu_1^1, \nu_1^2, \dots, \nu_1^l)$$

where ν_1^i denotes the first fundamental dominant weight of S_i. Again, we will still refer to such subgroups as "diagonal subgroups". For example, consider a subgroup $XZ = A_1B_2$ of $YZ = A_1^2B_2$ where $X = A_1$ is a simple diagonal subgroup of $Y = A_1^2$ with field twists 0 and r. Then we write $A_1B_2 \hookrightarrow A_1^2B_2$ via $(1, 1^{[r]}, 10)$.

Finally, we show how to combine all of these notations in the most general setting. Suppose that A is a diagonal subgroup of $Y = H_1 \dots H_n$ and B is a diagonal subgroup of $Z = J_1 \dots J_m$, where all the H_i (resp. J_i) are simple and of the same type H (resp. J) and H is not isomorphic to J. Suppose also that S is a semisimple group with no factors of type H or J. Then we naturally concatenate the notations above to give a notation for the subgroup ABS of YZS. For example, consider a subgroup $ABS = A_1^3B_2^2B_3$ of $YZS = A_1^4B_2^3B_3$ where A is a diagonal subgroup of Y via $(1_a, 1_b, 1_b^{[r]}, 1_c)$ and B is a diagonal subgroup of Z via $(10_a, 10_b^{[s]}, 10_b^{[t]})$. Then we write $A_1^3B_2^2B_3 \hookrightarrow A_1^4B_2^3B_3$ via $(1_a, 1_b, 1_b^{[r]}, 1_c, 10_a, 10_b^{[s]}, 10_b^{[t]}, 100)$.

When determining the different conjugacy classes of diagonal subgroups of a given group X we need to understand certain automorphisms of X induced from G. We define the following notation to describe the graph automorphisms of a subgroup X of G induced by $N_G(X)$. We let $\mathrm{Aut}_G(X) = N_G(X)/C_G(X)$ and $\mathrm{Inn}_G(X) = X/Z(X) \cong (XC_G(X))/C_G(X)$. Therefore $\mathrm{Aut}_G(X)/\mathrm{Inn}_G(X) \cong N_G(X)/(XC_G(X))$, the group of graph automorphisms of X induced by $N_G(X)$ and we let $\mathrm{Out}_G(X) = \mathrm{Aut}_G(X)/\mathrm{Inn}_G(X)$.

Now let G be of exceptional type. In Tables 1–11.5 we give an identification number to each of the conjugacy classes of G-irreducible connected subgroups arising in Theorems 5.1–9.1. The notation $G(\#a)$ (or simply a if G is clear from the context) means the G-irreducible subgroup corresponding to the ID number a. We set $G(\#0)$ to be G itself. Sometimes $G(\#a)$ will refer to infinitely many conjugacy classes of G-irreducible subgroups. This situation only occurs for diagonal subgroups, where the conjugacy class will depend on field twists $r_1, \dots, r_k$ and graph automorphisms $\theta_1, \dots, \theta_k$. Sometimes we refer to a subset of the conjugacy classes that $G(\#a)$ represents; we only do this when all graph automorphisms are trivial. These are thus described by an ordered set of field twists $r_1, \dots, r_k$ and are denoted by $G(\#a^{\{r_1,\dots,r_k\}})$. Let us give a concrete example to make this clearer. Consider $G_2(\#1)$, the conjugacy classes of diagonal subgroups $A_1 \hookrightarrow \bar{A}_1\tilde{A}_1$ via $(1^{[r]}, 1^{[s]})$ $(rs = 0;\ r \neq s)$ (see Table 1). Then the notation $G_2(\#1^{\{r,0\}})$ refers to the conjugacy classes with $s = 0$ and the notation $G_2(\#1^{\{1,0\}})$ refers to the single conjugacy class $A_1 \hookrightarrow \bar{A}_1\tilde{A}_1$ via $(1^{[1]}, 1)$.

In the tables in Section 11 we use a shorthand for $n^{\{s_1,\ldots,s_k\}}$ in certain situations. The notation $n^{\{\underline{0}\}}$ simply means that each s_i is equal to 0. The notation $n^{\{\delta_j\}}$ means $s_j = 1$ and $s_i = 0$ for all $i \neq j$.

We have chosen these identification numbers to be consistent with those given to G-irreducible subgroups of type A_1 in [**Tho16**, Tables 4–8]. After the first m identification numbers have accounted for the subgroups of type A_1 we then give the next set of identification numbers to the simple subgroups of rank at least 2 from [**Tho15**, Tables 3–7]. The next set is given to the remaining non-simple reductive, maximal connected subgroups and after this they are given in order.

In Tables 1–11.5 we also need a notation to be able to describe conjugacy classes of M-irreducible connected subgroups X of reductive, maximal connected subgroups M of G. Suppose that $M = M_1 M_2 \ldots M_r$. If all of the factors are simple classical algebraic groups then we define

$$V_M := V_{M_1}(\lambda_1) \otimes V_{M_2}(\lambda_1) \otimes \cdots \otimes V_{M_r}(\lambda_1)$$

and let $V_M \downarrow X$ be the usual restriction of the M-module V_M to X.

Now suppose that some M_i is of exceptional type. We do not wish to list composition factors for the action of X on a minimal module for M_i as this will make things more difficult to read. Instead, we use the fact that the projection of X to M_i will be M_i-irreducible (see Lemma 3.5) and therefore X has a unique ID number. In this case we give the ID number rather than the composition factors of the restriction of any module. So, if $M = M_1$ then we define the notation $V_M \downarrow X$ to simply be $M(\#a)$ where a is the ID number of the subgroup X of M. The final possibility is $M = M_1 M_2$ with at least one M_i of exceptional type. In this case, we combine the previous two notations by denoting $V_M \downarrow X = (V_{M_1} \downarrow X_1, V_{M_2} \downarrow X_2)$.

As above, let G be a simple exceptional algebraic group. Many of the G-irreducible connected subgroups have simple factors $\bar{X}$ generated by long root subgroups of G. In the following cases all subgroups of the given type are generated by long root subgroups of G and we will therefore omit the bar.

$$G = E_6,\ X = A_4, A_5, D_5, E_6.$$
$$G = E_7,\ X = A_4, A_5, A_6, A_7, D_5, D_6, E_6, E_7.$$
$$G = E_8,\ X = A_5, A_6, A_7, A_8, D_5, D_6, D_7, D_8, E_6, E_7, E_8.$$

Finally, we note the standard notation we will use for certain finite groups. A symmetric group acting on a finite set of size m will be denoted by S_m. A dihedral group of order $2n$ will be denoted by Dih_{2n}.

CHAPTER 3

Preliminaries

To prove our main theorems we require a number of preliminary results, which we record in this chapter. The first result is the starting point for our strategy. Recall that when we say a reductive, maximal connected subgroup we mean a reductive subgroup that is maximal among all connected subgroups.

THEOREM 3.1 ([**LS04**, Corollary 2]). *Let G be a simple exceptional algebraic group. Let M be a reductive, maximal connected subgroup of G. Then M is $Aut(G)$-conjugate to precisely one subgroup X as follows, where each isomorphism type of X denotes one G-conjugacy class of subgroups.*

G	X
G_2	$\bar{A}_2$, $\tilde{A}_2$ $(p=3)$, $\bar{A}_1\tilde{A}_1$, A_1 $(p \geq 7)$
F_4	B_4, C_4 $(p=2)$, $\bar{A}_1C_3$ $(p \neq 2)$, A_1G_2 $(p \neq 2)$, $\bar{A}_2\tilde{A}_2$, G_2 $(p=7)$, A_1 $(p \geq 13)$
E_6	$\bar{A}_1A_5$, $\bar{A}_2^3$, F_4, C_4 $(p \neq 2)$, A_2G_2, G_2 $(p \neq 7)$, A_2 $(p \geq 5)$
E_7	$\bar{A}_1D_6$, $\bar{A}_2A_5$, A_7, G_2C_3, A_1F_4, A_1G_2 $(p \neq 2)$, A_2 $(p \geq 5)$, A_1A_1 $(p \geq 5)$, A_1 $(p \geq 17)$, A_1 $(p \geq 19)$
E_8	D_8, $\bar{A}_1E_7$, $\bar{A}_2E_6$, A_8, $\bar{A}_4^2$, G_2F_4, F_4 $(p=3)$, B_2 $(p \geq 5)$, A_1A_2 $(p \geq 5)$, A_1 $(p \geq 23)$, A_1 $(p \geq 29)$, A_1 $(p \geq 31)$

REMARK. The class of maximal subgroups $F_4 < E_8$ when $p = 3$ were only discovered after the acceptance of this paper. We refer to [**CST**] for their construction.

Let G be a classical simple algebraic group, which we refer to as $\mathrm{Cl}(V) = \mathrm{SL}(V)$, $\mathrm{Sp}(V)$ or $\mathrm{SO}(V)$ for some finite-dimensional vector space V. We need to determine the G-conjugacy classes of reductive, maximal connected subgroups of G, when G is of small rank. Firstly, we need part of a theorem of Liebeck and Seitz concerning the maximal subgroups of classical algebraic groups. Let H be a subgroup of G. We introduce the following classes of subgroups (which is a subset of those from [**LS98**]).

Class $\mathcal{C}_1$: *Subspace stabilisers.* Here $H \in \mathcal{C}_1$ if $H = \mathrm{Stab}_G(W)$ where W is either a non-degenerate subspace of V, or $(G,p) = (\mathrm{SO}(V),2)$ and W is a non-singular subspace of dimension 1.

Class $\mathcal{C}_4$: *Tensor product subgroups.* Suppose that $V = V_1 \otimes V_2$ with $\dim V_i > 1$. Then $H \in \mathcal{C}_4$ if $H = \mathrm{Cl}(V_1) \circ \mathrm{Cl}(V_2)$ which acts naturally on V as follows:

$(g_1, g_2)(v_1 \otimes v_2) := (g_1 v_1) \otimes (g_2 v_2)$. The tensor product subgroups occurring are:

$$\mathrm{SL} \otimes \mathrm{SL} < \mathrm{SL}, \qquad \mathrm{Sp} \otimes \mathrm{SO} < \mathrm{Sp}\ (p \neq 2),$$
$$\mathrm{Sp} \otimes \mathrm{Sp} < \mathrm{SO}, \qquad \mathrm{SO} \otimes \mathrm{SO} < \mathrm{SO}\ (p \neq 2).$$

The following theorem can be immediately deduced from [**LS98**, Theorem 1].

THEOREM 3.2. *Let G be a classical simple algebraic group. Suppose that M is a reductive, maximal connected subgroup of G. Then one of the following holds:*
(i) *M belongs to $\mathcal{C}_1$;*
(ii) *M belongs to $\mathcal{C}_4$;*
(iii) *M is a simple algebraic group and $V \downarrow M$ is irreducible and restricted.*

In the following lemma we now apply this theorem to certain classical simple groups of small rank; the ones we treat are those which will arise in the work of Sections 5–9.

LEMMA 3.3. *Suppose that G is a classical group of type A_n for $n \in \{2,3,4,5,7,8\}$, B_n for $n \in \{2,3,4,5,6,7\}$, C_n for $n \in \{3,4\}$, or D_n for $n \in \{4,5,6,8\}$. Then the following table gives all G-conjugacy classes of reductive, maximal connected subgroups of G.*

Table 2: The maximal subgroups of certain low rank classical algebraic groups.

G	Max. sub. M	$V_G(\lambda_1) \downarrow M$	Comments
A_2	A_1 $(p \neq 2)$	2	
B_2	$\bar{A}_1^2$	$(1,1)+(0,0)$	
	$\tilde{A}_1^2$ $(p=2)$	$(2,0)+(0,2)$	
	A_1 $(p \geq 5)$	4	
A_3	B_2	01	
	A_1^2 $(p \neq 2)$	$(1,1)$	
B_3	$\bar{A}_3$	$010+000$ $(p \neq 2)$ or 010 $(p=2)$	
	$\bar{A}_1^2\tilde{A}_1$ $(p \neq 2)$	$(1,1,0)+(0,0,2)$	
	G_2	10	
	$\tilde{A}_1 B_2$ $(p=2)$	$(2,00)+(0,10)$	
C_3	$\bar{A}_1 C_2$	$(1,00)+(0,10)$	
	$A_1 A_1$ $(p \neq 2)$	$(2,1)$	
	$\tilde{A}_3$ $(p=2)$	010	
	G_2 $(p=2)$	10	
	A_1 $(p \geq 7)$	5	
A_4	B_2 $(p \neq 2)$	10	
B_4	$\bar{D}_4$	$\lambda_1 + 0$ $(p \neq 2)$ or λ_1 $(p=2)$	
	$\tilde{A}_1\bar{A}_3$ $(p \neq 2)$	$(2,000)+(0,010)$	
	$\bar{A}_1^2 B_2$ $(p \neq 2)$	$(1,1,00)+(0,0,10)$	
	A_1^2 $(p \neq 2)$	$(2,2)$	

	A_1 $(p \geq 11)$	8	
	$\tilde{A}_1 B_3$ $(p=2)$	$(2,000)+(0,100)$	
	B_2^2 $(p=2)$	$(10,00)+(00,10)$	
C_4	C_2^2	$(10,00)+(00,10)$	
	$\bar{A}_1 C_3$	$(1,000)+(0,100)$	
	A_1^3 $(p \neq 2)$	$(1,1,1)$	
	A_1 $(p \geq 11)$	7	
	$\tilde{D}_4$ $(p=2)$	λ_1	
D_4	B_3 (3 classes)	$100+000$ $(p \neq 2)$ or $000\|100\|000$ $(p=2)$ 001 001	classes are permuted by $\mathrm{Out}(D_4) \cong S_3$
	$A_1 B_2$ $(p \neq 2)$ (3 classes)	$(2,00)+(0,10)$ $(1,01)$ $(1,01)$	classes are permuted by $\mathrm{Out}(D_4) \cong S_3$
	$\bar{A}_1^4$	$(1,1,0,0)+(0,0,1,1)$	
	A_2 $(p \neq 3)$	11	involutory graph aut. of D_4 induces graph aut. of A_2
A_5	$A_1 A_2$	$(1,10)$	
	C_3	100	
	A_3 $(p \neq 2)$	010	
	A_2 $(p \neq 2)$	20	
B_5	$\bar{D}_5$	λ_1+0 $(p \neq 2)$ or λ_1 $(p=2)$	
	$\tilde{A}_1 \bar{D}_4$ $(p \neq 2)$	$(2,0)+(0,\lambda_1)$	
	$\bar{A}_1^2 B_3$ $(p \neq 2)$	$(1,1,000)+(0,0,100)$	
	$B_2 \bar{A}_3$ $(p \neq 2)$	$(10,000)+(00,010)$	
	A_1 $(p \geq 11)$	10	
	$\tilde{A}_1 B_4$ $(p=2)$	$(2,0)+(0,\lambda_1)$	
	$B_2 B_3$ $(p=2)$	$(10,000)+(00,100)$	
D_5	B_4	λ_1+0 $(p \neq 2)$ or $0\|\lambda_1\|0$ $(p=2)$	
	$A_1 B_3$ $(p \neq 2)$	$(2,000)+(0,100)$	
	$\bar{A}_1^2 \bar{A}_3$	$(1,1,000)+(0,0,010)$	
	B_2^2 $(p \neq 2)$	$(10,00)+(00,10)$	
	B_2 $(p \neq 2)$ (2 classes)	02 02	classes permuted by involutory graph aut.
B_6	$\bar{D}_6$	λ_1+0 $(p \neq 2)$ or λ_1 $(p=2)$	
	$\tilde{A}_1 \bar{D}_5$ $(p \neq 2)$	$(2,0)+(0,\lambda_1)$	
	$\bar{A}_1^2 B_4$ $(p \neq 2)$	$(1,1,0)+(0,0,\lambda_1)$	
	$B_2 \bar{D}_4$ $(p \neq 2)$	$(10,0)+(00,\lambda_1)$	

	$\bar{A}_3B_3$ $(p \neq 2)$	$(010, 000) + (000, 100)$	
	A_1 $(p \geq 13)$	12	
	B_2 $(p = 5)$	20	
	C_3 $(p = 3)$	010	
	$\tilde{A}_1B_5$ $(p = 2)$	$(2, 0) + (0, \lambda_1)$	
	B_2B_4 $(p = 2)$	$(10, 0) + (00, \lambda_1)$	
	B_3^2 $(p = 2)$	$(100, 000) + (000, 100)$	
D_6	B_5	$\lambda_1 + 0$ $(p \neq 2)$ or $0\|\lambda_1\|0$ $(p = 2)$	
	A_1B_4 $(p \neq 2)$	$(2, 0) + (0, \lambda_1)$	
	$\bar{A}_1^2\bar{D}_4$	$(1, 1, 0) + (0, 0, \lambda_1)$	
	B_2B_3 $(p \neq 2)$	$(10, 000) + (00, 100)$	
	$\bar{A}_3^2$	$(010, 000) + (000, 010)$	
	A_1C_3 (2 classes)	$(1, 100)$ $(1, 100)$	classes permuted by involutory graph aut.
A_7	C_4	λ_1	
	D_4 $(p \neq 2)$	λ_1	
	A_1A_3	$(1, 100)$	
B_7	$\bar{D}_7$	$\lambda_1 + 0$ $(p \neq 2)$ or λ_1 $(p = 2)$	
	$\tilde{A}_1\bar{D}_6$ $(p \neq 2)$	$(2, 0) + (0, \lambda_1)$	
	$\bar{A}_1^2B_5$ $(p \neq 2)$	$(1, 1, 0) + (0, 0, \lambda_1)$	
	$B_2\bar{D}_5$ $(p \neq 2)$	$(10, 0) + (00, \lambda_1)$	
	$\bar{A}_3B_4$ $(p \neq 2)$	$(010, 0) + (000, \lambda_1)$	
	$B_3\bar{D}_4$ $(p \neq 2)$	$(010, 0) + (000, \lambda_1)$	
	A_3 $(p \neq 2)$	101	
	A_1B_2 $(p \neq 2)$	$(2, 10)$	
	A_1 $(p \geq 17)$	14	
	$\tilde{A}_1B_6$ $(p = 2)$	$(2, 0) + (0, \lambda_1)$	
	B_2B_5 $(p = 2)$	$(10, 0) + (00, \lambda_1)$	
	B_3B_4 $(p = 2)$	$(100, 0) + (000, \lambda_1)$	
A_8	B_4 $(p \neq 2)$	λ_1	
	A_2^2	$(10, 10)$	
D_8	B_7	$\lambda_1 + 0$ $(p \neq 2)$ or $0\|\lambda_1\|0$ $(p = 2)$	
	A_1B_6 $(p \neq 2)$	$(2, 0) + (0, \lambda_1)$	
	$\bar{A}_1^2\bar{D}_6$	$(1, 1, 0) + (0, 0, \lambda_1)$	
	B_2B_5 $(p \neq 2)$	$(10, 0) + (00, \lambda_1)$	
	$\bar{A}_3\bar{D}_5$	$(010, 000) + (000, \lambda_1)$	
	B_3B_4 $(p \neq 2)$	$(100, 0) + (000, \lambda_1)$	
	$\bar{D}_4^2$	$(\lambda_1, 0) + (0, \lambda_1)$	

B_2^2 $(p \neq 2)$ (2 classes)	$(01, 01)$ $(01, 01)$	classes permuted by involutory graph aut.
A_1C_4 (2 classes)	$(1, \lambda_1)$ $(1, \lambda_1)$	classes permuted by involutory graph aut.
B_4 (2 classes)	λ_4 λ_4	classes permuted by involutory graph aut.

PROOF. We will give the details of how to apply Theorem 3.2 when G is of type A_n $(2 \leq n \leq 5)$ or of type D_4. The other types are similar. The strategy is to find all possible subgroups of G in $\mathcal{C}_1$, $\mathcal{C}_4$ and 3.2 (iii), and then check whether there are any containments amongst them.

Suppose that G is of type A_n. We apply Theorem 3.2, considering A_n as $\mathrm{SL}(V)$ where V is of dimension $n+1$ and equipped with the 0-form. Firstly, there are no subgroups in $\mathcal{C}_1$ since all subspaces are degenerate. Next, we consider $\mathcal{C}_4$. If $n+1$ is prime then there are no subgroups. For G of type A_3 we obtain the subgroup A_1^2 acting as $(1, 1)$ on V. This is only maximal when $p \neq 2$ because this subgroup is contained in B_2 when $p = 2$ (the subgroup B_2 comes from 3.2 (iii) as explained below). For G of type A_5 we obtain the subgroup A_1A_2 acting as $(1, 2)$.

Now consider subgroups from 3.2 (iii). We use [**Lüb01**] to find the simple groups S with an $(n+1)$-dimensional irreducible restricted representation. Any such representation will embed S into G so it remains to determine the maximality. When G is of type A_2 the only possibility for S is a subgroup A_1 when $p \neq 2$ acting as the symmetric square of its natural representation. Since there are no other possible subgroups it follows that S is maximal. Now let G be of type A_3. The possibilities for S are a subgroup B_2 acting as $V_{B_2}(01)$ and a subgroup A_1 $(p \geq 5)$ acting as $V_{A_1}(4)$. We claim the subgroup A_1 is contained in B_2. This follows since for any group of type A_n $(n \geq 3)$, the subgroup A_1 acting as $V_{A_1}(n)$ is contained in $C_{(n+1)/2}$ (n odd) or $B_{n/2}$ (n even) because it fixes a symplectic (n odd) or orthogonal (n even) form. Therefore only the subgroup B_2 is maximal.

Let G be of type A_4. Then the only possibility for S is a subgroup B_2 acting as $V_{B_2}(10)$. Since there are no other possible subgroups this B_2 is maximal. Finally, let G be of type A_5. Then the possibilities for S are C_3 acting via $V_{C_3}(100)$, A_3 acting via $V_{A_3}(010)$, A_2 $(p \neq 2)$ acting via $V_{A_2}(20)$ and A_1 $(p \geq 7)$. The subgroup A_1 $(p \geq 7)$ is contained in C_3 by the argument in the previous paragraph. Also, the subgroup A_3 is contained in C_3 when $p = 2$ $(\mathrm{SO}_6 < \mathrm{Sp}_6)$. There are no further containments.

Now let G be of type D_4, which we consider as SO_8 with natural module V. The subgroups in $\mathcal{C}_1$ are B_3 (SO_7), A_1B_2 $(p \neq 2)$ $(\mathrm{SO}_3\mathrm{SO}_5)$ and A_1^4 $(\mathrm{SO}_4\mathrm{SO}_4 < \mathrm{SO}_8)$. The tensor decomposition $V = V_4 \otimes V_2$ (with V_i of dimension i) gives two D_4-conjugacy classes of maximal subgroups A_1B_2 $(\mathrm{Sp}_2 \otimes \mathrm{Sp}_4)$ when $p \neq 2$. Indeed, there is only one GO_8-conjugacy class but this splits in the index two subgroup SO_8 because any element $g \in \mathrm{GO}_8 \setminus \mathrm{SO}_8$ normalising A_1B_2 centralises it (since the outer automorphism group of A_1B_2 is trivial) and is hence contained in the centre of GO_8. However, the centre of GO_8 is contained in SO_8 so no such g exists and $N_{\mathrm{GO}_8}(A_1B_2) < \mathrm{SO}_8$. It follows that there is an involution in the automorphism group of D_4 which swaps the two SO_8-conjugacy classes. Furthermore, they are both mapped to the subgroup A_1B_2 from $\mathcal{C}_1$ by appropriate triality automorphisms since neither is centralised by any triality automorphisms ([**GLS98**, Table 4.7.1]).

Now we consider simple groups with an 8-dimensional irreducible restricted representation from [**Lüb01**]. The group A_2 has such a representation when $p \neq 3$, namely the adjoint module $V_{A_2}(11)$. Since it is the adjoint module for A_2, we know A_2 preserves the Killing form on it. The Killing form is a bilinear symmetric form and therefore $V_{A_2}(11)$ is an orthogonal module and $A_2 < D_4$ when $p > 3$. When $p = 2$ the proof of [**Kle87**, Proposition 2.3.3] shows that A_2 preserves a quadratic form on $V_{A_2}(11)$ and thus A_2 is a subgroup of D_4. Moreover, there is an element of $\mathrm{GO}_8 \setminus \mathrm{SO}_8$ acting as a graph automorphism of A_2, again by the proof of [**Kle87**, Proposition 2.3.3]. Therefore, $N_{\mathrm{GO}_8}(A_2) \not< \mathrm{SO}_8$ and there is just one D_4-conjugacy class of maximal subgroup A_2. We also note that we have shown that $A_2.2 < D_4.2$ but $A_2.2 \not< D_4$. The module $V_{A_1}(7)$ is a self-dual, restricted, 8-dimensional representation for A_1 when $p \geq 11$. However, this module is symplectic and therefore there is no maximal subgroup A_1 of D_4. The group B_3 has a self-dual, restricted, 8-dimensional, representation $V_{B_3}(001)$. This is indeed an orthogonal representation by [**KL90**, Proposition 5.4.9] (the argument holds for algebraic groups also) and hence yields a subgroup B_3 of D_4. Since the outer automorphism group of B_3 is trivial, it follows as before that there are two D_4-conjugacy classes and they are swapped by an involution in the automorphism group of D_4; they are mapped to the subgroup B_3 from $\mathcal{C}_1$ under triality automorphisms. Finally, when $p = 2$ the group C_3 has an 8-dimensional self-dual irreducible representation obtained by the special isogeny map from C_3 to B_3. Therefore this module is also orthogonal, but the image of this Sp_6 inside D_4 is a B_3, otherwise a triality automorphism conjugates a subgroup C_3 to one of type B_3, which is absurd since they are not even isomorphic as algebraic groups. □

For the following lemmas let G be a semisimple connected algebraic group. We state some elementary results about G-irreducible subgroups.

LEMMA 3.4 ([**LT04**, Lemma 2.1]). *If X is a G-irreducible connected subgroup of G, then X is semisimple and $C_G(X)$ is finite.*

LEMMA 3.5 ([**Tho15**, Lemma 3.6]). *Suppose that a G-irreducible subgroup X is contained in K_1K_2, a commuting product of connected non-trivial subgroups K_1, K_2 of G. Then X has non-trivial projection to both K_1 and K_2. Moreover, each projection is a K_i-irreducible subgroup.*

LEMMA 3.6 ([**LT04**, Lemma 2.2]). *Suppose that G is a classical simple algebraic group, with natural module $V = V_G(\lambda_1)$. Let X be a semisimple connected subgroup of G. If X is G-irreducible then one of the following holds:*

(i) *$G = A_n$ and X is irreducible on V;*
(ii) *$G = B_n, C_n$ or D_n and $V \downarrow X = V_1 \perp \ldots \perp V_k$ with the V_i all non-degenerate, irreducible and inequivalent as X-modules;*
(iii) *$G = D_n$, $p = 2$, X fixes a non-singular vector $v \in V$, and X is a G_v-irreducible subgroup of $G_v = B_{n-1}$.*

The next lemma and corollary are used in the proofs of Theorems 5.1 to 9.1 to show that an M-irreducible subgroup is G-irreducible, where M is a reductive, maximal connected subgroup of a simple exceptional algebraic group G.

LEMMA 3.7 ([**Tho15**, Lemma 3.8]). *Let X be a semisimple connected subgroup of G and let V be a G-module. Suppose that X does not have the same composition factors as any semisimple L'-irreducible connected subgroup H of the same type as*

X for some proper Levi subgroup L of G. If X is of type B_n and $p = 2$ then assume further that there is no subgroup H of type C_n with $H \leq L'$ and L-irreducible, for some Levi subgroup L of G, such that there is an isogeny $\phi : X \to H$ inducing a mapping which takes the composition factors of $V \downarrow X$ to those of $V \downarrow H$. Then X is G-irreducible.

COROLLARY 3.8 ([**Tho15**, Corollary 3.9]). *Suppose that $X < G$ is semisimple and $L(G) \downarrow X$ has no trivial composition factors. Then X is G-irreducible.*

The following two well known results will be used throughout the proof of Theorem 1 when showing that a G-irreducible subgroup contained in a maximal subgroup M_1 is conjugate to a G-irreducible subgroup of a maximal subgroup M_2.

LEMMA 3.9 ([**LS12**, Lemma 11.13]). *Let G be an adjoint simple algebraic group of exceptional type and let $s \in G$ be an element of prime order $r \neq p$. Then $C_G(s)^\circ$ is semisimple if and only if $C_G(s)$ is listed below.*

G	$C_G(s)$	r
G_2	$\bar{A}_1\tilde{A}_1$, $\bar{A}_2$	2, 3 (resp.)
F_4	$\bar{A}_1C_3$, B_4, $\bar{A}_2\tilde{A}_2$	2, 2, 3 (resp.)
E_6	$\bar{A}_1A_5$, $\bar{A}_2^3.3$	2, 3 (resp.)
E_7	$\bar{A}_1D_6$, $A_7.2$, $\bar{A}_2A_5$	2, 2, 3 (resp.)
E_8	$\bar{A}_1E_7$, D_8, $\bar{A}_2E_6$, A_8, $\bar{A}_4^2$	2, 2, 3, 3, 5 (resp.)

LEMMA 3.10 ([**LS94**, Section 4]). *Let G be a simple algebraic group of exceptional type and X be a subgroup of type A_1 or A_2. If X is generated by long root subgroups of G then $C_G(X)^\circ$ is given in the table below.*

G	$C_G(\bar{A}_1)^\circ$	$C_G(\bar{A}_2)^\circ$
G_2	$\tilde{A}_1$	1
F_4	C_3	$\tilde{A}_2$
E_6	A_5	$\bar{A}_2^2$
E_7	D_6	A_5
E_8	E_7	E_6

CHAPTER 4

Strategy for the proofs of Theorems 5.1–9.1

To prove Theorem 1 we prove Theorems 5.1–9.1 in Sections 5–9, respectively. In this chapter we describe the strategy used in proving those theorems.

Let G be a simple exceptional algebraic group over an algebraically closed field of characteristic p. Suppose that X is a G-irreducible connected subgroup of G. Then X is contained in a maximal connected subgroup M of G. Since X is G-irreducible, M is reductive. Furthermore, X is M-irreducible as any parabolic subgroup of M is contained in a parabolic subgroup of G by the Borel-Tits Theorem [**BT71**, Théorème 2.5]. Therefore, X is M-irreducible in some reductive, maximal connected subgroup M of G and the following strategy will find all such X. Not only do we wish to classify all G-irreducible connected subgroups of G, we also want to describe the lattice structure of connected overgroups. This last point explains why the strategy we describe below is different to that used in [**Tho15**] and [**Tho16**].

Take the first reductive, maximal connected subgroup M from Theorem 3.1 (the ordering is chosen to make the proof and resulting tables easier to follow; of course, one could use any ordering). We iterate the following schematic process, explaining each stage below. Find all reductive, maximal connected subgroups of M; these subgroups will be denoted by M_1 in each case. All such subgroups will clearly be M-irreducible. Now find all of the reductive, maximal connected subgroups of each M_1, in turn. At this point it is not necessarily the case that all such subgroups will be M-irreducible and we only wish to keep the irreducible ones. Continuing in this way will lead to a set containing every M-irreducible subgroup; we want to consider the G-conjugacy classes of these subgroups and so we specify when a class is repeated in our list. Moreover, for a given representative of a G-conjugacy class of irreducible subgroups we know all of its connected overgroups and thus understand the lattice structure of the irreducible connected subgroups of M.

To find the reductive, maximal connected subgroups of a subgroup of M we repeatedly use Theorem 3.1 and Lemma 3.3. To check whether a subgroup X of M is M-irreducible we first note that by Lemma 3.5, the projection of X to each simple factor of M must be irreducible. We then use Lemma 3.6 for classical factors of M; if M has a factor of exceptional type then this will have rank less than that of G and so we may use Theorem 1 inductively, since we will prove this in ascending order of rank. We also need to consider G-conjugacy rather than just M-conjugacy of all of the M-irreducible connected subgroups. There are a large number of cases where G does not fuse any M-classes together. Given two M-conjugacy classes of irreducible connected subgroups, with representatives X_1 and X_2, say, it is easy to check that X_1 is not G-conjugate to X_2 by comparing the composition factors of their actions on the adjoint module. If the composition factors are the same then we will give an argument showing either that X_1 is G-conjugate to X_2, or that we are in the case of the exception given in Corollary 1 with $G = E_8$, $p = 3$, and X_1 and X_2 are of type A_2. Many of the arguments are based on information from [**Car72**] on $\mathrm{Out}_G(H)$ for certain maximal rank subgroups H of G. We give each G-conjugacy class of M-irreducible connected subgroups a unique identification number n, and the class (or representative subgroup) is denoted by $G(\#n)$ (see Section 2).

We now consider the second reductive, maximal connected subgroup of G from Theorem 3.1. In the proofs and tables, this will also be denoted by M, but to be clear in our explanation in this paragraph we denote it by N so as to be able to differentiate between the first and second reductive, maximal connected subgroups. We use the same method as above to find the lattice structure of G-conjugacy classes of N-irreducible connected subgroups but with an important extra step. For each N-irreducible connected subgroup X we need to determine whether X is conjugate to an M-irreducible connected subgroup of M. To show that this does not happen is straightforward. Indeed, we can compare the composition factors of X on $L(G)$ with those of each M-irreducible subgroup of the same type as X. This routine but tedious check will not be explicitly mentioned. When X has the same composition factors as an M-irreducible subgroup Y then we need to prove that X is conjugate to Y, since we also wish to prove Corollary 1 (when X is simple this has already been proved in [**Tho15**, Corollary 1], [**Tho16**, Corollary 3]). Proving that X and Y are conjugate requires results from Section 3 such as Lemmas 3.9, 3.10 and some ad-hoc methods. The main task is showing that X is contained in some conjugate of M. The fact that X and Y are conjugate then follows from the fact that composition factors on $L(G)$ determine conjugacy of M-irreducible subgroups. For example, suppose G is of type E_7. If a subgroup X of N has an $\bar{A}_1$ factor generated by long root subgroups of G then X will be contained in some conjugate of $M = \bar{A}_1 D_6$ by Lemma 3.10. For a specific example, consider $N = G_2C_3$ and $X = \bar{A}_1 A_1 C_3 = E_7(\#42)$.

We repeat this process until we have considered all of the reductive, maximal connected subgroups from Theorem 3.1. The information obtained from this method is displayed in Tables 1–11.5, except for the M-irreducible subgroups which are G-reducible. We explained the notation used in these tables in Section 2, and we explain how to read the tables at the start of Section 11. In particular, when the identification number n for a subgroup X is written in italics it means that X is listed elsewhere in the table and so should be discounted if one wants exactly one conjugate of each G-irreducible connected subgroup, as in Theorem 1.

We now describe the final part of the strategy: determining whether an M-irreducible connected subgroup is G-irreducible. Let X be an M-irreducible connected subgroup of G. If X is simple then we use [**Tho15**] and [**Tho16**] to check if X is irreducible. If X contains a simple G-irreducible subgroup then X is of course itself G-irreducible. In both of these cases we will not explicitly mention this in the proofs. Now suppose that X is non-simple and contains no simple G-irreducible subgroup. Then to prove that X is G-irreducible we use Lemma 3.7 and Corollary 3.8, for which we require the composition factors for the action of X on the minimal or adjoint module. These can be found by restricting the composition factors of M to X and the composition factors for the G-irreducible ones can be found in Section 12. To apply Lemma 3.7 we also need the composition factors for the action of the Levi subgroups of G on the minimal and adjoint modules. These can be found in Section 13. In most cases, a non-simple M-irreducible connected subgroup is G-irreducible. Corollary 3 lists the connected subgroups X (simple or not) for which X is irreducible in every reductive, maximal connected overgroup, yet G-reducible. To prove that X is G-reducible requires different methods and we explain these as and when we use them.

We modify the above approach when considering diagonal subgroups. Suppose that X is an M-irreducible connected subgroup of M of the form A^nB for some $n \geq 2$ with A and B of different types. Then X has maximal diagonal subgroups of the form $A^{n-1}B$. We do not want to then consider all of their maximal subgroups including those of the form $A^{n-2}B$. Instead, we list all diagonal subgroups of X immediately and do not list their subgroups in the table. Doing this significantly reduces the size of the tables, and it does not mean that we miss any M-irreducible subgroups. It does however mean that some additional combinatorial work is required to recover the lattice of overgroups of certain diagonal subgroups. In light of the large number of diagonal subgroups, we often produce a complete set of non-conjugate classes of diagonal subgroups of an irreducible subgroup X in a supplementary table. This allows for easier reading of the tables referenced in Theorem 1.

There is another situation where we deviate slightly from the above approach, but only to alter the order in which we study the subgroups in question. Let us explain this with an example. Suppose that $M = \mathrm{SO}_8$ of type D_4. When $p \neq 2$ there is a maximal connected subgroup $\mathrm{SO}_5\mathrm{SO}_3$. However, when $p = 2$ this subgroup still exists and is still SO_8-irreducible but is now contained in SO_7. In view of this we use the same identification number n for both groups but use $n\mathrm{a}$ for the subgroup when $p \neq 2$ and $n\mathrm{b}$ for the subgroup when $p = 2$. Moreover, we wish to study the subgroups of $n\mathrm{a}$ and $n\mathrm{b}$ together. In the tables, if we first arrive at the subgroup labelled by $n\mathrm{b}$ we postpone listing its subgroups until we come to the subgroup labelled by $n\mathrm{a}$.

This generalises to any situation where a subgroup X occurs for $p \neq m$ somewhere in the lattice and elsewhere when $p = m$. In particular, we use this only if the composition factors on both the minimal and adjoint module for $n\mathrm{b}$ are the same as those for $n\mathrm{a}$ when $p = m$.

CHAPTER 5

Irreducible subgroups of G_2

In this chapter we deduce Theorem 1 when G is of type G_2. It follows immediately from Theorem 3.1 that the only non-simple G_2-irreducible connected subgroup is the maximal subgroup $A_1\tilde{A}_1$. The simple G_2-irreducible connected subgroups of rank at least 2 are given in [**Tho15**, Lemma 3.3]. The G_2-irreducible subgroups of type A_1 are given in [**Tho16**, Theorem 2], where it is also proved that $A_1 < A_2$ embedded via the representation $V_{A_1}(2)$ is conjugate to $A_1 \hookrightarrow A_1\tilde{A}_1$ via $(1,1)$ when $p \neq 2$. Combining these results we deduce the following theorem, concluding the case where G is of type G_2.

THEOREM 5.1. *Let X be a G_2-irreducible connected subgroup of G_2. Then X is conjugate to exactly one subgroup of Table 1 and each subgroup in Table 1 is G_2-irreducible. Moreover, Table 1 gives the lattice structure of the irreducible connected subgroups of G_2.*

CHAPTER 6

Irreducible subgroups of F_4

In this chapter we use the strategy described in Section 4 to prove Theorem 1 when G is of type F_4. In doing this we prove the following theorem.

THEOREM 6.1. *Let X be an F_4-irreducible connected subgroup of F_4. Then X is conjugate to exactly one subgroup of Table* 11.2 *and each subgroup in Table* 11.2 *is F_4-irreducible. Moreover, Table* 11.2 *gives the lattice structure of the irreducible connected subgroups of F_4.*

As mentioned in Section 4, throughout the proof we will use [**Tho15**, Theorem 3.4] and [**Tho16**, Theorem 3], which classify the simple F_4-irreducible connected subgroups, without reference. In addition, we will implicitly use the fact that any subgroup that contains an F_4-irreducible subgroup is itself F_4-irreducible.

By Theorem 3.1, the reductive, maximal connected subgroups of F_4 are B_4, C_4 $(p = 2)$, $\bar{A}_1C_3$ $(p \neq 2)$, A_1G_2 $(p \neq 2)$, $\bar{A}_2\tilde{A}_2$, G_2 $(p = 7)$ and A_1 $(p \geq 13)$. The maximal connected subgroup A_1 contains no proper irreducible connected subgroups and so requires no further consideration. In the following sections we consider each of the remaining reductive, maximal connected subgroups M in turn.

6.1. $M = B_4$ $(F_4(\#12))$

Here we will only treat the case $p \neq 2$; the $p = 2$ case will follow from Section 6.2 below, where we consider the subgroups of C_4, by applying an exceptional graph automorphism of F_4. The reductive, maximal connected subgroups of B_4 are given by Lemma 3.3 and the possibilities are $\bar{D}_4$, $\tilde{A}_1\bar{A}_3$, $\bar{A}_1^2B_2$, A_1^2 and A_1 $(p \geq 11)$. There are no proper non-simple M-irreducible connected subgroups contained in either A_1^2 or A_1 $(p \geq 11)$. We consider the remaining maximal connected subgroups M_1 in the following sections.

6.1.1. $M_1 = \bar{D}_4$ **$(F_4(\#13))$.** From Lemma 3.3, the reductive, maximal connected subgroups of $\bar{D}_4$ are B_3, A_1B_2 (3 conjugacy classes of each), A_2 $(p \neq 3)$ and A_1^4. We have $\mathrm{Out}_{F_4}(\bar{D}_4) \cong S_3$ by [**Car72**, Table 8] and therefore there is only one F_4-conjugacy class of maximal subgroups of type B_3 and A_1B_2. The subgroup B_3 is F_4-reducible, since it is a Levi subgroup of F_4 (recalling $p \neq 2$). The subgroup A_1B_2 is a diagonal subgroup of $\bar{A}_1^2B_2 = F_4(\#28)$ since there is only one SO_9-conjugacy class of subgroups acting as $\mathrm{SO}_3\mathrm{SO}_5$ on the natural 9-dimensional module. Therefore A_1B_2 is conjugate to $F_4(\#40^{\{0\}})$ and no further consideration is given to it here. The maximal subgroup A_2 is conjugate to $F_4(\#19)$. It remains to consider the M-irreducible connected subgroups of $\bar{A}_1^4$, all of which are diagonal. From [**Car72**, Table 8], we see $\mathrm{Out}_{F_4}(\bar{A}_1^4) \cong S_4$, acting naturally on the four $\bar{A}_1$ factors. The conjugacy classes of diagonal subgroups are hence as listed in Table 11.2 with the following exception: the subgroup $\bar{A}_1^2\tilde{A}_1 \hookrightarrow \bar{A}_1^4$ via $(1_a, 1_b, 1_c, 1_c)$ is not M-irreducible (and hence none of its subgroups is) as it is contained in the F_4-reducible subgroup B_3. This explains the ($r \neq 0$) condition on $F_4(\#33)$ and subsequent conditions on $F_4(\#34)$, $F_4(\#35)$ and $F_4(\#1)$.

6.1.2. $M_1 = \tilde{A}_1\bar{A}_3$ **($F_4(\#27)$).** The reductive, maximal connected subgroups of $\tilde{A}_1\bar{A}_3$ are of the form $\tilde{A}_1Y$ where Y is a reductive, maximal connected subgroup of $\bar{A}_3$. From Lemma 3.3 we see the possibilities for Y are B_2 and $\tilde{A}_1^2$ (the subgroups of type A_1 are generated by short root subgroups of B_4 and hence of F_4). The subgroup $\tilde{A}_1B_2$ is contained in $\bar{A}_1^2B_2$ and conjugate to $F_4(\#40^{\{0\}})$. The diagonal subgroups of $\tilde{A}_1^3$ follow from the fact that $\mathrm{Out}_{F_4}(\tilde{A}_1^3) \cong S_3$, acting naturally on the three $\tilde{A}_1$ factors. Note that the subgroup $A_1\tilde{A}_1 \hookrightarrow \tilde{A}_1^3$ via $(1_a, 1_a, 1_b)$ is M-reducible by Lemma 3.6 since it acts as $(2,0)^2 + (0,2)$ on $V_{B_4}(\lambda_1)$.

6.1.3. $M_1 = \bar{A}_1^2B_2$ **($F_4(\#28)$).** The reductive, maximal connected subgroups of $\bar{A}_1^2B_2$ are either of the form $\bar{A}_1^2Y$ where Y is a reductive, maximal connected subgroup of B_2, or a diagonal subgroup $A_1B_2 \hookrightarrow \bar{A}_1^2B_2$ via $(1_a, 1_a^{[r]}, 10)$. By Lemma 3.3, the possibilities for Y are $\bar{A}_1^2$ and A_1 ($p \geq 5$) (recalling our assumption that $p \neq 2$). The subgroup $\bar{A}_1^4$ is conjugate to $F_4(\#32)$ since there is only one class of such subgroups in B_4. The diagonal subgroups of $X = \bar{A}_1^2A_1$ ($p \geq 5$) are as in Table 11.2, noting that the normaliser of X contains an involution swapping the two $\bar{A}_1$ factors. As explained in Section 4, we do not consider the reductive, maximal connected subgroups of $A_1B_2 \hookrightarrow \bar{A}_1^2B_2$ separately.

This completes the $M = B_4$ case.

6.2. $M = C_4$ ($p = 2$) ($F_4(\#14)$)

By Lemma 3.3, the reductive, maximal connected subgroups of C_4 when $p = 2$ are B_2^2, $\bar{A}_1C_3$ and $\tilde{D}_4$ (this subgroup $\tilde{D}_4$ is generated by short root subgroups of F_4 because it is the image of $\bar{D}_4$ under the exceptional graph automorphism of F_4). The subgroup $\bar{A}_1C_3$ is a maximal subgroup of F_4 when $p \neq 2$. We therefore consider the subgroups of $\bar{A}_1C_3$ when $p = 2$ in the $\bar{A}_1C_3$ section below (as indicated by the ID number 24b in Table 11.2). The remaining maximal connected subgroups M_1 are considered in the following sections.

6.2.1. $M_1 = B_2^2$ **($F_4(\#30)$).** The only reductive, maximal connected subgroups of B_2 when $p = 2$ are $\bar{A}_1^2$ and $\tilde{A}_1^2$. Almost everything follows from this, by working out the conjugacy classes of diagonal C_4-irreducible subgroups of $\bar{A}_1^2\tilde{A}_1^2$, $\bar{A}_1^4$ and $\tilde{A}_1^4$. We note that there is only one F_4-conjugacy class of $\bar{A}_1^4$ and that it is contained in $\bar{D}_4 < B_4$ even when $p = 2$. Applying the exceptional graph automorphism of F_4 shows that $\mathrm{Out}_{F_4}(\bar{A}_1^4) \cong S_4$ implies that $\mathrm{Out}_{F_4}(\tilde{A}_1^4) \cong S_4$. The diagonal C_4-irreducible subgroups of $\tilde{A}_1^4$ now follow. The normalisers of both $\bar{A}_1^2$ and $\tilde{A}_1^2$ in B_2 contain an involution swapping the two A_1 factors and hence $\mathrm{Out}_{F_4}(\bar{A}_1^2\tilde{A}_1^2) \cong S_2 \times S_2$. Again, the diagonal C_4-irreducible subgroups now follow.

6.2.2. $M_1 = \tilde{D}_4$ **($F_4(\#15)$).** By Lemma 3.3, the reductive, maximal connected subgroups of $\tilde{D}_4$ when $p = 2$ are B_3 (3 conjugacy classes), $\tilde{A}_1^4$ and A_2. Since $\mathrm{Out}_{F_4}(\tilde{D}_4) \cong S_3$, it follows that there is only one F_4-conjugacy class of maximal subgroups of type B_3. Therefore Lemma 3.6 shows that B_3 is C_4-reducible since one of the $\tilde{D}_4$ classes acts as $000|100|000$ on $V_{\tilde{D}_4}(\lambda_1)$. The subgroup A_2 is conjugate to $F_4(\#18^{\{1,0\}})$. Finally, consider $\tilde{A}_1^4$. This subgroup is contained in B_2^2 because there is only one C_4-conjugacy class of subgroups of type A_1^4 acting as $(1,1,0,0)+(0,0,1,1)$ on $V_{C_4}(\lambda_1)$. Therefore $\tilde{A}_1^4$ is conjugate to $F_4(\#51)$.

This completes the C_4 $(p = 2)$ case.

6.3. $M = \bar{A}_1 C_3$ $(F_4(\#24))$

The subgroup $\bar{A}_1 C_3$ is maximal when $p \neq 2$ and contained in C_4 when $p = 2$. We consider both cases in this section. By Lemma 3.3, the reductive, maximal connected subgroups of C_3 are $\bar{A}_1 B_2$, $A_1 A_1$ $(p \neq 2)$, $\tilde{A}_3$ $(p = 2)$, G_2 $(p = 2)$ and A_1 $(p \geq 7)$. This yields the reductive, maximal connected subgroups of $\bar{A}_1 C_3$, as listed in Table 11.2.

By [**LS94**, p.333, Table 2], we have $C_{F_4}(B_2) = \bar{A}_1^2$ $(p \neq 2)$ and B_2 $(p = 2)$. Thus, the subgroup $\bar{A}_1^2 B_2$ has already been considered in Sections 6.1.3 and 6.2.1, respectively and in particular, is conjugate to $F_4(\#28)$. The only thing noteworthy about the diagonal subgroups of $\bar{A}_1 A_1 A_1$ is that $X = \bar{A}_1 A_1 \hookrightarrow \bar{A}_1 A_1 A_1$ via $(1_a, 1_b, 1_b)$ is contained in $\bar{A}_1^2 B_2$ when $p \geq 5$ since $2 \otimes 1 = 3 + 1$ and it follows that X is conjugate to $F_4(\#42^{\{0\}})$. When $p = 3$, we have $2 \otimes 1 = T(3) = 1|3|1$ and hence X is M-reducible by Lemma 3.6.

The only reductive, maximal connected subgroup of $\tilde{A}_3$ when $p = 2$ is B_2, which is C_3-reducible by Lemma 3.6. Therefore $\bar{A}_1 \tilde{A}_3$ contains no proper F_4-irreducible subgroups. However, since it has maximal rank it is clearly F_4-irreducible.

Finally, we consider the subgroups contained in $\bar{A}_1 G_2$ $(p = 2)$. The G_2-irreducible subgroups are given by Theorem 5.1. The maximal connected subgroup $A_2 < G_2$ is C_3-reducible by Lemma 3.6 and so $\bar{A}_1 \tilde{A}_2$ is M-reducible. The subgroup $\bar{A}_1 \tilde{A}_1 A_1 < \bar{A}_1 G_2$ is also contained in $\bar{A}_1^2 \tilde{A}_1^2 < \bar{A}_1^2 B_2 < C_4$ and conjugate to $F_4(\#44^{\{0,1\}})$. Indeed, the subgroup $\tilde{A}_1 A_1 < G_2 < C_3$ is conjugate to $\tilde{A}_1 A_1 \hookrightarrow \tilde{A}_1^2 \bar{A}_1 < C_3$ via $(1_a, 1_b, 1_b^{[1]})$ since both act as $(1,1)+(0,2)$ on $V_{C_3}(\lambda_1)$.

6.4. $M = A_1 G_2$ $(p \neq 2)$ $(F_4(\#25))$

The lattice structure of M-irreducible connected subgroups of M follows from Theorem 5.1. The G_2 factor of M is contained in $\bar{D}_4$ as seen from the construction in [**Sei91**, 3.9] and therefore subgroups generated by long root subgroups of G_2 are generated by long root subgroups of F_4. It follows that $A_1 \bar{A}_1 A_1$ is conjugate to $F_4(\#57)$, since $C_{F_4}(\bar{A}_1)^\circ = C_3$ by Lemma 3.10.

6.5. $M = \bar{A}_2 \tilde{A}_2$ $(F_4(\#26))$

By Lemma 3.3, the only reductive, maximal connected subgroup of A_2 is A_1 $(p \neq 2)$. The reductive, maximal connected subgroups of M are thus as in Table 11.2. We note that $\bar{A}_2 A_1$ $(p \neq 2)$ is conjugate to $F_4(\#64) < A_1 G_2$, since the $\bar{A}_2 < G_2$ is generated by long root subgroups of F_4 and $C_{F_4}(\bar{A}_2) = \tilde{A}_2$ by Lemma 3.10. It subsequently follows that $A_1 A_1 < \bar{A}_2 \tilde{A}_2$ is conjugate to $F_4(\#62^{\{0,0\}})$ since $A_1 A_1$ is a subgroup of $\bar{A}_2 A_1$.

6.6. $M = G_2$ $(p = 7)$ $(F_4(\#16))$

The lattice structure of M-irreducible subgroups is given by Theorem 5.1. We show that all reductive, maximal connected subgroups of M are conjugate to subgroups already considered. For the maximal connected subgroups A_2 and A_1 this follows from the proofs of [**Tho15**, Theorem 3.4] and [**Tho16**, Theorem 3]. Now consider $X = A_1A_1$; we claim that X is conjugate to $Y = F_4(\#61^{\{0,0\}}) < \bar{A}_1C_3$. To do this it suffices to show that A_1A_1 is contained in $\bar{A}_1C_3$ as comparing composition factors on V_{26} shows that X is then conjugate to Y. We know by Lemma 3.9 that X is the centraliser in G_2 of a semisimple element of order 2, call it t. Again by Lemma 3.9, we know that the centraliser in F_4 of t is B_4 or $\bar{A}_1C_3$ and by [**LS99**, Proposition 1.2] the trace on V_{26} is -6 or 2, respectively. We calculate the trace of t on V_{26} from the restriction $V_{26} \downarrow X = (2,2) + (1,1) + (1,3) + (0,4)$, noting that the element t can be seen as minus the identity in both A_1 factors and hence has trace 2 on V_{26}. Therefore X is contained in $\bar{A}_1C_3$, proving the claim.

This completes the proof of Theorem 6.1.

CHAPTER 7

Irreducible subgroups of $G = E_6$

In this chapter we use the strategy described in Section 4 to prove the following theorem.

THEOREM 7.1. *Let X be an E_6-irreducible connected subgroup of E_6. Then X is Aut(E_6)-conjugate to exactly one subgroup of Table* 11.3 *and each subgroup in Table* 11.3 *is E_6-irreducible. Moreover, Table* 11.3 *gives the lattice structure of the irreducible connected subgroups of E_6.*

Recall that throughout the proof we will use [**Tho15**, Theorem 1] and [**Tho16**, Theorem 4], which classify the simple E_6-irreducible connected subgroups, without reference.

By Theorem 3.1, the reductive, maximal connected subgroups of E_6 are $\bar{A}_1A_5$, $\bar{A}_2^3$, A_2G_2, F_4, C_4 $(p \neq 2)$, G_2 $(p \neq 7)$ and A_2 $(p \geq 5)$. The only irreducible connected subgroup contained in A_2 is A_1 $(p \neq 2)$. Since this is a simple subgroup, the maximal connected subgroup A_2 $(p \geq 5)$ requires no further consideration. In the following sections we consider each of the remaining reductive, maximal connected subgroups M in turn.

7.1. $M = \bar{A}_1A_5$ $(E_6(\#24))$

The reductive, maximal connected subgroups of A_5 are C_3, A_1A_2, A_3 $(p \neq 2)$ and A_2 $(p \neq 2)$ by Lemma 3.3. Therefore the reductive, maximal connected subgroups of M are as in Table 11.3, and we consider them in the following sections.

7.1.1. $M_1 = \bar{A}_1C_3$ $(E_6(\#27))$**.** Using Lemma 3.3, we find that the reductive, maximal connected subgroups of $\bar{A}_1C_3$ are $\bar{A}_1^2C_2$, $\bar{A}_1A_1A_1$ $(p \neq 2)$, $\bar{A}_1A_1$ $(p \geq 7)$, $\bar{A}_1A_3$ $(p = 2)$ and $\bar{A}_1G_2$ $(p = 2)$. Since the subgroup $\bar{A}_1C_2 < C_3$ acts reducibly on $V_{C_3}(100)$, it follows that $\bar{A}_1^2C_2$ is M-reducible.

The diagonal subgroups of $\bar{A}_1A_1A_1$ $(p \neq 2)$ and $\bar{A}_1A_1$ $(p \geq 7)$ are easily seen to be as in Table 11.3, noting that $\bar{A}_1A_1 \hookrightarrow \bar{A}_1A_1A_1$ via $(1_a, 1_b, 1_b)$ is M-reducible by Lemma 3.6. Neither $X = \bar{A}_1A_3$ $(p = 2)$ nor $Y = \bar{A}_1G_2$ $(p = 2)$ has a proper M-irreducible connected subgroup, by Lemma 3.6.

It remains to prove that X and Y are E_6-irreducible. Suppose that X is E_6-reducible. Then by Lemma 3.7, there exists a subgroup Z of type A_1A_3 contained L-irreducibly in a Levi subgroup L, such that X and Z have the same composition factors on V_{27}. It follows that L' has type D_5 or A_1A_3. From Table 12.3, the dimensions of the X-composition factors on V_{27} are $14, 12, 1$, whereas Table 13.3 shows that D_5 and A_1A_3 have composition factors of dimensions $16, 10, 1$ and $8, 6, 4^2, 2^2, 1$, respectively. Thus X and Z do not have the same composition factors on V_{27}, a contradiction. Hence X is E_6-irreducible.

The same argument applies to Y since the composition factors of $V_{27} \downarrow Y$ also have dimensions $14, 12, 1$ and only a Levi subgroup of type D_5 contains an irreducible subgroup of type A_1G_2 when $p = 2$.

7.1.2. $M_1 = \bar{A}_1A_1A_2$ ($E_6(\#28)$)**.** The only reductive, maximal connected subgroup of A_2 is A_1 ($p \neq 2$). Therefore the lattice structure of M_1-irreducible subgroups is as given in Table 11.3 along with the subgroup $Y = A_1A_2 \hookrightarrow \bar{A}_1A_1A_2$ via $(1, 1, 10)$ when $p = 2$. There is only one A_5-conjugacy class of subgroups A_1A_1 acting as $(2, 1)$ on $V_{A_5}(\lambda_1)$ and so $\bar{A}_1A_1A_1$ is contained in $\bar{A}_1C_3$ and thus conjugate to $E_6(\#31)$.

It remains to prove that when $p = 2$ the subgroup $X = A_1A_2 \hookrightarrow \bar{A}_1A_1A_2$ via $(1^{[r]}, 1^{[s]}, 10)$ $(rs = 0; r \neq s)$ is E_6-irreducible and that Y, defined as above, is E_6-reducible. To prove that X is E_6-irreducible we use Lemma 3.7. Suppose that Z is an L-irreducible subgroup A_1A_2 of a Levi subgroup L, such that X and Z have the same composition factors on V_{27}. Since L' contains an irreducible subgroup A_1A_2 it follows from Lemma 3.6 that L' is of type A_5, $\bar{A}_1\bar{A}_2^2$, $\bar{A}_1^2A_2$ or $\bar{A}_1A_2$. The X-composition factors of V_{27} are $(1^{[r]} \otimes 1^{[s]}, 10)/(2^{[s]}, 10)/(0, 02)/(0, 10)^2$. Now it follows that Z is not contained in A_5 since $V_{27} \downarrow A_5 = \lambda_1^2/\lambda_4$. Similarly, since Z has a composition factor of dimension 12, we find that Z is not a subgroup of $\bar{A}_1\bar{A}_2^2$, $\bar{A}_1^2A_2$ or $\bar{A}_1A_2$. This is a contradiction and thus X is E_6-irreducible.

Now consider Y. It is shown in Section 7.3 below that M_1 is conjugate to $A_2\bar{A}_1A_1 < A_2G_2$. When $p = 2$, the subgroup $A_1 \hookrightarrow \bar{A}_1A_1 < G_2$ via $(1, 1)$ is G_2-reducible by Theorem 5.1. Therefore Y is A_2G_2-reducible and thus E_6-reducible.

7.1.3. $M_1 = \bar{A}_1A_3$ ($p \neq 2$) ($E_6(\#29\text{a})$) **and** $\bar{A}_1A_2$ ($p \neq 2$) ($E_6(\#30)$)**.** There are no proper M-irreducible connected subgroups of either $\bar{A}_1A_3$ ($p \neq 2$) or $\bar{A}_1A_2$ ($p \neq 2$) since any proper connected subgroup of A_3 or A_2 is A_5-reducible by Lemma 3.6. It remains to prove that they are both E_6-irreducible. The dimensions of the composition factors on V_{27} are $15, 12$ for both subgroups, as seen from Table 11.3. An easy application of Lemma 3.7, as in the previous sections, shows that both are E_6-irreducible.

7.2. $M = \bar{A}_2^3$ ($E_6(\#25)$)

The only reductive, maximal connected subgroup of A_2 is A_1 ($p \neq 2$). We have $\mathrm{Out}_{\mathrm{Aut}(E_6)}(\bar{A}_2^3) \cong S_3 \times S_2$ where the central involution acts as a graph automorphism on each A_2 factor; a two-cycle in the S_3 direct factor swaps two of the A_2 factors whilst also inducing a graph automorphism on them; a three-cycle in the S_3 direct factor acts naturally as a three-cycle on the three A_2 factors. The lattice structure of the M-irreducible subgroups then follows. It remains for us to show that $X = A_1A_2 \hookrightarrow A_1\bar{A}_2^2$ via $(1, 10, 01)$ is conjugate to $Y = E_6(\#37^{\{0,0\}})$ when $p \neq 2$. Let $Z = E_6(\#28) = \bar{A}_1A_1A_2$ so $Y \hookrightarrow Z$ via $(1, 1, 10)$. It suffices to prove that Y is contained in $\bar{A}_2^3$. It is shown in Section 7.3 below that Z is conjugate to $A_2\bar{A}_1A_1 < A_2G_2$. Therefore Y is conjugate to $A_2A_1 \hookrightarrow A_2\bar{A}_1A_1 < A_2G_2$ via $(10, 1, 1)$. By Theorem 5.1 this implies that Y is also conjugate to $A_2A_1 < A_2\bar{A}_2 < A_2G_2$. Using Lemma 3.10, we see that $A_2\bar{A}_2$ is contained in $\bar{A}_2^3$, and hence Y is contained in $\bar{A}_2^3$, as required.

7.3. $M = A_2G_2$ ($E_6(\#26)$)

By Lemma 3.3, the only proper irreducible connected subgroup of A_2 is A_1 ($p \neq 2$); the G_2-irreducible connected subgroups are given by Theorem 5.1. Using this we find that the lattice structure of M-irreducible connected subgroups is as given in Table 11.3, noting that the G_2 factor is contained in a Levi subgroup D_4 and hence the long root subgroups of G_2 are long root subgroups of E_6.

Since $C_{E_6}(\bar{A}_2) = \bar{A}_2^2$ by Lemma 3.10, the subgroup $A_2\bar{A}_2$ is contained in $\bar{A}_2^3$ and comparing composition factors shows that it is conjugate to $E_6(\#40^{\{0\}})$. Similarly, the subgroup $A_2\bar{A}_1A_1$ is conjugate to $E_6(\#28) < \bar{A}_1A_5$. All other conjugacies follow from these two facts.

7.4. $M = F_4$ ($E_6(\#7)$)

Theorem 6.1 gives the lattice structure of the F_4-irreducible subgroups. The maximal subgroup F_4 of E_6 is the centraliser of a standard graph automorphism of E_6 by [**Sei91**, Theorem 15.1]. Therefore the maximal subgroup B_4 of F_4 is contained in a D_5 Levi subgroup. It also follows that the maximal connected subgroup $\bar{A}_2A_2$ is conjugate to $E_6(\#39^{\{0\}})$.

The subgroup C_4 is a maximal subgroup of E_6 when $p \neq 2$; we therefore consider the subgroups of the maximal subgroup C_4 ($p = 2$) in the next section. Similarly, the subgroup G_2 is a maximal subgroup of E_6 when $p \neq 7$; we therefore consider the subgroups of the maximal subgroup G_2 ($p = 7$) in Section 7.6.

The maximal subgroup $\bar{A}_1C_3$ ($p \neq 2$) is contained in $\bar{A}_1A_5$ since $C_{E_6}(\bar{A}_1)^\circ = A_5$ by Lemma 3.10 and hence $\bar{A}_1C_3$ is conjugate to $E_6(\#27)$. Finally, we consider the maximal subgroup $X = A_1G_2$ ($p \neq 2$). By [**LS94**, p.333, Table 3], we have $C_{E_6}(G_2)^\circ = A_2$ and thus $X < A_2G_2$. It follows that X is conjugate to $E_6(\#47)$.

7.5. $M = C_4$ ($E_6(\#8)$)

The subgroup C_4 is maximal when $p \neq 2$ and contained in F_4 when $p = 2$. We consider both cases in this section. Lemma 3.3 yields the reductive, maximal connected subgroups of C_4. They are B_2^2, $\bar{A}_1C_3$, A_1^3 ($p \neq 2$), A_1 ($p \geq 11$) and D_4 ($p = 2$). The subgroup B_2^2 is E_6-reducible by Lemma 3.4 since $C_{E_6}(B_2)^\circ = B_2T_1$ by [**LS94**, p.333, Table 2]. The subgroup $\bar{A}_1C_3$ is conjugate to $E_6(\#27)$, by the argument given in the previous section.

Next, we prove that the maximal connected subgroup $X = A_1^3$ ($p \neq 2$) is conjugate to $Y = E_6(\#44) < \bar{A}_2^3$. It suffices to show that Y is contained in C_4. Consider the standard graph automorphism of E_6, call it τ and let w_0 be the longest word of the Weyl group. Then $w_0 = -\tau$ and so $t := \tau w_0$ acts by inversion on a maximal torus of E_6. Therefore t induces a graph automorphism on each A_2 factor of $\bar{A}_2^3$. Thus $Y < C_{E_6}(t)^\circ$ since the irreducible subgroup A_1 in a subgroup A_2 is centralised by the graph automorphism of A_2. Finally, we check that $\dim(C_{L(E_6)}(t)) = 36$ and so $\dim(C_{E_6}(t)) = 36$ (by [**Bor91**, 9.1], since t is semisimple). Therefore, $C_{E_6}(t)^\circ = C_4$ by [**GLS98**, Table 4.3.1].

Finally, we consider the subgroups of the maximal subgroup D_4 ($p = 2$). In this case M is contained in F_4 since $p = 2$ and so we need only consider the F_4-irreducible subgroups contained in D_4, which are A_2 and $\bar{A}_1^4$. Moreover, Theorem 6.1 shows that A_2 is a subgroup of $\bar{A}_2A_2 < F_4$ and hence conjugate to $E_6(\#15^{\{1\}})$. It also shows that $\bar{A}_1^4$ is a subgroup of B_2^2 and thus E_6-reducible.

7.6. $M = G_2$ ($E_6(\#10)$)

The subgroup G_2 is maximal when $p \neq 7$ and contained in F_4 when $p = 7$. In this section we consider both cases. Theorem 5.1 gives the lattice structure of M-irreducible connected subgroups. We need only consider the maximal connected subgroup $X = A_1A_1$. We claim that if $p \neq 2$ then X is conjugate to $E_6(\#34^{\{0,0\}})$ but when $p = 2$ it is E_6-reducible.

First suppose that $p \neq 2$. It suffices to prove that X is contained in $\bar{A}_1A_5$. By Lemma 3.9, X is the centraliser in G_2 of a semisimple element of order 2, call this t. Moreover, since X contains a simple E_6-irreducible subgroup of type A_1 it is E_6-irreducible. Therefore, the centraliser in E_6 of t is $\bar{A}_1A_5$ by Lemma 3.9.

Now let $p = 2$. To prove that X is E_6-reducible we consider the action of X on $L(E_6)$. By [**LS04**, Table 10.1], we have $L(E_6) \downarrow G_2 = 11 + 01$. In Table 12.1, the composition factors of $V_{G_2}(01) \downarrow X$ are given and moreover, $V_{G_2}(01) \downarrow X = ((0,0)|((2,0)+(0,2))|(0,0)) + (1,3)$. Therefore, X fixes a non-trivial vector of $L(E_6)$. By [**Sei91**, Lemma 1.3], it follows that X is contained in either a parabolic subgroup, or $\bar{A}_1A_5$ or $\bar{A}_2^3$. The $\bar{A}_1A_5$ and $\bar{A}_2^3$ sections show that neither $\bar{A}_1A_5$ nor $\bar{A}_2^3$ contains an E_6-irreducible subgroup A_1A_1 when $p = 2$. Therefore X is E_6-reducible, as claimed.

This completes the proof of Theorem 7.1.

CHAPTER 8

Irreducible subgroups of $G = E_7$

In this chapter we prove Theorem 1 when G is of type E_7 by proving the following theorem.

THEOREM 8.1. *Let X be an E_7-irreducible connected subgroup of E_7. Then X is conjugate to exactly one subgroup of Table* 11.4 *and each subgroup in Table* 11.4 *is E_7-irreducible. Moreover, Table* 11.4 *gives the lattice structure of the irreducible connected subgroups of E_7.*

As in the previous sections we consider each of the reductive, maximal connected subgroups of E_7 in turn. The simple E_7-irreducible connected subgroups are classified in [**Tho15**, Theorem 2] and [**Tho16**, Theorem 5] and we will use these without reference throughout. By Theorem 3.1, the reductive, maximal connected subgroups of E_7 are $\bar{A}_1 D_6$, $\bar{A}_2 A_5$, A_7, $G_2 C_3$, $A_1 F_4$, $A_1 G_2$ $(p \neq 2)$, $A_1 A_1$ $(p \geq 5)$, A_2 $(p \geq 5)$ and A_1 (2 classes, $p \geq 17, 19$).

There are no proper non-simple irreducible subgroups of $A_1 A_1$ $(p \geq 5)$, A_2 $(p \geq 5)$, A_1 $(p \geq 17)$ or A_1 $(p \geq 19)$ so these require no examination. We treat the remaining cases in the following sections.

8.1. $M = \bar{A}_1 D_6$ $(E_7(\#30))$

Using Lemma 3.3 we find that the reductive, maximal connected subgroups of M are $\bar{A}_1^3 \bar{D}_4$, $\bar{A}_1 A_1 B_4$ $(p \neq 2)$, $\bar{A}_1 B_2 B_3$ $(p \neq 2)$, $\bar{A}_1 \bar{A}_3^2$, $\bar{A}_1 B_5$ and $\bar{A}_1 A_1 C_3$ (2 classes). We consider each of these in the following sections.

8.1.1. $M_1 = \bar{A}_1^3 \bar{D}_4$ $(E_7(\#36))$**.** We first note that $\mathrm{Out}_{E_7}(M_1) \cong S_3$ where the S_3 acts simultaneously as the outer automorphism group of both $\bar{A}_1^3$ and $\bar{D}_4$. Therefore, by Lemma 3.3, the E_7-conjugacy classes of reductive, maximal connected subgroups of M_1 are $\bar{A}_1^7$, $\bar{A}_1^3 B_3$, $\bar{A}_1^3 A_1 B_2$ $(p \neq 2)$, $\bar{A}_1^3 A_2$ $(p \neq 3)$ and maximal diagonal subgroups. As explained in Section 4, we do not explicitly consider the subgroup lattice of the maximal diagonal subgroups and we write down all classes of diagonal subgroups of M_1 in the same place as the reductive, maximal connected subgroups in Table 11.4.

All M_1-irreducible subgroups of $\bar{A}_1^7$ are diagonal and are given in Table 11.6. They follow in the same way that the diagonal subgroups of type A_1 are found in the proof of [**Tho16**, Theorem 5]. Indeed, we have chosen the same isomorphism of $\mathrm{PSL}(2,7) \cong \mathrm{Out}_{E_7}(\bar{A}_1^7)$ with a subgroup of S_7 acting on the seven factors, namely that the generators are mapped to $(1,2,3)(5,6,7)$ and $(2,4)(3,5)$.

By Lemmas 3.3 and 3.6, the M-irreducible, reductive, maximal connected subgroups of $\bar{A}_1^3 B_3$ are $\bar{A}_1^5 A_1$ $(p \neq 2)$, $\bar{A}_1^3 G_2$, $\bar{A}_1^3 A_1 B_2$ $(p = 2)$ and maximal diagonal subgroups. The diagonal subgroups follow from noting that $\mathrm{Out}_{E_7}(\bar{A}_1^3 B_3) \cong S_2$ where the involution swaps the second and third $\bar{A}_1$ factors. The subgroup $\bar{A}_1^3 A_1 B_2$

is a maximal connected subgroup of $\bar{A}_1^3\bar{D}_4$ when $p \neq 2$ and we will consider it below (in the notation of Table 11.4, the subgroup $\bar{A}_1^3A_1B_2$ is $E_7(\#45\text{b})$). The subgroup $\bar{A}_1^5A_1$ is contained in $\bar{A}_1^7$ since the sixth A_1 factor is a subgroup of $\bar{A}_1^2$ diagonally embedded via $(1,1)$.

Now consider $X = \bar{A}_1^3G_2$. The diagonal M-irreducible subgroups of X follow from Lemma 3.6 and the fact that $\text{Out}_{E_7}(X) \cong S_3$ acting naturally on the three $\bar{A}_1$ factors. Theorem 5.1 yields the remaining X-irreducible connected subgroups. By Lemma 3.6, the subgroup $\bar{A}_1^3\bar{A}_2$ is M-reducible. We also see that the subgroup $\bar{A}_1A_1 < G_2 < \bar{D}_4$ is a diagonal subgroup of $\bar{A}_1^4 < \bar{D}_4$ via $(1_a, 1_b, 1_b, 1_b)$ and hence $\bar{A}_1^4A_1$ is a subgroup of $\bar{A}_1^7$. The subgroup $\bar{A}_1^3A_2$ $(p = 3)$ is a maximal connected subgroup of $\bar{A}_1^3\bar{D}_4$ when $p \neq 3$ and we will consider the subgroups for all p below. Finally, the diagonal subgroups of $\bar{A}_1^3A_1$ $(p \geq 7)$ follow in the same way as the diagonal subgroups of X.

Next, we consider the subgroup $X = \bar{A}_1^3A_1B_2$ which is a maximal connected subgroup of M_1 when $p \neq 2$ and contained in $\bar{A}_1^3B_3$ when $p = 2$. The diagonal subgroups follow from the fact that $\text{Out}_{E_7}(X) \cong S_2$ where the involution swaps the second and third $\bar{A}_1$ factors. Using Lemma 3.3, we see the reductive, maximal connected subgroups of X are $\bar{A}_1^5A_1$, $\bar{A}_1^3A_1^3$ $(p = 2)$ and $\bar{A}_1^3A_1A_1$ $(p \geq 5)$. The first subgroup is contained in $\bar{A}_1^7$ and conjugate to $E_7(\#49^{\{0\}})$. The diagonal subgroups of $Y_1 = \bar{A}_1^3A_1^3$ are found in Table 11.13 and are deduced from the fact that $\text{Out}_{E_7}(Y_1) \cong S_2 \times S_3$, where the S_3 direct factor acts naturally on A_1^3 and the central involution swaps the second and third $\bar{A}_1$ factors. Similarly, the diagonal subgroups of $Y_2 = \bar{A}_1^3A_1A_1$ can be found in Table 11.12. Here they follow from the fact that $\text{Out}_{E_7}(Y_2) \cong S_2$, where the involution swaps the second and third $\bar{A}_1$ factors. However, the normaliser of $\bar{A}_1^3A_1 \hookrightarrow Y_2$ via $(1_a, 1_b, 1_c, 1_d, 1_d)$ contains an S_3 acting naturally on the three $\bar{A}_1$ subgroups. Indeed, since a subgroup A_1 acting as $4+2$ on $V_{\bar{D}_4}(\lambda_1)$ is contained in both a $\bar{D}_4$-irreducible A_2 and A_1B_2 it is centralised by both a triality and involutory automorphism of $\bar{D}_4$.

Finally, consider $\bar{A}_1^3A_2$ which is a maximal connected subgroup of M_1 when $p \neq 3$ and contained in $\bar{A}_1^3G_2$ when $p = 3$. Other than diagonal subgroups, the only reductive, maximal connected subgroup is $\bar{A}_1^3A_1$ $(p \neq 2)$. When $p \geq 5$ this is a subgroup of $\bar{A}_1^3A_1A_1 = E_7(\#87)$ and when $p = 3$ the maximal subgroup A_1 of $A_2 < G_2$ is also contained in $\bar{A}_1A_1 < G_2$ and hence $\bar{A}_1^3A_1$ is a subgroup of $\bar{A}_1^7$.

It remains to prove that when $p = 2$ the diagonal subgroups $E_7(\#184)$ are E_7-irreducible. Let $X = A_1A_2 \hookrightarrow \bar{A}_1^3A_2 = E_7(\#46)$ via $(1, 1^{[r]}, 1^{[s]}, 10)$ $(0 < r < s)$ and suppose that X is E_7-reducible. From Table 12.4, we have $V_{56} \downarrow \bar{A}_1^3A_2 = (1,1,1,00)/(1,0,0,11)/(0,1,0,11)/(0,0,1,11)$. Therefore, X has three 16-dimensional composition factors on V_{56}. By Lemma 3.7, there exists a subgroup Z of type A_1A_2 contained L-irreducibly in a Levi subgroup L, such that X and Z have the same composition factors on V_{56}. Since Z has three 16-dimensional composition factors on V_{56}, it follows from Table 13.4 that L' is of type A_1D_5. Using Lemma 3.6, we see that $L' = A_1D_5$ does not contain an L'-irreducible subgroup A_1A_2, a contradiction. Hence X is E_7-irreducible.

8.1.2. $M_1 = \bar{A}_1A_1B_4$ **$(E_7(\#37))$.** In this section we consider $\bar{A}_1A_1B_4$ both when it is a maximal connected subgroup of M $(p \neq 2)$ and when it is contained in $\bar{A}_1B_5$ $(p = 2)$. By Lemma 3.3, the reductive, maximal connected subgroups of $\bar{A}_1A_1B_4$ are those listed in Table 11.4. The subgroups $\bar{A}_1A_1\bar{D}_4$ and $\bar{A}_1^3A_1B_2$ $(p \neq 2)$ are both contained in $\bar{A}_1^3\bar{D}_4$, as seen by considering their action on V_M.

We now consider $\bar{A}_1 A_1^2 \bar{A}_3$, which is maximal connected when $p \neq 2$ and contained in $\bar{A}_1 A_1^2 B_3$ when $p = 2$. The reductive, maximal connected subgroups of $\bar{A}_3$ are B_2 and A_1^2 ($p \neq 2$). However, when $p = 2$ the subgroup $\bar{A}_1 A_1^2 B_2$ is M-reducible by Lemma 3.6. When $p \neq 2$ the subgroup $\bar{A}_1 A_1^2 B_2$ is contained in $\bar{A}_1^3 \bar{D}_4$ and thus conjugate to $E_7(\#91^{\{0\}})$. The diagonal subgroups of $Y = \bar{A}_1 A_1^4$ follow from $\mathrm{Out}_{E_7}(Y) \cong S_4$, acting naturally on the last four factors. The diagonal M-irreducible subgroups of $\bar{A}_1 A_1^2 \bar{A}_3$, as listed in Table 12.4, follow from noting that there is an involution in the Weyl group of D_6 swapping the second and third A_1 factors, and the fact that the field twists corresponding to the second and third factors must be distinct for the subgroup to be M-irreducible.

Now consider $X = \bar{A}_1 A_1 A_1^2$ ($p \neq 2$). Since $N_{B_4}(A_1^2)$ contains an involution swapping the two A_1 factors we have $\mathrm{Out}_{E_7}(X) \cong S_2$. Also, the subgroup $A_1 \hookrightarrow A_1^2 < B_4$ via $(1,1)$ is contained in $\bar{D}_4$ when $p \geq 5$ and is B_4-reducible when $p = 3$. Indeed, this follows from Lemma 3.6, since $2 \otimes 2 = 4 + 2 + 0$ when $p \geq 5$ and $2 \otimes 2 = (0|4|0) + 2$ when $p = 3$. The non-simple diagonal subgroups of X follow from this and are found in Table 11.15.

Now consider $X = \bar{A}_1 A_1 B_2^2$ ($p = 2$). The reductive, maximal connected subgroups of B_2 when $p = 2$ are $\bar{A}_1^2$ and A_1^2. The maximal connected subgroup $\bar{A}_1^3 A_1 B_2$ is contained in $\bar{A}_1^3 \bar{D}_4$ and the diagonal subgroups of X follow easily, noting that there is an involution in $N_{B_4}(B_2^2)$ swapping the B_2 factors and that $\bar{A}_1 A_1 B_2 \hookrightarrow X$ via $(1,1,10,10)$ is $\bar{A}_1 D_6$-reducible by Lemma 3.6. That leaves us to consider the subgroups of $Y = \bar{A}_1 A_1^3 B_2$.

The diagonal subgroups of Y follow as usual, noting that $\mathrm{Out}_M(Y) \cong S_3$, acting naturally on the three conjugate A_1 factors. The non-diagonal reductive, maximal connected subgroups of Y are $\bar{A}_1^3 A_1^3$ and $\bar{A}_1 A_1^5$. The first subgroup is contained in $\bar{A}_1^3 \bar{D}_4$ and has already been considered. Let $Z = \bar{A}_1 A_1^5$. Then $\mathrm{Out}_M(Z) \cong S_5$ acting naturally on the last five A_1 factors, and the diagonal M-irreducible subgroups follow using Lemma 3.6.

Finally, it remains to consider $X = \bar{A}_1 A_1^2 B_3$ ($p = 2$). The reductive, maximal connected subgroups of B_3 when $p = 2$ are $A_1 B_2$, $\bar{A}_3$ and G_2. The maximal subgroup $\bar{A}_1 A_1^3 B_2$ is contained in $\bar{A}_1 A_1 B_2^2$, as may be seen by considering its action on V_M and thus has already been considered. The diagonal irreducible subgroups of X are as given in Table 11.4, noting that there is an involution in the Weyl group of D_6 swapping the second and third A_1 factors. The maximal subgroup $\bar{A}_1 A_1^2 \bar{A}_3$ was considered above and we are left to consider $Y = \bar{A}_1 A_1^2 G_2$.

The reductive, maximal connected subgroups of G_2 when $p = 2$ are A_2 and $\bar{A}_1 A_1$, by Theorem 5.1. The subgroup A_2 is B_3-reducible by Lemma 3.6 and hence $\bar{A}_1 A_1^2 A_2$ is M-reducible. The subgroup $\bar{A}_1 A_1$ is contained in $A_1 \bar{A}_1^2 < A_1 B_2 < B_3$ and it follows that $\bar{A}_1^2 A_1^2 A_1$ is contained in $\bar{A}_1^3 \bar{D}_4$ and conjugate to $E_7(\#134^{\{0,0\}})$. Finally, the diagonal irreducible subgroups of Y follow in the same way as those of X.

8.1.3. $M_1 = \bar{A}_1 B_2 B_3$ **(**$E_7(\#38)$**).** The subgroup $\bar{A}_1 B_2 B_3$ is maximal when $p \neq 2$ and contained in $\bar{A}_1 B_5$ when $p = 2$. We consider both cases in this section. By Lemma 3.3, the reductive, maximal connected subgroups of $\bar{A}_1 B_2 B_3$ are those listed in Table 11.4. The subgroups $\bar{A}_1^3 B_3$ and $\bar{A}_1^3 A_1 B_2$ ($p \neq 2$) are both contained in $\bar{A}_1^3 \bar{D}_4$, as seen by considering their action on V_M. Similarly, the subgroups $\bar{A}_1 A_1^2 B_3$ ($p = 2$) and $\bar{A}_1 A_1 B_2^2$ ($p = 2$) are both contained in $\bar{A}_1 A_1 B_4$. We now consider the three remaining reductive, maximal connected subgroups.

First, let $X = \bar{A}_1A_1B_3$ ($p \geq 5$). The reductive, maximal connected subgroups again follow from Lemma 3.3. The subgroups that require further comment are $Y_1 = \bar{A}_1A_1\bar{A}_3$ and $Y_2 = \bar{A}_1A_1G_2$. First note that the diagonal subgroups of Y_1 and Y_2 follow in the same way as the diagonal subgroups of X.

The reductive, maximal connected subgroups of $\bar{A}_3$ when $p \geq 5$ are B_2 and A_1^2 by Lemma 3.3. The subgroup $\bar{A}_1A_1B_2$ of Y_1 is M-reducible by Lemma 3.6. The subgroup $\bar{A}_1A_1A_1^2$ of Y_1 is contained in $\bar{A}_1^3\bar{D}_4$ and conjugate to $E_7(\#102^{\{0\}})$.

Now consider Y_2. The G_2-irreducible subgroups are given by Theorem 5.1. The subgroup $\bar{A}_2 < G_2$ is B_3-reducible by Lemma 3.6. The subgroup $\bar{A}_1A_1$ is contained in $\bar{A}_1^2A_1 < B_3$ and therefore $\bar{A}_1^2A_1A_1$ is contained in $\bar{A}_1^3\bar{D}_4$ and conjugate to $E_7(\#103^{\{0,0\}})$. The diagonal subgroups of $\bar{A}_1A_1A_1$ ($p \geq 7$) are clear and finish the study of the subgroups of X.

Next, we consider $\bar{A}_1B_2\bar{A}_3$. The reductive, maximal connected subgroups $\bar{A}_1^3\bar{A}_3$ and $\bar{A}_1B_2^2$ are both M-reducible. The other reductive, maximal connected subgroups of $\bar{A}_1B_2\bar{A}_3$ are listed in Table 11.4 and have been considered previously, as seen from their action on V_M.

Finally, we consider $X = \bar{A}_1B_2G_2$. All of the reductive, maximal connected subgroups of the form $\bar{A}_1YG_2$, where Y is a reductive, maximal connected subgroup of B_2, have been considered above and the same is true for $\bar{A}_1B_2\bar{A}_1A_1$. The subgroup $\bar{A}_1B_2\bar{A}_2$ is M-reducible by Lemma 3.6. That leaves us to consider $\bar{A}_1B_2A_2$ ($p = 3$) and $\bar{A}_1B_2A_1$ ($p \geq 7$). By Theorem 5.1, the maximal subgroup $A_1 < A_2 < G_2$ when $p = 3$ is contained in $\bar{A}_1A_1$ and hence we have already considered all of the reductive, maximal connected subgroups of $\bar{A}_1B_2A_2$ ($p = 3$). The only reductive, maximal connected subgroups of $\bar{A}_1B_2A_1$ ($p \geq 7$) we have not already considered are diagonal subgroups, which are listed in Table 11.4.

8.1.4. $M_1 = \bar{A}_1\bar{A}_3^2$ **($E_7(\#39)$).** By [**Car72**, Table 10], we have $\mathrm{Out}_{E_7}(M_1) \cong S_2 \times S_2$, where one generator swaps the two $\bar{A}_3$ factors and another generator induces a graph automorphism on both of them. The diagonal irreducible subgroups of M_1 are therefore as in Table 11.4. The reductive, maximal connected subgroups of $\bar{A}_3$ are B_2 and A_1^2 ($p \neq 2$) and so the non-diagonal reductive, maximal connected subgroups of $\bar{A}_1\bar{A}_3^2$ are $\bar{A}_1B_2\bar{A}_3$ and $\bar{A}_1A_1^2\bar{A}_3$. The first subgroup is contained in $\bar{A}_1B_2B_3$ and the second is contained in $\bar{A}_1A_1B_4$ and hence both have already been considered.

8.1.5. $M_1 = \bar{A}_1B_5$ **($E_7(\#40)$).** As before, we use Lemma 3.3 to find that the reductive, maximal connected subgroups of $\bar{A}_1B_5$ are as listed in Table 11.4, as well as the M-reducible subgroup $\bar{A}_1\bar{D}_5$. The subgroups $\bar{A}_1A_1\bar{D}_4$ and $\bar{A}_1^3B_3$ are contained in $\bar{A}_1^3\bar{D}_4$, and the subgroup $\bar{A}_1B_2\bar{A}_3$ is contained in $\bar{A}_1B_2B_3$. The subgroups $\bar{A}_1A_1B_4$ and $\bar{A}_1B_2B_3$ are both maximal subgroups of M when $p \neq 2$; we considered their subgroups in previous sections.

The only subgroup left to consider is $\bar{A}_1A_1 = E_7(\#267)$ ($p \geq 11$). All proper irreducible connected subgroups are diagonal and as listed in Table 11.4.

8.1.6. $M_1 = \bar{A}_1A_1C_3$ **($E_7(\#41)$).** We note that there are two classes of $\bar{A}_1A_1C_3$ and they are distinguished by their action on V_{56}, for example, as given in Table 12.4.

By Lemma 3.3, the reductive, maximal connected subgroups of C_3 are $\bar{A}_1B_2$, A_1A_1 ($p \neq 2$), A_1 ($p \geq 7$), A_3 ($p = 2$) and G_2 ($p = 2$). The subgroup $\bar{A}_1^2A_1B_2$ is contained in $\bar{A}_1^3\bar{D}_4$, as seen from its action on V_M; there are two classes of such subgroups, namely $E_7(\#90^{\{0,0\}})$ and $E_7(\#92^{\{0,0\}})$. By considering composition factors on V_{56}, we see that $\bar{A}_1^2 A_1 B_2$ is conjugate to $E_7(\#90^{\{0,0\}})$.

The irreducible subgroups of $X = \bar{A}_1 A_1 A_1 A_1$ ($p \neq 2$) are all diagonal and these can be found in Table 11.17. We note that if $Y_1 \hookrightarrow X$ via $(1_a, 1_b, 1_b, 1_c)$ then Y_1 acts as $(1,3,1) + (1,1,1)$ on V_M when $p \geq 5$ and as $(1, T(3), 1)$ on V_M when $p = 3$. Therefore, Y_1 is contained in $\bar{A}_1^3 A_1 A_1$ and conjugate to $E_7(\#119^{\{0,0,0,0\}})$ when $p \geq 5$, but Y_1 is M-reducible by Lemma 3.6 when $p = 3$. Similarly, if $Y_2 \hookrightarrow X$ via $(1_a, 1_b, 1_c, 1_c)$, then Y_2 is conjugate to $E_7(\#117^{\{0,0,0,0\}})$ when $p \geq 5$, but M-reducible when $p = 3$. Finally, if $Y_3 \hookrightarrow X$ via $(1_a, 1_b, 1_c, 1_b)$, then Y_3 acts as $(1,2,2) + (1,2,0)$ and is conjugate to $E_7(\#207^{\{0,0\}})$.

The diagonal subgroups of $X = \bar{A}_1 A_1 A_1$ ($p \geq 7$) are easily seen to be as in Table 11.4. Since $1 \otimes 5 = 6 + 4$, it follows that $Y \hookrightarrow X$ via $(1_a, 1_b, 1_b)$ is conjugate to $E_7(\#262^{\{0,0\}})$.

By Lemma 3.6, the only $\bar{A}_1 A_1 C_3$-irreducible subgroups contained in $\bar{A}_1 A_1 A_3$ ($p = 2$) are diagonal subgroups. Again by Lemma 3.6, the only $\bar{A}_1 A_1 C_3$-irreducible subgroups contained in $\bar{A}_1 A_1 G_2$ ($p = 2$) are $\bar{A}_1 A_1 A_1 A_1$ and diagonal subgroups. The subgroup $A_1 A_1 < G_2$ acts on $V_{C_3}(\lambda_1)$ as $(1,1) + (0,2)$ and is hence a subgroup of $\bar{A}_1 B_2 < C_3$. Therefore, we have already considered $\bar{A}_1 A_1 A_1 A_1 < \bar{A}_1 A_1 G_2$.

Finally, we need to prove that when $p = 2$ the diagonal subgroups $E_7(\#289)$ are E_7-irreducible. Let $X = A_1 A_3 \hookrightarrow \bar{A}_1 A_1 A_3 = E_7(\#270)$ via $(1^{[r]}, 1^{[s]}, 100)$ ($rs = 0$) and suppose that X is E_7-reducible. From Table 12.4, we have $V_{56} \downarrow \bar{A}_1 A_1 A_3 = (1,1,010)/(0,3,000)/(0,1,101)$. Therefore, X has a 28-dimensional composition factor on V_{56}. By Lemma 3.7, there exists a subgroup Z of type $A_1 A_3$ contained L-irreducibly in a Levi subgroup L, such that X and Z have the same composition factors on V_{56}. Since Z has a 28-dimensional composition factor on V_{56}, it follows from Table 13.4 that L' is of type D_6. Using Lemma 3.3, we find that there are two conjugacy classes of D_6-irreducible subgroups of type $A_1 A_3$ when $p = 2$ both acting as $(1,010)$ on $V_{D_6}(\lambda_1)$. The composition factors of Z on V_{56} are thus $(1,010)^2/(1,101)/(3,000)$ or $(1,010)^2/(2,010)/(0,200)/(0,010)^2/(0,002)$. In both cases Z does not have the same composition factors as X on V_{56}, which is a contradiction. Hence X is E_7-irreducible.

8.1.7. $M_1 = \bar{A}_1 A_1 C_3$ **($E_7(\#42)$).** The reductive, maximal connected subgroups of $E_7(\#42)$ follow in the same way as those of $E_7(\#41)$ in the previous section. We note that the subgroup $\bar{A}_1^2 A_1 B_2$ is conjugate to $E_7(\#92^{\{0,0\}})$, which can be seen by considering its composition factors on V_{56}. There is only one M-conjugacy class of $\bar{A}_1 A_1 A_1 A_1$ acting as $(1,1,2,1)$ on $(1, \lambda_1)$, since the graph automorphism of D_6 swaps the second and fourth A_1 factors.

We claim that $X = A_1 C_3 \hookrightarrow \bar{A}_1 A_1 C_3$ via $(1,1,100)$ is E_7-reducible when $p = 2$. It then follows that both $A_1 A_3 \hookrightarrow \bar{A}_1 A_1 A_3$ via $(1,1,100)$ and $A_1 G_2 \hookrightarrow \bar{A}_1 A_1 G_2$ via $(1,1,10)$ are E_7-reducible. To prove the claim we first show that $\bar{A}_1 A_1 C_3$ is contained in the maximal subgroup $G_2 C_3$. Indeed, the C_3 factor of $G_2 C_3$ is contained in $A_5 < D_6$ and is generated by long root subgroups of E_7. By [**LS94**, p.333, Table 3], the centraliser of such a subgroup C_3 is either $\bar{A}_1 A_1$ or G_2. By considering the composition factors on V_{56}, we see that the C_3 factor of $E_7(\#42)$ has connected centraliser G_2 and therefore $\bar{A}_1 A_1 C_3 < G_2 C_3$. Now, by Theorem 5.1, the subgroup $A_1 \hookrightarrow \bar{A}_1 A_1 < G_2$ via $(1,1)$ is G_2-reducible when $p = 2$ and hence X is $G_2 C_3$-reducible, proving the claim.

Finally, we prove the diagonal subgroups $X = E_7(\#298)$ are E_7-irreducible when $p = 2$. The dimensions of the X-composition factors of V_{56} are $24, 12, 6^2, 4^2$. Using a similar argument to that given for $E_7(\#289)$ in the previous section we find that no subgroup of a Levi factor has the same composition factors as X on V_{56}. Therefore X is E_7-irreducible by Lemma 3.7.

This completes the case $M = \bar{A}_1 D_6$.

8.2. $M = \bar{A}_2 A_5$ ($E_7(\#31)$)

By Lemma 3.3, the reductive, maximal connected subgroups of A_5 are A_2A_1, C_3, A_3 ($p \neq 2$) and A_2 ($p \neq 2$). Similarly, by Lemma 3.3, the only reductive, maximal connected subgroup of $\bar{A}_2$ is A_1 ($p \neq 2$). It follows that the reductive, maximal connected subgroups of $\bar{A}_2 A_5$ are as in Table 11.4. We consider these in the following sections. Note that all reductive, maximal connected subgroups of $A_1 A_5$ ($p \neq 2$) are contained in $\bar{A}_2 X$ for some reductive, maximal connected subgroups X of A_5. We therefore give them no further consideration.

8.2.1. $M_1 = \bar{A}_2 A_2 A_1$ **($E_7(\#300)$).** The M-irreducible subgroups contained in M_1 are straightforward to find and given in Table 11.4, along with the subgroup $X = A_2 A_1 \hookrightarrow M_1$ via $(10, 01, 1)$ when $p = 3$. We need to show that X is conjugate to $E_7(\#184^{\{0,0,0\}}) < \bar{A}_1 D_6$ when $p \neq 3$ and E_7-reducible when $p = 3$. We also show that $Y = A_1 A_1 A_1 < \bar{A}_2 A_2 A_1$ is conjugate to $E_7(\#275^{\{0,0\}}) < \bar{A}_1 D_6$.

We first note that $\bar{A}_2 A_2 A_1$ is a subgroup of $A_1 F_4$. Indeed, consider the maximal subgroup $\bar{A}_2 A_2$ of F_4. The $\bar{A}_2$ factor generated by long root subgroups of F_4 is generated by long root subgroups of E_7 and so $A_1 \bar{A}_2 A_2 < \bar{A}_2 C_{E_7}(\bar{A}_2)^\circ = \bar{A}_2 A_5$, by Lemma 3.10. Theorem 6.1 shows that $Z = A_2 \hookrightarrow \bar{A}_2 A_2 < F_4$ via $(10, 01)$ is conjugate to $A_2 < \bar{D}_4$ embedded via $V_{A_2}(11)$ when $p \neq 3$ but Z is F_4-reducible when $p = 3$. It follows that $X < \bar{D}_4 C_{E_7}(\bar{D}_4)^\circ = \bar{A}_1^3 \bar{D}_4 < \bar{A}_1 D_6$ when $p \neq 3$ and X is $A_1 F_4$-reducible when $p = 3$. Comparing composition factors when $p \neq 3$ shows that X is conjugate to $E_7(\#184^{\{0,0,0\}})$, as required. Also by Theorem 6.1, the irreducible subgroup $A_1 A_1 < \bar{A}_2 A_2 < F_4$ is contained in $\bar{A}_1 C_3$. Therefore, $Y < \bar{A}_1 C_3 A_1 < F_4 A_1$. Comparing composition factors shows that Y is conjugate to $E_7(\#275^{\{0,0\}})$.

When $p = 2$ the diagonal subgroups $E_7(\#307)$ and $E_7(\#308)$ contain no proper E_7-irreducible subgroups and so we need to prove that they are E_7-irreducible. From Table 12.4, the composition factors of $E_7(\#300) = \bar{A}_2 A_2 A_1$ acting on V_{56} are $(10, 10, 1)/(01, 01, 1)/(00, 11, 1)/(00, 00, 3)$. Let X be the diagonal subgroup embedded via $(10, 10, 1)$. We prove that X is E_7-irreducible; the other cases are similar and easier. Suppose that X is E_7-reducible. By Lemma 3.7, there exists a subgroup Z of type $A_1 A_2$ contained L-irreducibly in a Levi subgroup L, such that X and Z have the same composition factors on V_{56}. The Z-composition factors of V_{56} are thus $(3, 00)/(1, 20)/(1, 11)/(1, 10)^2/(1, 02)/(1, 01)^2$; in particular, the dimensions of the composition factors are $16, 6^6, 4$. Using Table 13.4 we see that L' has type D_6, $A_1 A_5$ or A_5. Using Lemma 3.6 we find all of the L'-irreducible subgroups of type $A_1 A_2$. If L' has type D_6 then Z acts as $((0, 00)|(2, 00)|(0, 00)) + (0, 11)$ on $V_{D_6}(\lambda_1)$. Therefore Z has a trivial composition factor on V_{56}, a contradiction. If L' has type $A_1 A_5$ then Z is contained in the subgroup $A_1 A_1 A_2$ where the $A_1 A_2 < A_5$ acts on $V_{A_5}(\lambda_1)$ as $(1, 10)$. Therefore the Z-composition factors of V_{56} are $(1^{[r]} \otimes 1^{[s]}, 10)/(1^{[r]} \otimes 1^{[s]}, 01)/(3^{[s]}, 00)/(1^{[s]}, 11)/(1^{[s]}, 10)/(1^{[s]}, 01)$ for some r, s with $rs = 0$.

These are not the same as the X-composition factors of V_{56} for any r, s, which is a contradiction. Finally, suppose that L' is of type A_5, of which there are two such conjugacy classes. Then Z acts as $(1, 10)$ on $V_{A_5}(\lambda_1)$. Using the previous case and Table 13.4 we see that the Z-composition factors of V_{56} are not the same as the X-composition factors. This final contradiction shows that X is E_7-irreducible.

8.2.2. $M_1 = \bar{A}_2C_3$ ($E_7(\#301)$). The only irreducible subgroup of $\bar{A}_2$ is A_1 when $p \neq 2$. We claim that A_1C_3 is contained in $\bar{A}_1D_6$. To prove this, we first note that by [**LS94**, p.333, Table 3], the centraliser in E_7 of the C_3 factor is G_2 and therefore $\bar{A}_2C_3$ is contained in G_2C_3. By Theorem 5.1, the irreducible subgroup A_1 of $\bar{A}_2 < G_2$ is contained in $\bar{A}_1A_1 < G_2$. It follows that A_1C_3 is therefore contained in $\bar{A}_1A_1C_3 < \bar{A}_1D_6$ and by considering composition factors we conclude that A_1C_3 is conjugate to $E_7(\#294^{\{0,0\}})$.

From Lemma 3.3, the reductive, maximal connected subgroups of C_3 are A_1B_2, A_1A_1 ($p \neq 2$), A_1 ($p \geq 7$), A_3 ($p = 2$) and G_2 ($p = 2$). The action of A_1B_2 on $V_{A_5}(\lambda_1)$ is reducible and hence $\bar{A}_2A_1B_2$ is M-reducible by Lemma 3.6. We also note that the maximal subgroup A_1A_1 ($p \neq 2$) acts as $(2, 1)$ on $V_{A_5}(\lambda_1)$ and therefore $\bar{A}_2A_1A_1$ is conjugate to $E_7(\#306)$.

The subgroup $\bar{A}_2A_3$ is a maximal connected subgroup of M when $p \neq 2$ and so we consider the subgroups of $\bar{A}_2A_3$ ($p = 2$) in the next section.

Finally, consider $X = \bar{A}_2G_2$ ($p = 2$). The reductive, maximal connected subgroups of G_2 when $p = 2$ are M-reducible by Lemma 3.6 and there are no reductive, maximal connected subgroups of $\bar{A}_2$ when $p = 2$. Thus there are no proper M-irreducible subgroups of X. It remains to prove that X is E_7-irreducible. From Table 12.4, we have $V_{56} \downarrow X = (10, 10)/(01, 10)/(00, 20)/(00, 10)^2/(00, 00)^2$. As in previous cases, it is straightforward to show that there are no Levi subgroups with an irreducible subgroup A_2G_2 having the same composition factors as X on V_{56}. Therefore Lemma 3.7 implies that X is E_7-irreducible.

8.2.3. $M_1 = \bar{A}_2A_3$ ($E_7(\#302)$). The subgroup $\bar{A}_2A_3$ is a maximal connected subgroup of M when $p \neq 2$ and contained in $\bar{A}_2C_3$ when $p = 2$. In this section we consider both cases.

Both reductive, maximal connected subgroups of A_3 act reducibly on $V_{A_3}(010)$ and hence on $V_{A_5}(\lambda_1)$. Therefore, the only proper irreducible subgroup to consider is $X = A_1A_3$ ($p \neq 2$), where the A_1 factor is irreducibly embedded in A_2. We need to prove that X is E_7-irreducible when $p \neq 2$ and that M_1 is E_7-irreducible when $p = 2$.

First we consider M_1 when $p = 2$. Suppose that M_1 is E_7-reducible. By Lemma 3.7, there exists a subgroup Z of type A_2A_3 contained L-irreducibly in a Levi subgroup L, such that X and Z have the same composition factors on V_{56}. It follows that L' has no simple factors of type A_1 and has rank at least 5. Therefore L' is of type A_2A_3. Using Table 12.4 and Table 13.4, we see that the composition factors of the Levi subgroup A_2A_3 are not the same as the composition factors of M_1 on V_{56}. This contradiction proves that M_1 is E_7-irreducible.

Now we consider X. When $p \geq 5$, the action of X on $L(E_7)$ has no trivial composition factors and so X is E_7-irreducible by Corollary 3.8. When $p = 3$, Table 12.4 shows that $V_{56} \downarrow X = (2, 010)^2/(0, 200)/(0, 002)$. As in previous cases, it is straightforward to show that there are no Levi subgroups with a subgroup A_1A_3

having the same composition factors as X on V_{56}. Therefore Lemma 3.7 implies that X is E_7-irreducible.

8.2.4. $M_1 = \bar{A}_2 A_2$ $(p \neq 2)$ $(E_7(\#303))$. The only non-simple proper irreducible connected subgroup to consider is $X = A_1 A_2 = E_7(\#314)$, since the irreducibly embedded subgroup A_1 of the second A_2 factor acts reducibly on $V_{A_5}(\lambda_1)$. Since X contains no proper E_7-irreducible subgroups we need to show that it is E_7-irreducible.

When $p \geq 5$ we see from Table 12.4 that there are no trivial X-composition factors of $L(E_7)$ and hence X is E_7-irreducible by Corollary 3.8. Now let $p = 3$ and suppose that X is E_7-reducible. From Table 12.4 we have $V_{56} \downarrow X = (2,20)/(2,02)/(0,30)/(0,11)^2/(0,03)$. By Lemma 3.7, there exists a subgroup Z of type $A_1 A_2$ contained L-irreducibly in a Levi subgroup L, such that X and Z have the same composition factors on V_{56}. The dimensions of the Z-composition factors of V_{56} are $18^2, 7^2, 3^2$. Using Table 13.4, we deduce that there is no Levi subgroup containing such a subgroup Z. Therefore X is E_7-irreducible.

This completes the case $M = \bar{A}_2 A_5$.

8.3. $M = A_7$ $(E_7(\#22))$

From Lemma 3.3, we find that the reductive, maximal connected subgroups of A_7 are C_4, D_4 $(p \neq 2)$ and $A_1 A_3$. The subgroup C_4 is E_7-reducible and hence we do not need to consider any of its subgroups.

Next we consider D_4 $(p \neq 2)$. By [**CLSS92**, Lemma 2.15], we have $\mathrm{Out}_{E_7}(D_4) \cong S_3$. It follows that the only reductive, maximal connected subgroup of D_4 which is A_7-irreducible is A_2 when $p \geq 5$, acting as 11 on $V_{A_7}(\lambda_1)$. Such a subgroup A_2 has already been considered in [**Tho15**] and shown to be contained in $\bar{A}_2 A_5$. By comparing composition factors, we find that it is conjugate to $E_7(\#24)$. The only proper irreducible connected subgroup of A_2 is A_1. Such a subgroup A_1 acts as $4+2$ on $V_{A_7}(\lambda_1)$ and is therefore M-reducible.

We now show that $X = A_1 A_3$ is contained in $\bar{A}_2 A_5$. Let Y be the A_3 factor of X. Then Y acts as 100^2 on $V_{A_7}(\lambda_1)$. Thus Y is contained in $\bar{A}_3^2$, a Levi subgroup of A_7. By [**LS94**, p.333, Table 2], we have $C_{E_7}(\bar{A}_3)^\circ = \bar{A}_1 \bar{A}_3$ and it follows that $\bar{A}_3^2 < C_{E_7}(\bar{A}_1)^\circ = D_6$. By considering the composition factors of Y on V_{56} it follows that Y acts as 010^2 on $V_{D_6}(\lambda_1)$ and is hence contained in an A_5-Levi subgroup of D_6, acting as 010 on $V_{A_5}(\lambda_1)$. There are two E_7-conjugacy classes of A_5-Levi subgroups and again by considering the composition factors of Y, it follows that Y is contained in an A_5-Levi whose connected centraliser is $\bar{A}_2$. Therefore $X < \bar{A}_2 A_5$.

When $p \neq 2$, we find that X is conjugate to $E_7(\#313)$, as may be seen by considering their composition factors on V_{56}. When $p = 2$, there are no $\bar{A}_2 A_5$-irreducible subgroups of type $A_1 A_3$ contained in $\bar{A}_2 A_5$. Therefore X is $\bar{A}_2 A_5$-reducible and hence G-reducible.

8.4. $M = G_2 C_3$ $(E_7(\#32))$

The reductive, maximal connected subgroups of G_2 are given by Theorem 5.1 and the reductive, maximal subgroups of C_3 are given by Lemma 3.3. The lattice of M-irreducible subgroups now easily follows with further use of Lemma 3.3 and Lemma 3.6. It remains for us to show the claimed conjugacies between subgroups of

M and subgroups of previously considered reductive, maximal connected subgroups of E_7.

The G_2 factor of G_2C_3 is contained in a Levi subgroup D_4 and hence the subgroups of G_2 generated by long root subgroups of G_2 are generated by long root subgroups of E_7. It follows that $\bar{A}_2C_3$ is conjugate to $E_7(\#301) < \bar{A}_2A_5$ and $\bar{A}_1A_1C_3$ is conjugate to $E_7(\#42) < \bar{A}_1D_6$. By Theorem 5.1, the irreducible subgroup A_1 of $A_2 < G_2$ ($p = 3$) is contained in $\bar{A}_1A_1$ and so when $p = 3$, the subgroup A_1C_3 is conjugate to $E_7(\#294^{\{1,0\}})$. The subgroup $G_2\bar{A}_1B_2$ is contained in $\bar{A}_1D_6$ and conjugate to $E_7(\#253)$.

We need to prove that some diagonal subgroups of $X = A_1A_1A_1 = E_7(\#323)$ ($p \geq 7$) have also been seen previously. Firstly, the subgroup A_1A_1 embedded via $(1_a, 1_b, 1_b)$ is contained in $A_1\bar{A}_1B_2$ since $2 \otimes 1 = 3 + 1$. Secondly, we claim the subgroup $Y = A_1A_1$ embedded via $(1_a, 1_a, 1_b)$ is contained in $\bar{A}_1D_6$ also. To prove this, first consider the maximal connected subgroup $A_1A_1G_2$ of A_1F_4. Then $C_{E_7}(G_2)^\circ = C_3$ and hence $A_1A_1G_2 < G_2C_3$. By considering composition factors on V_{56} we see that the A_1 factor of A_1F_4 is the third A_1 factor of X. Appealing to Theorem 6.1 shows that $A_1 \hookrightarrow A_1A_1 < A_1G_2 < F_4$ via $(1,1)$ is conjugate to $A_1 \hookrightarrow \bar{A}_1A_1 < \bar{A}_1C_3 < F_4$ via $(1,1)$ since both have identification number $F_4(\#8^{\{0,0\}})$. Therefore Y is contained in $A_1\bar{A}_1C_3$, which is a subgroup of $\bar{A}_1D_6$ by Lemma 3.10.

8.5. $M = A_1F_4$ ($E_7(\#33)$)

Theorem 6.1 gives the lattice structure of the M-irreducible subgroups of M. By [**LS94**, p.333, Table 3], $C_{E_7}(B_4)^\circ = \bar{A}_1A_1$ and hence A_1B_4 is conjugate to $E_7(\#190^{\{0,0\}}) < \bar{A}_1D_6$. Similarly, $A_1\bar{A}_1C_3$ is a subgroup of $\bar{A}_1D_6$ and $A_1\bar{A}_2A_2$ is a subgroup of $\bar{A}_2A_5$. In Section 8.4, we also proved that $A_1A_1G_2$ is contained in G_2C_3. We are therefore left to consider A_1C_4 ($p = 2$), A_1G_2 ($p = 7$) and A_1A_1 ($p \geq 13$). The diagonal subgroups of A_1A_1 ($p \geq 13$) follow immediately.

From Theorem 6.1, we see that the M-irreducible maximal connected subgroups of A_1C_4 are A_1D_4, $A_1B_2^2$ and $A_1\bar{A}_1C_3$. The second subgroup is contained in A_1B_4 and the third subgroup is contained in $\bar{A}_1D_6$ as well. Hence both have already been considered. The M-irreducible maximal connected subgroups of A_1D_4 are $A_1A_1^4$ and A_1A_2. The former is a subgroup of A_1B_4 and the latter a subgroup of $A_1\bar{A}_2A_2$. Hence both have already been considered.

Similarly, using Theorem 6.1 we see that all F_4-irreducible connected subgroups of G_2 ($p = 7$) are contained in either $\bar{A}_2A_2$ or $\bar{A}_1C_3$ and so we have already considered all M-irreducible subgroups of A_1G_2 ($p = 7$).

8.6. $M = A_1G_2$ ($p \neq 2$) ($E_7(\#34)$)

The lattice structure of M-irreducible subgroups follows from Theorem 5.1. We claim that A_1A_2 (where the factor A_2 is generated by long root subgroups of G_2) is contained in $\bar{A}_2A_5$ and conjugate to $E_7(\#309^{\{0,0\}})$. To prove this, let X be the A_2 factor. The G_2 factor of M is contained in A_6, acting on $V_{A_6}(\lambda_1)$ as $V_{G_2}(10)$, by the construction in [**Sei91**, 3.12]. It follows that X acts as $10 + 01 + 00$ on $V_{A_6}(\lambda_1)$ and therefore X is contained in a Levi subgroup $\bar{A}_2^2$ of A_6 embedded via $(10, 01)$. Furthermore, X is contained in A_5 acting as 10^2, since $\bar{A}_2^2$ acts as $(10,00)+(00,01)$ on $V_{A_5}(\lambda_1)$. We now have $C_{E_7}(X)^\circ \geq \bar{A}_2A_1$ and since $X\bar{A}_2A_1$ is E_7-irreducible (subgroup $E_7(\#300)$) this must be an equality. Therefore $A_1X < \bar{A}_2C_{E_7}(\bar{A}_2)^\circ = \bar{A}_2A_5$ and comparing composition factors finishes the proof of the claim.

Now consider $X = A_1A_1A_1$. The projection of X to G_2 is the centraliser in G_2 of a semisimple involution. Therefore X centralises a semisimple involution t of E_7 and thus X is contained in $\bar{A}_1D_6$ or A_7 by Lemma 3.9. Since X contains a simple E_7-irreducible subgroup, X is E_7-irreducible. As A_7 does not contain an E_7-irreducible subgroup of type $A_1A_1A_1$ it follows that X is contained in $\bar{A}_1D_6$. By considering the X-composition factors of V_{56}, we see that X is conjugate to $E_7(\#206^{\{0,0\}})$.

The only reductive, maximal connected subgroup of A_1A_2 ($p = 3$) is A_1A_1 and by Theorem 5.1 this is a subgroup of $A_1A_1A_1$ and has thus already been considered.

This completes the proof of Theorem 8.1.

CHAPTER 9

Irreducible subgroups of $G = E_8$

In this chapter we prove Theorem 1 when G is of type E_8; we classify the E_8-irreducible connected subgroups thus proving the following theorem.

THEOREM 9.1. *Let X be an E_8-irreducible connected subgroup of E_8. Then X is conjugate to exactly one subgroup of Table* 11.5 *and each subgroup in Table* 11.5 *is E_8-irreducible. Moreover, Table* 11.5 *gives the lattice structure of the irreducible connected subgroups of E_8.*

As in the previous sections we consider each of the reductive, maximal connected subgroups of E_8 in turn. The simple connected E_8-irreducible subgroups are classified in [**Tho15**, Theorem 3] and [**Tho16**, Theorem 6] and we will use these without reference throughout. By Theorem 3.1, the reductive, maximal connected subgroups of E_8 are D_8, $\bar{A}_1E_7$, $\bar{A}_2E_6$, A_8, $\bar{A}_4^2$, G_2F_4, B_2 ($p \geq 5$), A_1A_2 ($p \geq 5$) and A_1 (3 classes, $p \geq 23, 29, 31$).

We remind the reader that throughout the proof we will only make reference to the E_8-irreducibility of an M-irreducible subgroup when it does not properly contain an E_8-irreducible subgroup. Since the simple E_8-irreducible connected subgroups are known we only need to consider a small number of cases.

9.1. $M = D_8$ ($E_8(\#43)$)

The reductive, maximal connected subgroups of M are given in Lemma 3.3. They are $\bar{A}_1^2D_6$, $\bar{D}_4^2$, $\bar{A}_3D_5$, B_7, A_1B_6 ($p \neq 2$), B_2B_5 ($p \neq 2$), B_3B_4 ($p \neq 2$), B_2^2 ($p \neq 2$) (2 classes), A_1C_4 (2 classes) and B_4 (2 classes). In the cases where there are two conjugacy classes these are distinguished by their action on $L(E_8)$, which is given in Table 12.5.

Firstly we prove that two of the maximal connected subgroups are E_8-reducible when $p = 2$. In [**Tho15**, Lemma 7.4], it is shown that one of the classes of B_4 subgroups is E_8-reducible, denoted $B_4(\ddagger)$. In the current notation this is subgroup $E_8(\#46)$ for which $p \neq 2$. We also claim one of the classes of A_1C_4 is E_8-reducible when $p = 2$. Let $A_1C_4(\ddagger)$ denote the subgroup A_1C_4 with composition factors $(4,0)/(2,\lambda_2)/(2,0)^2/(0,\lambda_2)^2/(0,\lambda_4)/(0,0)^2$ on $V_{D_8}(\lambda_7)$ when $p = 2$ (this is subgroup $E_8(\#116)$ when $p \neq 2$). By [**LS94**, p.333, Table 3], the connected centraliser of the C_4 factor of $A_1C_4(\ddagger)$ is $\bar{A}_1U_5$, where U_5 is a 5-dimensional connected unipotent subgroup. Therefore $X < U_5\bar{A}_1C_4$, which by the Borel-Tits Theorem [**BT71**, Théorème 2.5] is contained in a parabolic subgroup of E_8. Therefore X is E_8-reducible.

There are a very large number of E_8-irreducible subgroups contained in D_8. We therefore suppress many of the routine parts of constructing the lattice of irreducible connected subgroups since we have been more explicit in the proofs of the previous theorems, especially the case $\bar{A}_1D_6 < E_7$. For example, the use of Lemma 3.3 to find the lattice of D_8-irreducible connected subgroups is mainly omitted, as are most of the considerations of when two subgroups contained in different reductive, maximal connected subgroups of D_8 are conjugate. Our use of Lemma 3.6 to remove any D_8-reducible classes inside reductive, maximal connected subgroups of D_8 will also be implicit.

When finding the E_8-conjugacy classes of diagonal subgroups contained in $X < D_8$ we give only the required information concerning $\mathrm{Out}_{E_8}(X)$. On many occasions $\mathrm{Out}_{E_8}(X) \cong \mathrm{Out}_{D_8}(X)$; in such cases we will not be explicit.

9.1.1. $M_1 = \bar{A}_1^2 D_6$ $(E_8(\#107))$**.** By [**Car72**, Table 11], we have $\mathrm{Out}_{E_8}(M_1) \cong S_2$, where the involution acts simultaneously as a graph automorphism of $\bar{A}_1^2$ and D_6. Up to E_8-conjugacy we may therefore choose to take just one representative of each non-conjugate pair of D_6-irreducible subgroups which are conjugate in $D_6.2$. This leads to the reductive, maximal connected subgroups of $\bar{A}_1^2 D_6$ given in Table 11.5. We will consider each of them in turn.

9.1.1.1. $M_2 = \bar{A}_1^4 \bar{D}_4$ $(E_8(\#117))$. By [**Car72**, Table 11], we have $\mathrm{Out}_{E_8}(M_2) \cong S_4$ acting naturally on the four A_1 factors and inducing the full outer automorphism group of $\bar{D}_4$. Therefore, we have only one E_8-conjugacy class of each of the subgroups $\bar{A}_1^4 B_3$ and $\bar{A}_1^4 A_1 B_2$ contained in M_2. Moreover, the stabiliser of B_3 and $A_1 B_2$ under the action of S_4 is Dih_8. Therefore $\mathrm{Out}_{E_8}(\bar{A}_1^4 B_3) \cong \mathrm{Out}_{E_8}(\bar{A}_1^4 A_1 B_2) \cong \mathrm{Dih}_8$. The irreducible subgroups G_2 and A_2 are normalised by all outer automorphisms of $\bar{D}_4$ and so $\mathrm{Out}_{E_8}(\bar{A}_1^4 G_2) \cong \mathrm{Out}_{E_8}(\bar{A}_1^4 A_2) \cong S_4$. Finding most of the diagonal subgroups contained in M_2 now follows; we give further details below where needed.

From [**LS04**, Table 10.3] we see $\mathrm{Out}_{E_8}(\bar{A}_1^8) \cong \mathrm{AGL}_3(2)$. The action on the eight factors is described in the proof of [**Tho16**, Theorem 6] and the non-simple diagonal subgroups follow using the same method as for the simple ones.

We note two intricacies in finding the diagonal subgroups of $E_8(\#189) = \bar{A}_1^4 A_1 A_1$ $(p \geq 5)$. There are two classes of pairs $\bar{A}_1^2 < \bar{A}_1^4$, which may be represented by the first and second A_1 factors and the first and third A_1 factors, for example. When taking a diagonal subgroup, say X, in the first and second A_1 factors via $(1,1)$ it is then conjugate to the fifth A_1 factor, say Y. It then follows that there is an involution in D_8 that swaps X and Y. This leads to the condition "if $r = s$ then $r < t$" in the definition of $E_8(\#213)$ in Table 11.29. The second thing to note is that a diagonal subgroup of the fifth and sixth A_1 factors via $(1,1)$ is conjugate to a subgroup A_1 contained in $A_2 < \bar{D}_4$. Indeed, the irreducible subgroup A_1 of A_2 acts as $4+2$ on $V_{\bar{D}_4}(\lambda_1)$. Since this subgroup A_2 is centralised by a triality automorphism of $\bar{D}_4$ it follows that $\mathrm{Out}_{E_8}(E_8(\#211^{\{0,0\}})) \cong S_4$ acting naturally on the first four A_1 factors.

Finally, we consider the subgroup $E_8(\#190) = \bar{A}_1^4 A_1^3 < \bar{A}_1^4 A_1 B_2$ when $p = 2$. Since $\mathrm{Out}_{\bar{D}_4}(A_1^3) \cong S_3$ it follows that $\mathrm{Out}_{E_8}(E_8(\#190)) \cong S_4 \times S_3$, where the S_4 direct factor acts naturally on the first four factors and the S_3 direct factor acts naturally on the final three factors.

9.1.1.2. $M_2 = \bar{A}_1^2 B_5$ $(E_8(\#118))$. Since B_5 is centralised by a graph automorphism of D_6 we have $\mathrm{Out}_{E_8}(M_2) \cong S_2$. This is enough to determine the classes of diagonal subgroups of all subgroups of M_2.

9.1.1.3. $M_2 = \bar{A}_1^2 \bar{A}_3^2$ $(E_8(\#119))$. Here $\mathrm{Out}_{E_8}(M_2) \cong S_2 \times S_2$ by [**Car72**, Table 11], with an involution swapping the two $\bar{A}_3$ factors and another involution simultaneously swapping the two $\bar{A}_1$ factors whilst fixing one $\bar{A}_3$ factor and acting as a graph automorphism on the other. The diagonal subgroups of M_2 now follow.

Because $\mathrm{Out}_{\bar{A}_3}(A_1^2) \cong S_2$ it follows that $\mathrm{Out}_{E_8}(\bar{A}_1^2 A_1^2 \bar{A}_3) \cong S_2 \times S_2$, containing an involution swapping the first and second $\bar{A}_1$ factors whilst acting as a graph automorphism of the $\bar{A}_3$ factor and another involution swapping the third and fourth A_1 factors. For the diagonal subgroup $X = A_1^3 \bar{A}_3 \hookrightarrow \bar{A}_1^2 A_1^2 \bar{A}_3$ via $(1_a, 1_a, 1_b, 1_c, 100)$ we have $\mathrm{Out}_{E_8}(X) \cong S_3$ acting naturally on the three A_1 factors. Finally, note that $\mathrm{Out}_{E_8}(\bar{A}_1^2 A_1^4) \cong S_2 \times S_4$ and taking a diagonal subgroup $X = A_1^5$ via $(1_a, 1_a, 1_b, 1_c, 1_d, 1_e)$ yields $\mathrm{Out}_{E_8}(X) \cong S_5$.

It remains to prove that the diagonal subgroups $E_8(\#373)$ are irreducible when $p = 2$ as they do not properly contain any E_8-irreducible subgroups. Let $X = E_8(\#373) = A_1 \bar{A}_3 \hookrightarrow \bar{A}_1^2 A_1^2 \bar{A}_3 = E_8(\#355) = Y$ via $(1^{[r]}, 1^{[s]}, 1^{[t]}, 1^{[u]}, 100)$ where $rt = 0$, $r < s$ and $t < u$. From Table 12.5, we have

$$\begin{aligned} L(E_8) \downarrow Y = {} & (2,0,0,0,000)/(1,1,2,0,000)/(1,1,0,2,000)/(1,1,0,0,010)/ \\ & (1,1,0,0,000)^2/(1,0,1,1,100)/(1,0,1,1,001)/(0,2,0,0,000)/ \\ & (0,1,1,1,100)/(0,1,1,1,001)/(0,0,2,2,000)/(0,0,2,0,010)/ \\ & (0,0,2,0,000)^2/(0,0,0,2,010)/(0,0,0,2,000)^2/(0,0,0,0,101)/ \\ & (0,0,0,0,010)^2/(0,0,0,0,000)^6. \end{aligned}$$

Therefore X, has at most seven trivial composition factors and at least two composition factors of dimension 32. Looking for a contradiction we suppose that X is E_8-reducible. By Lemma 3.7, there exists a subgroup Z of type $A_1 A_3$ contained L-irreducibly in a Levi subgroup L, such that X and Z have the same composition factors on $L(E_8)$. By considering the number of trivial composition factors on $L(E_8)$ of each Levi subgroup that contains an irreducible subgroup $A_1 A_3$ (using Table 13.5 and Lemma 3.6), it follows that L' has type E_7, $A_1 E_6$, D_7, D_6, $A_1 D_5$, A_7, $A_1 A_5$ or A_3^2.

Suppose that L' has type E_7. Then Theorem 8.1 shows that Z is conjugate to $E_7(\#m)$ for $m =$ 194, 265, 266, 289, or 298. Using the restriction $L(E_8) \downarrow E_7 = V_{E_7}(\lambda_1)/V_{E_7}(\lambda_7)^2/V_{E_7}(0)^4$ from Table 13.5 and the composition factors given in Table 12.4, we calculate that Z does not have the same composition factors as X on $L(E_8)$, a contradiction.

Now suppose that L' has type $A_1 E_6$. Then by Theorem 7.1 the projection of Z to E_6 is conjugate to $E_6(\#29)$. As before, we use the composition factors of the action of $A_1 E_6$ on $L(E_8)$ from Table 13.5 and the composition factors of $E_6(\#29)$ acting on V_{27} and $L(E_6)$ from Table 12.3 to see that Z does not have the same composition factors as X on $L(E_8)$.

In the remaining cases all of the simple factors of L' are classical. We therefore use Lemma 3.6 to find the irreducible subgroups of type $A_1 A_3$. Firstly, suppose that L' has type D_7. Then by Lemma 3.6, Z acts on $V_{D_7}(\lambda_1)$ as $((0,000)|((2,000) + (2^{[r]},000) + (2^{[s]},000))|(0,000)) + (0,010)$ with $0 < r < s$ since Z is D_7-irreducible. Using the composition factors of $L(E_8) \downarrow D_7$ given in Table 13.5, we calculate that Z does not have the same composition factors as X on $L(E_8)$, a contradiction. Next suppose that L' has type D_6 or $A_1 D_5$. Then the action of Z on λ_1 (respectively $(0, \lambda_1)$) has two trivial composition factors. It then follows that Z has more than seven trivial composition factors on $L(E_8)$, a contradiction.

Now suppose that L' has type A_7. Then using Lemma 3.6 we find that there is a unique conjugacy class of A_7-irreducible subgroups of type $A_1 A_3$; they act as $(1, 100)$ on $V_{A_7}(\lambda_1)$. Using the restriction $L(E_8) \downarrow A_7$ from Table 13.5 we find that

Z is not contained in A_7. Finally, suppose that L' has type A_1A_5 or A_3^2. Using the composition factors of $L(E_8) \downarrow L'$ given in Table 13.5, it follows that Z has at most one composition factor of dimension at least 32. This is a contradiction since X has at least two such composition factors.

9.1.1.4. $M_2 = \bar{A}_1^2A_1B_4$ (E_8(#120)) *or* $\bar{A}_1^2B_2B_3$ (E_8(#121)). Since the subgroups A_1B_4 and B_2B_3 of D_6 are centralised by a graph automorphism of D_6 we have $\mathrm{Out}_{E_8}(M_2) \cong S_2$ with the involution swapping the two $\bar{A}_1$ factors. We note that as in previous discussions the subgroup $X = A_1 \hookrightarrow \bar{A}_1^2 < D_8$ via $(1,1)$ is conjugate to $Y = A_1 < D_8$ acting via $2 + 0^{13}$ on $V_{D_8}(\lambda_1)$. This is enough to determine the classes of diagonal subgroups of all subgroups of M_2.

9.1.1.5. $M_2 = \bar{A}_1^2A_1C_3$ (E_8(#122)). Firstly we note that $\mathrm{Out}_{E_8}(M_2)$ is trivial, as can be seen from the M_2-composition factors of $L(E_8)$. The diagonal subgroups of M_2 therefore follow. We also note $X = A_1\bar{A}_1C_3 \hookrightarrow M_2$ via $(1_a, 1_b, 1_a, 100)$ is contained in the E_8-reducible subgroup A_1C_4 when $p = 2$ and hence X is E_8-reducible.

Next, we consider $X = \bar{A}_1^2A_1^2A_1 = E_8$(#558) $(p \neq 2)$. A graph automorphism of D_6 swaps the third and fourth A_1 factors and hence $\mathrm{Out}_{E_8}(\bar{A}_1^2A_1^2A_1) \cong S_2$ where the involution simultaneously swaps the two $\bar{A}_1$ factors as well as swapping the third and fourth A_1 factors. The subgroup $A_1^2A_1 < D_6$ acts as $(1,1,2)$ on $V_{D_6}(\lambda_1)$. Therefore the subgroups $\bar{A}_1^2A_1A_1 \hookrightarrow X$ via $(1_a, 1_b, 1_c, 1_c, 1_d)$, $(1_a, 1_b, 1_c, 1_d, 1_c)$ or $(1_a, 1_b, 1_c, 1_d, 1_d)$ are contained in previously considered subgroups of $\bar{A}_1^2D_6$.

It remains to prove that when $p = 2$ the diagonal subgroups $X = E_8$(#614) $= A_1A_3 \hookrightarrow \bar{A}_1^2A_1A_3 = E_8$(#560) $= Y$ via $(1^{[r]}, 1^{[s]}, 1^{[t]}, 100)$ $(rst = 0; r \neq t)$ are E_8-irreducible. From Table 12.5, we have

$$\begin{aligned} L(E_8) \downarrow Y = {} & (2,0,0,000)/(1,1,1,010)/(1,0,3,000)/(1,0,1,101)/(0,2,0,000)/ \\ & (0,1,2,010)/(0,1,0,200)/(0,1,0,010)^2/(0,1,0,002)/(0,0,2,101)/ \\ & (0,0,2,000)/(0,0,0,101)^2/(0,0,0,020)/(0,0,0,000)^4. \end{aligned}$$

It follows that X has exactly 4 trivial composition factors as well as composition factors of dimensions 56 and 48. A routine use of Lemma 3.7 as before shows that X is E_8-irreducible.

9.1.2. $M_1 = \bar{D}_4^2$ **(**E_8**(#108)).** It easily follows from [**LS04**, Table 10.3] that $\mathrm{Out}_{E_8}(M_1) \cong S_2 \times S_3$, where the central involution swaps the two $\bar{D}_4$ factors, and the S_3 direct factor acts simultaneously as the group of graph automorphisms of each $\bar{D}_4$ factor. In particular, there is just one E_8-conjugacy class of each of the subgroups $B_3\bar{D}_4$ and $A_1B_2\bar{D}_4$ contained in M_1. It then follows that there are two E_8-conjugacy classes of each of the subgroups B_3^2, $B_3A_1B_2$ and $A_1^2B_2^2$. We make some remarks about them now.

Firstly, one class of subgroups B_3^2 acts with composition factors $(W(100), 000)/(000, W(100))/(000, 000)^2$ on $V_{D_8}(\lambda_1)$ and is hence D_8-reducible by Lemma 3.6. The other class is D_8-irreducible, namely E_8(#623). One class of subgroups $A_1B_2B_3$ acts with composition factors $(W(2), 00, 000)/(0, W(10), 000)/(0, 00, W(100))/(0, 00, 000)$ on $V_{D_8}(\lambda_1)$ and is therefore D_8-irreducible if and only if $p \neq 2$. Moreover, when $p \neq 2$ this subgroup is contained in $\bar{A}_1^2D_6$ and conjugate to E_8(#533$^{\{0\}}$). Similarly, one class of subgroups $A_1^2B_2^2$ acts with composition factors $(W(2), 0, 00, 00)/(0, W(2), 00, 00)/(0, 0, W(10), 00)/(0, 0, 00, W(10))$ on $V_{D_8}(\lambda_1)$ and is therefore D_8-irreducible if and only if $p \neq 2$. This class is denoted E_8(#677) and in this case

$\mathrm{Out}_{E_8}(A_1^2B_2^2) = \mathrm{Out}_{D_8}(A_1^2B_2^2) \cong S_2 \times S_2$ where one involution swaps the two A_1 factors and another involution swaps the two B_2 factors. The second class of subgroups $A_1^2B_2^2$ is denoted $E_8(\#676)$ and in that case $\mathrm{Out}_{E_8}(A_1^2B_2^2) \cong S_2$ where the involution simultaneously swaps the two A_1 factors and the two B_2 factors.

Most of the classes of diagonal subgroups contained in M_1 follow immediately. We will point out those that are not entirely obvious. First, consider $X = B_3A_1^3 = E_8(\#636)$. In this case $\mathrm{Out}_{\bar{D}_4}(A_1^3) \cong S_3$ and so $\mathrm{Out}_{E_8}(X) \cong S_3$. Next, we consider the subgroup $Y = A_1^4B_2 = E_8(\#638) < B_3A_1^3$. We note that $\mathrm{Out}_{E_8}(Y) \cong S_3$, with the S_3 acting naturally on the last three A_1 factors. This yields the diagonal subgroups of Y. Finally, we consider the subgroup $Z = A_1^6 < B_3A_1^3$. Here we have $\mathrm{Out}_{E_8}(Z) \cong S_3 \times S_3 \times S_2$, where the first S_3 direct factor acts naturally on the first three A_1 factors, the second S_3 direct factor acts naturally on the final three A_1 factors and the central involution swaps the first three A_1 factors with the last three A_1 factors.

We also note that $\mathrm{Out}_{E_8}(A_2^2) \cong S_2 \times S_2$, where one involution simultaneously acts as a graph automorphism of both A_2 factors and another involution swaps the two A_2 factors.

It remains to prove that the diagonal subgroups $X = E_8(\#673) = A_1A_2 \hookrightarrow A_1^3A_2 = E_8(\#670) = Y$ via $(1, 1^{[r]}, 1^{[s]}, 10)$ $(0 < r < s)$ are E_8-irreducible when $p = 2$. To do this we use a standard application of Lemma 3.7. The X-composition factors of $L(E_8)$ can be found from those of $L(E_8) \downarrow Y$, which are given in Table 12.5. We note that there are six trivial X-composition factors of $L(E_8)$ as well as two 64-dimensional X-composition factors. We omit the details of checking that no Levi subgroup L contains an L'-irreducible subgroup of type A_1A_2 having the same composition factors as X on $L(E_8)$.

9.1.3. $M_1 = \bar{A}_3D_5$ **(**$E_8(\#109)$**) or** B_7 **(**$E_8(\#47)$**) or** A_1B_6 **(**$E_8(\#110)$**) or** B_2B_5 **(**$E_8(\#111)$**) or** B_3B_4 **(**$E_8(\#112)$**).** There is nothing for us to explicitly note here. In particular, all outer automorphisms acting on subgroups contained in M_1 are induced by elements of D_8.

9.1.4. $M_1 = B_2^2$ **(maximal if** $p \neq 2$**, contained in** B_4 **if** $p = 2$**) (**$E_8(\#113)$**).** The D_8-irreducible subgroups contained in M_1 are straightforward to determine using Lemma 3.3. When $p = 2$ the subgroup A_1^4 acting as $(1,1,1,1)$ on $V_{D_8}(\lambda_1)$ is also contained in $A_1C_4 = E_8(\#115)$. Indeed, there are just two D_8-classes of such A_1^4 subgroups and these can be distinguished by their composition factors on $L(E_8)$, as given in Table 12.5. We consider A_1^4 as a subgroup of $E_8(\#115)$ in Section 9.1.6.

9.1.5. $M_1 = B_2^2$ **(**$p \neq 2$**) (**$E_8(\#114)$**).** We note that this subgroup B_2^2 is contained in the E_8-reducible maximal subgroup B_4 of D_8 when $p = 2$ and is therefore E_8-reducible itself. The subgroups of M_1 follow in a similarly straightforward manner to those of $E_8(\#113)$ in Section 9.1.4.

9.1.6. $M_1 = A_1C_4$ **(**$E_8(\#115)$**).** By Lemma 3.3, the reductive, maximal connected subgroups of M_1 are $A_1\bar{A}_1C_3$, $A_1B_2^2$, A_1^4 $(p \neq 2)$, A_1A_1 $(p \geq 11)$ and A_1D_4 $(p = 2)$. The subgroup $A_1\bar{A}_1C_3$ is contained in $\bar{A}_1^2A_1C_3 = E_8(\#122)$ and by considering composition factors of $L(E_8)$ we find that it is conjugate to $E_8(\#564^{\{0,0\}})$. The maximal connected subgroup $A_1B_2^2$ acts as $(1,01,00) + (1,00,01)$ on $V_{D_8}(\lambda_1)$ and is thus a subgroup of either $E_8(\#676)$ or $E_8(\#677)$, both of which are of type

$A_1^2B_2^2$. It again follows from considering the composition factors of $L(E_8)$ that $A_1B_2^2$ is conjugate to $E_8(\#708^{\{0\}})$.

The maximal subgroup A_1^4 ($p \neq 2$) is still E_8-irreducible when $p = 2$, but it is now contained in $A_1^2B_2 < A_1B_3 < A_1D_4$. The diagonal subgroups of A_1^4 follow easily since $\mathrm{Out}_{D_8}(A_1^4) \cong S_4$. The diagonal subgroups of the maximal subgroup A_1A_1 ($p \geq 11$) are clear.

Next, we consider the maximal subgroup $M_2 = A_1D_4$. We note that since $\mathrm{Out}_{C_4}(D_4) \cong S_2$ we have $\mathrm{Out}_{D_8}(M_2) \cong S_2$, where the involution acts as an involutory graph automorphism of the D_4 factor. There are thus two classes of maximal subgroups of type A_1B_3. One such class acts as $(1,000)|(1,100)|(1,000)$ on $V_{D_8}(\lambda_1)$ and is thus D_8-reducible by Lemma 3.6 and hence E_8-reducible. The subgroup $A_1A_1^4$ of M_2 is contained in $A_1B_2^2 < M_1$ and hence conjugate to $E_8(\#649^{\{0,0\}})$.

Finally, we prove that $X = A_1A_2 = E_8(\#872)$ is E_8-irreducible when $p = 2$. From Table 12.5, we have

$$\begin{aligned} L(E_8) \downarrow X = {} & (3,11)/(2,30)/(2,11)/(2,03)/(2,00)^2/(1,30)^2/(1,22)/ \\ & (1,03)^2/(1,00)^4/(0,30)^2/(0,22)/(0,11)^2/(0,03)^2/(0,00)^4. \end{aligned}$$

Suppose that X is E_8-reducible. By Lemma 3.7, there exists a subgroup Z of type A_1A_2 contained L-irreducibly in a Levi subgroup L, such that X and Z have the same composition factors on $L(E_8)$. By considering the number of trivial composition factors on $L(E_8)$ of each Levi subgroup that contains an irreducible subgroup A_1A_2 (using Table 13.5 and Lemma 3.6) it follows that L' has type E_7, A_1E_6, D_7, A_2D_5, A_2D_4 or $A_1^2A_2^2$. Suppose that L' has type E_7. Then Theorem 8.1 shows that Z is conjugate to $E_7(\#m)$ for $m = 184, 307$, or 308. Using the restriction $L(E_8) \downarrow E_7 = V_{E_7}(\lambda_1)/V_{E_7}(\lambda_7)^2/V_{E_7}(0)^4$ and the composition factors given in Table 12.4, we calculate that Z does not have the same composition factors as X on $L(E_8)$, a contradiction. Now suppose that L' has type A_1E_6. Then by Theorem 7.1 the projection of Z to E_6 is conjugate to $E_6(\#m)$ for $m = 13, 14, \ldots, 23$ or 37. As before, we calculate that Z does not have the same composition factors as X on $L(E_8)$. Suppose that L' has type D_7. Then by Lemma 3.6, Z acts on $V_{D_7}(\lambda_1)$ as $((0,00)|((2,00) + (2^{[r]},00))|(0,00)) + (0,11)$ with $r \neq 0$, since Z is D_7-irreducible. Now, because $V_{D_7}(\lambda_1)$ occurs as composition factor of $L(E_8)$ with multiplicity two, it follows that Z has at least six trivial composition factors on $L(E_8)$, which is a contradiction. Now suppose that L' has type A_2D_5 or A_2D_4. Then the A_2 factor of Z is conjugate to $\bar{A}_2$ and so in all cases the composition factors of Z are not the same as those of X. Finally, suppose that L' has type $A_1^2A_2^2$. Then L' has only four composition factors of dimension at least 16 on $L(E_8)$ and all four of them have dimension 18. Since X has four composition factors of dimension 18 and two composition factors of dimension 16, it follows that X and Z do not have the same composition factors on $L(E_8)$. This final contradiction shows that X is E_8-irreducible.

9.1.7. $M_1 = A_1C_4$ $(p \neq 2)$ $(E_8(\#116))$. Notice that $X = A_1B_2^2 < M_1$ acts on $V_{D_8}(\lambda_1)$ as $(1,01,00)+(1,00,01)$ and is thus a subgroup of $E_8(\#677) = A_1^2B_2^2 < \bar{D}_4^2$. By considering the X-composition factors of $L(E_8)$ we find that X is conjugate to a D_8-reducible subgroup acting as $(2,00,00)^2 + (0,10,00) + (0,00,10)$. The rest of the lattice structure follows as for $E_8(\#115)$ in Section 9.1.6.

9.1.8. $M_1 = B_4$ $(E_8(\#45))$**.** The reductive, maximal connected subgroups of M_1 are given by Lemma 3.3. It is then a routine calculation to find the action of each subgroup on $V_{D_8}(\lambda_1) \downarrow B_4 = V_{B_4}(\lambda_4)$, which is also required when finding the composition factors on V_{26} of the subgroups of the maximal subgroup B_4 of F_4.

Firstly, the maximal connected subgroup D_4 of M_1 acts as $\lambda_3 + \lambda_4$ on $V_{D_8}(\lambda_1)$. There are two D_8-conjugacy classes of D_4 subgroups acting in such a way. By restricting the action of M_1 on $L(E_8)$ to D_4 we find the composition factors of D_4 on $L(E_8)$. This shows that D_4 is conjugate to $E_8(\#49)$.

The maximal connected subgroup A_1A_3 acts as $(1,100) + (1,001)$ on $V_{D_8}(\lambda_1)$. Therefore it is D_8-reducible by Lemma 3.6. The maximal connected subgroup B_2^2 $(p = 2)$ acts as $(01,01)$ and is maximal when $p \neq 2$. Therefore it is given the identification number $E_8(\#113\text{b})$ and has been considered previously. The action of the maximal subgroup $A_1^2B_2$ $(p \neq 2)$ on $V_{D_8}(\lambda_1)$ is $(1,1,01)$ and therefore $A_1^2B_2$ is conjugate to $E_8(\#709^{\{0\}})$.

Next, we consider the maximal connected subgroup A_1^2. When $p \geq 5$ it acts as $(3,1) + (1,3)$ on $V_{D_8}(\lambda_1)$ and is conjugate to $E_8(\#707^{\{0,0,0,0\}})$. When $p = 3$ it acts with composition factors $(3,1)/(1,1)^2/(1,3)$ and is therefore D_8-reducible by Lemma 3.6. Finally, the maximal subgroup A_1 $(p \geq 11)$ acts as $10 + 4$ and is therefore conjugate to $E_8(\#12^{\{0,0\}})$.

9.1.9. $M_1 = B_4$ $(p \neq 2)$ $(E_8(\#46))$**.** As for $E_8(\#45)$, we find the action of the maximal subgroups of M_1 on $V_{D_8}(\lambda_1)$. In this case the subgroups D_4, A_1A_3, $A_1^2B_2$ and A_1^2 $(p = 3)$ are all conjugate to D_8-reducible subgroups. When $p \geq 5$, the maximal subgroup A_1^2 is conjugate to $E_8(\#693^{\{0,0,0,0\}})$. As in the previous case, the maximal subgroup A_1 $(p \geq 11)$ is conjugate to $E_8(\#12^{\{0,0\}})$.

This completes the analysis of the case $M = D_8$.

9.2. $M = \bar{A}_1E_7$ $(E_8(\#102))$

In this section we find the lattice of E_8-irreducible subgroups contained in M. Theorem 8.1 allows us to immediately write down the lattice of M-irreducible subgroups. The subgroup $\bar{A}_1^2D_6 < \bar{A}_1E_7$ is conjugate to $E_8(\#107) < D_8$ and so any subgroup of $\bar{A}_1^2D_6$ has already been considered.

The remainder of this section deals with the question of whether an M-irreducible subgroup is E_8-irreducible.

There is one M-irreducible subgroup that we prove is E_8-reducible and hence all of its subgroups are thus E_8-reducible. Consider the subgroup $X = A_1F_4 \hookrightarrow \bar{A}_1A_1F_4 = E_8(\#881)$ via $(1,1,\lambda_1)$ when $p = 2$. By [**LS94**, p.333, Table 3] the connected centraliser of the F_4 factor is G_2 and hence $\bar{A}_1A_1F_4 < G_2F_4$. Theorem 5.1 implies that the projection of X to G_2 is G_2-reducible since $p = 2$. Therefore X is G_2F_4-reducible and thus E_8-reducible.

Now we need to prove that each subgroup of M in Table 11.5 is E_8-irreducible. Most subgroups either contain a simple E_8-irreducible subgroup or have already been considered as subgroups of D_8. The remaining cases are $E_8(\#n)$ for $n = 879, 889, 897, 898$ when $p = 2$ and $n = 911$, $913, 914, 915, 916$ when $p \neq 2$.

We start by considering the cases where $p \neq 2$. If $p \geq 5$ then the action of each subgroup on $L(E_8)$ has no trivial composition factors and hence all of the subgroups are E_8-irreducible by Corollary 3.8. Since $E_8(\#913)$ is only A_1E_7-irreducible for $p \geq 5$, it requires no further consideration.

Now let $p = 3$ and $X = E_8(\#911) = A_1A_3 \hookrightarrow \bar{A}_1A_1A_3 = E_8(\#910) = Y$ via $(1^{[r]}, 1^{[s]}, 100)$. Suppose that X is E_8-reducible. By Lemma 3.7, there exists a subgroup Z of type A_1A_3 contained L-irreducibly in a Levi subgroup L, such that X and Z have the same composition factors on $L(E_8)$. From Table 12.5 we see that

$$L(E_8) \downarrow Y = (2,0,000)/(1,2,010)^2/(1,0,200)/(1,0,002)/(0,4,000)/ \\ (0,2,101)^2/(0,2,000)/(0,0,101)/(0,0,020)/(0,0,000)^2.$$

Therefore, X and Z have only two trivial composition factors on $L(E_8)$ and so L' has at most two trivial composition factors on $L(E_8)$. Using the restrictions in Table 13.5, we find that L' has one of the following types: A_1E_6, D_7, A_2D_5, A_7, A_3A_4, A_1A_6, $A_1A_2A_4$, $A_1^2A_4$, or A_3^2. Lemma 3.6 immediately rules out L' having type A_3A_4, A_1A_6, $A_1A_2A_4$ or $A_1^2A_4$ since such a Levi subgroup contains no irreducible subgroup A_1A_3. We will now take the other five types and rule them out in turn. Suppose that L' has type A_1E_6. Then one L'-composition factor of $L(E_8)$ is isomorphic to $(1,0)$. Hence Z has a 2-dimensional composition factor on $L(E_8)$, a contradiction. Now suppose that L' has type A_7. Then from Lemma 3.6 we deduce that Z acts on $V_{A_7}(\lambda_1)$ as $(1,100)$. It is straightforward to calculate the Z-composition factors of $L(E_8)$ from the L'-composition factors given in Table 13.5. These are not the same as those of X, a contradiction. Now suppose that L' has type A_2D_5 or D_7. Then the A_3 factor of Z is conjugate to a Levi subgroup $\bar{A}_3$, acting as $010 + 000^4$ on $V_{D_5}(\lambda_1)$ and $010 + 000^8$ on $V_{D_7}(\lambda_1)$. In particular, there is no Z-composition factor of $L(E_8)$ isomorphic to $(0,020)$. Therefore the composition factors of X and Z are not the same on $L(E_8)$, a contradiction. Finally, suppose that L' has type A_3^2. Then by Table 13.5, the largest dimension of any composition factor of $L(E_8) \downarrow L'$ is 24. This is a contradiction since X has a 45-dimensional composition factor on $L(E_8)$, namely $(2^{[s]}, 101)$.

A similar argument applies to show that $X = E_8(\#n)$ is E_8-irreducible for $n = 914$, 915 and 916. In particular, there are only two trivial X-composition factors of $L(E_8)$.

We now consider the cases where $p = 2$. Firstly, since $\bar{A}_1A_7 = E_8(\#879)$ has rank 8 it is clearly G-irreducible. Now let $X = \bar{A}_1\bar{A}_2A_3 = E_8(\#889)$. Suppose that X is E_8-reducible. By Lemma 3.7, there exists a subgroup Z of type $A_1A_2A_3$ contained L-irreducibly in a Levi subgroup L, such that X and Z have the same composition factors on $L(E_8)$. From Table 12.5 we find that X has just two trivial composition factors on $L(E_8)$. It thus follows that L' is of type A_2D_5. Using Lemma 3.6, we deduce that the A_1A_3 factor of Z is contained in the D_5 factor of A_2D_5 and acts as $(1^{[r]} \otimes 1^{[s]}, 000) + (0,010)$ on $V_{D_5}(\lambda_1)$. From this we calculate the Z-composition factors of $L(E_8)$ and see that they are not the same as those of X, a contradiction. Therefore X is E_8-irreducible.

Finally, we need to consider the cases where X is $E_8(\#897)$ or $E_8(\#898)$ and so $X = A_1A_2 \hookrightarrow \bar{A}_1\bar{A}_2A_1A_2 = E_8(\#887) = Y$ via $(1^{[r]}, 10^{[t]}, 1^{[s]}, 10^{[u]})$ $(rs = tu = 0)$ or $(1^{[r]}, 10^{[t]}, 1^{[s]}, 01^{[u]})$ $(rs = tu = 0;\ t \neq u)$, respectively. Using the composition factors of $L(E_8) \downarrow Y$ given in Table 12.5, we find that X has at most six trivial composition factors on $L(E_8)$. The case where X has exactly six occurs when X is embedded via $(1, 10^{[1]}, 1, 10)$. We prove that X is E_8-irreducible in this case; the others are similar and easier. Suppose that X is E_8-reducible. By Lemma 3.7, there exists a subgroup Z of type A_1A_2 contained L-irreducibly in a Levi subgroup L, such that X and Z have the same composition factors on $L(E_8)$. Restricting the Y-composition factors of $L(E_8)$ to X yields

$$L(E_8) \downarrow X = (4,00)/(2,30)^2/(2,11)^2/(2,03)^2/(2,00)^4/(0,30)^4/(0,22)^3/ \\ (0,11)^4/(0,03)^4/(0,00)^6.$$

By considering the number of trivial composition factors on $L(E_8)$ of each Levi subgroup containing an irreducible subgroup A_1A_2 (using Table 13.5 and Lemma 3.6), it follows that L' has type E_7, A_1E_6, D_7, A_2D_5, A_2D_4, A_1A_5, $A_1A_2A_3$ or $A_1^2A_2^2$. Suppose that L' has type E_7 or A_1E_6. Then Theorem 8.1 implies that Z is conjugate to $E_7(\#m)$ for $m = 184$, 307, or 308 and Theorem 7.1 implies that the projection of Z to E_6 is conjugate to $E_6(\#m)$ for $m = 13, 14, \ldots, 23$ or 37. Using the restrictions in Tables 12.3, 12.4 and 13.5 we calculate the Z-composition factors of $L(E_8)$; in all cases these are not the same as the X-composition factors, a contradiction. Now suppose that L' has type D_7. Lemma 3.6 implies that Z acts on $V_{D_7}(\lambda_1)$ as $((0,00)|((2,00)+(2^{[r]},00))|(0,00))+(0,11)$ with $r \neq 0$, since Z is D_7-irreducible. From Table 13.5, the D_7-composition factor $V_{D_7}(\lambda_1)$ of $L(E_8)$ has multiplicity two. Hence $(2,00)$ and $(2^{[r]},00)$ both occur as Z-composition factors of $L(E_8)$ with multiplicity at least two. Comparing with the X-composition factors we see that $r=0$, a contradiction. Suppose that L' has type A_2D_5, A_2D_4 or $A_1A_2A_3$. Then the A_2 factor of Z is conjugate to $\bar{A}_2$ and so in all cases the composition factors of Z are not the same as those of X. For the next case let L' have type A_1A_5. Then L' has only five composition factors of dimension at least 16 on $L(E_8)$: four of them have dimension 20 and the other has dimension 23. Since X has four composition factors of dimension 18 and two composition factors of dimension 16, it follows that X and Z do not have the same composition factors on $L(E_8)$. Finally, we suppose that L' has type $A_1^2A_2^2$. Then L' has only four composition factors of dimension at least 16 on $L(E_8)$ and all four of them have dimension 18. As in the previous case, it follows that X and Z do not have the same composition factors on $L(E_8)$. This final contradiction proves that X is E_8-irreducible.

This completes the analysis of the E_8-irreducible subgroups contained in $M = \bar{A}_1E_7$.

9.3. $M = \bar{A}_2E_6$ ($E_8(\#103)$)

Theorem 7.1 and Lemma 3.3 allow us to write down the lattice of M-irreducible subgroups. Firstly, we note that the maximal subgroup $\bar{A}_2\bar{A}_1A_5$ is contained in $\bar{A}_1E_7$ and conjugate to $E_8(\#878)$. We will now consider subgroups of $\bar{A}_2Y$ for the remaining reductive, maximal connected subgroups Y of E_6. In doing this we will consider all of the subgroups of $A_1E_6 = E_8(\#981)$ ($p \neq 2$) in the following sections and thus give it no further consideration.

9.3.1. $M_1 = \bar{A}_2^4$ **($E_8(\#975)$).** As in [**Tho15**, Section 7.3] we fix a conjugacy class representative of M_1 by giving the M_1-composition factors of $L(E_8)$ in Table 12.5. The diagonal subgroups follow from the action of $\mathrm{Out}_{E_8}(M_1) \cong \mathrm{GL}(2,3)$ (by [**LS04**, Table 10.3]) on the four A_2 factors.

We prove that $A_1^4 < M_1$ ($p \neq 2$) is contained in D_8 and conjugate to $E_8(\#873)$. Theorem 7.1 implies that $A_1^3 < \bar{A}_2^3 < E_6$ is contained in the maximal subgroup C_4 of E_6. When we consider $\bar{A}_2C_4$ in Section 9.3.3, we prove that A_1C_4 is conjugate to $E_8(\#116)$. Hence A_1^4 is contained in $E_8(\#116)$ and conjugate to $E_8(\#873)$.

9.3.2. $M_1 = \bar{A}_2F_4$ $(E_8(\#976))$**.** We first note M_1 is a subgroup of G_2F_4. Thus, by Theorem 5.1, the subgroup A_1F_4 $(p \neq 2)$ is conjugate to $A_1F_4 \hookrightarrow \bar{A}_1A_1F_4 = E_8(\#881) < \bar{A}_1E_7$ via $(1, 1, \lambda_1)$. Therefore we need only consider subgroups of the form $\bar{A}_2X$ where X is an E_6-irreducible connected subgroup of F_4.

By Theorem 7.1, we have that $\bar{A}_2\bar{A}_1C_3$ is a subgroup of $\bar{A}_2\bar{A}_1A_5$ and $\bar{A}_2^2A_2$ is a subgroup of $\bar{A}_2^4$. We will consider $\bar{A}_2C_4$ $(p = 2)$ in Section 9.3.3, and similarly we consider $\bar{A}_2G_2$ $(p = 7)$ in Section 9.3.5.

9.3.3. $M_1 = \bar{A}_2C_4$ $(E_8(\#977))$**.** The subgroup M_1 is maximal if $p \neq 2$ and contained in $\bar{A}_2F_4$ when $p = 2$. Firstly, we prove that $X = A_1C_4$ $(p \neq 2)$ is contained in D_8 and conjugate to $Y = A_1C_4 = E_8(\#116)$. Indeed, since $p \neq 2$, [**LS96**, Table 8.1] shows that there are exactly two classes of subgroups of type C_4 in E_8; they are contained in the two classes of subgroups of type A_7. In particular, the C_4 factor of Y is contained in E_6 and the connected centraliser is $\bar{A}_2$. It follows that Y is contained in $\bar{A}_2C_4$ and thus conjugate to X.

We now need to consider subgroups of the form $X = \bar{A}_2Y$ where Y is an E_6-irreducible connected subgroup of C_4. If Y is contained in $\bar{A}_1C_3$ or A_1^3 then X is contained in $\bar{A}_2\bar{A}_1A_5$ or $\bar{A}_2^4$, respectively. If $Y = A_1$ $(p \geq 11)$ then X requires no further consideration. The final case is $Y = D_4$ $(p = 2)$. In this case the only E_6-irreducible subgroup of Y is A_2. This subgroup A_2 is contained in $\bar{A}_2^3$, again by Theorem 7.1.

9.3.4. $M_1 = \bar{A}_2A_2G_2$ $(E_8(\#978))$**.** Theorem 7.1 shows that $\bar{A}_2A_1G_2$ is contained in $\bar{A}_2F_4$. Similarly, $\bar{A}_2A_2\bar{A}_1A_1$ and $\bar{A}_2A_2\bar{A}_2$ are contained in $\bar{A}_2\bar{A}_1A_5$ and $\bar{A}_2^4$, respectively.

The diagonal subgroups of M_1 follow by noting that $\bar{A}_2A_2$ is a maximal subgroup of F_4 and thus there is an involution in E_8 acting as a graph automorphism on both A_2 factors. Moreover, by Theorem 6.1 the subgroup $A_2G_2 \hookrightarrow M_1$ via $(10, 10, 10)$ is G_2F_4-reducible when $p = 3$. Note that looking at Theorem 6.1 would seem to suggest the subgroup $A_2G_2 \hookrightarrow M_1$ via $(10, 01, 10)$ is the G_2F_4-reducible subgroup. However, the given M_1-composition factors of $L(E_8)$ in Table 12.5 show that a graph automorphism of the second A_2 factor has been introduced.

The non-diagonal subgroups of $\bar{A}_2A_2A_2$ $(p = 3)$ have all been covered previously and the simple diagonal subgroups are given by [**Tho16**, Lemma 7.11]. The non-simple diagonal subgroups follow in the same way. Similarly, the non-diagonal subgroups of $\bar{A}_2A_2A_1$ $(p \geq 7)$ have already been considered and the diagonal subgroups follow in the same way as the diagonal subgroups of M_1.

9.3.5. $M_1 = \bar{A}_2G_2$ $(E_8(\#979))$**.** The subgroup M_1 is maximal if $p \neq 7$ and contained in $\bar{A}_2F_4$ if $p = 7$. By Theorem 7.1, except for A_2 $(p = 3)$, all reductive, maximal connected subgroups of the G_2 factor are contained in a previously considered maximal connected subgroup of E_6. It remains to consider subgroups of $\bar{A}_2A_2$ $(p = 3)$. We postpone this until Section 9.3.6.

9.3.6. $M_1 = \bar{A}_2A_2$ $(p \neq 2)$ $(E_8(\#980))$**.** The subgroup M_1 is maximal if $p \geq 5$ and contained in $\bar{A}_2G_2$ if $p = 3$. The non-diagonal reductive, maximal connected subgroups of M_1 are $\bar{A}_2A_1$ and A_1A_2. Theorem 7.1 implies that $\bar{A}_2A_1$ is contained in $\bar{A}_2\bar{A}_1A_5$ and hence the subgroup A_1A_1 of A_1A_2 is also contained in $\bar{A}_2\bar{A}_1A_5$.

The subgroup $X = A_1A_2 = E_8(\#1027)$ when $p = 5$ is the only subgroup that does not properly contain an E_8-irreducible subgroup. The composition factors of

$L(E_8) \downarrow X$ are given in Table 12.5. Since there are no trivial composition factors, Corollary 3.8 shows that X is E_8-irreducible.

This completes the case $M = \bar{A}_2E_6$.

9.4. $M = A_8$ ($E_8(\#62)$)

By Lemma 3.3, the only reductive, maximal connected subgroups of A_8 are B_4 ($p \neq 2$) and A_2^2. The subgroup B_4 is contained in D_8 and so its non-simple subgroups require no consideration. The subgroup A_2^2 was studied in [**Tho16**, Section 7.2]. In particular, [**Tho16**, Lemma 7.10] shows that A_2^2 is E_8-irreducible and conjugate to $E_8(\#668)$ when $p \neq 3$ but E_8-reducible when $p = 3$.

9.5. $M = \bar{A}_4^2$ ($E_8(\#104)$)

By Lemma 3.3, the only reductive, maximal connected subgroup of $\bar{A}_4$ is B_2 ($p \neq 2$). The only $\bar{A}_4$-irreducible maximal connected subgroup of B_2 is A_1 ($p \geq 5$) embedded via the representation of high weight 4. The final thing to note is that the subgroup B_2^2 of $\bar{A}_4^2$ is contained in D_8. Indeed, this is shown in [**LS96**, p.63].

9.6. $M = G_2F_4$ ($E_8(\#105)$)

The irreducible connected subgroups of G_2 and F_4 are given by Theorems 5.1 and 6.1, respectively. This allows us to write down all of the M-irreducible connected subgroups when noting the following details.

Firstly, the G_2 factor is contained in a Levi subgroup D_4 by [**Sei91**, 3.16] and thus $\bar{A}_1A_1F_4$ is contained in $\bar{A}_1E_7$ and $\bar{A}_2F_4$ is contained in $\bar{A}_2E_6$. Similarly, the subgroups G_2B_4, $G_2\bar{A}_1C_3$ and $G_2\bar{A}_2A_2$ have all been considered before in the D_8, $\bar{A}_1E_7$ and $\bar{A}_2E_6$ cases, respectively.

We now consider the remaining maximal connected subgroups of M in turn. Once we have considered A_2F_4 ($p = 3$) and A_1F_4 ($p \geq 7$) we will only consider subgroups of the form G_2X where X is an F_4-irreducible subgroup, since all others will have been covered. In particular, we will not make any further mention of G_2A_1 ($p \geq 13$).

9.6.1. $M_1 = A_2F_4$ ($p = 3$) ($E_8(\#1030)$)**.** By Theorem 5.1, the irreducible subgroup A_1 of the A_2 factor is contained in $\bar{A}_1A_1$. We have hence already considered the maximal connected subgroup A_1F_4 of M_1 and need only consider subgroups of the form A_2X where X is an F_4-irreducible subgroup. Similarly, A_2B_4, $A_2\bar{A}_1C_3$ and $A_2\bar{A}_2A_2$ have already been considered and so it remains to examine $A_2A_1G_2$.

Theorem 6.1 implies that the only reductive, maximal connected subgroup of A_1G_2 not previously considered is A_1A_2 (where the subgroup A_2 of G_2 is generated by short root subgroups of G_2). This yields $A_2^2A_1$, since the G_2 factor of M and the G_2 factor of A_1G_2 are conjugate and there exists an involution in the normaliser of $G_2A_1G_2$ swapping the two G_2 factors by the construction in [**Sei91**, p.39]. It remains to consider the diagonal subgroups of $A_2^2A_1$. Each subgroup G_2 contains an element inducing a graph automorphism of A_2 and combined with the involution which swaps the A_2 subgroups we obtain the classes of diagonal subgroups $E_8(\#1038)$.

9.6.2. $M_1 = A_1F_4$ $(p \geq 7)$ $(E_8(\#1031))$. We need to consider the subgroups $A_1A_1G_2$, A_1A_1 $(p \geq 13)$ and A_1G_2 $(p = 7)$ since the other reductive, maximal connected subgroups of M_1 have already been considered. Firstly, let $M_2 = A_1A_1G_2 = E_8(\#1039)$. The diagonal subgroups of M_2 are as listed in Table 11.5. We use Theorem 6.1 to prove that $X = A_1G_2 \hookrightarrow M_2$ via $(1,1,10)$ is conjugate to $E_8(\#952^{\{0,0\}})$. We may consider X as a subgroup of $G_2A_1A_1 < G_2A_1G_2 < G_2F_4$ by swapping the two G_2 factors. Theorem 6.1 shows that $A_1 \hookrightarrow A_1A_1 = F_4(\#66) < A_1G_2 < F_4$ via $(1,1)$ is conjugate to $A_1 \hookrightarrow \bar{A}_1A_1 = F_4(\#58) < \bar{A}_1C_3 < F_4$ where the subgroup A_1 of C_3 is maximal. Therefore X is a subgroup of $\bar{A}_1G_2C_3 < \bar{A}_1E_7$ and comparing composition factors shows that X is indeed conjugate to $E_8(\#952^{\{0,0\}})$.

Now, using Theorem 6.1, we need only consider the reductive, maximal connected subgroup A_1A_1 of $A_1G_2 < F_4$, where the subgroup A_1 of G_2 is maximal. This yields $A_1^2A_1 = E_8(\#1042)$ since the G_2 factors are conjugate, as shown in the previous section. The classes of diagonal subgroups of $A_1^2A_1$ then follow. We claim that $X_1 \hookrightarrow A_1^2A_1$ via $(1_a,1_a,1_b)$ and $X_2 \hookrightarrow A_1^2A_1$ via $(1_a,1_b,1_b)$ are contained in $\bar{A}_1E_7$. To prove the claim for X_1, we consider the involution $t \in N_{E_8}(A_1^2A_1)$ swapping the first two A_1 factors. The subgroup X_1 is centralised by t and a routine check shows that the full centraliser of t in E_8 is $\bar{A}_1E_7$ rather than D_8. The claim for X_2 follows from Theorem 6.1. Indeed, as above we have $A_1 \hookrightarrow A_1A_1 < A_1G_2 < F_4$ via $(1,1)$ is contained in $\bar{A}_1C_3$ and thus X_2 is contained in $\bar{A}_1G_2C_3 < \bar{A}_1E_7$, as claimed.

The irreducible subgroups of A_1A_1 $(p \geq 13)$ are all simple and thus require no further consideration. Finally, we use Theorem 6.1 to see that all subgroups of A_1G_2 $(p = 7)$ have already been considered.

9.6.3. $M_1 = G_2C_4$ $(p = 2)$ $(E_8(\#1032))$. We need only consider the maximal connected subgroup G_2D_4 since $G_2\bar{A}_1C_3$ and $G_2B_2^2$ are contained in $\bar{A}_1E_7$ and D_8, respectively. Theorem 6.1 implies that the subgroups of the D_4 factor are contained in B_4 or $\bar{A}_2A_2$ and so have already been considered.

9.6.4. $M_1 = G_2^2A_1$ $(p \neq 2)$ $(E_8(\#1033))$. As noted above, the G_2 factors of M_1 are E_8-conjugate and furthermore, there is an involution t in $\mathrm{Out}_{E_8}(G_2^2A_1)$ that swaps the two factors. We need only consider the diagonal subgroups of M_1 since any other reductive, maximal connected subgroup of M_1 has been considered above. The diagonal subgroups of M_1 are as in Table 11.5; we note that only when $p = 3$ does there exist a special isogeny of G_2 yielding the subgroups $E_8(\#1048)$, with the notation as in Section 2. Finally, we note that $X = G_2A_1 \hookrightarrow M_1$ via $(10,10,1)$ is centralised by t and thus contained in $\bar{A}_1E_7$, as above. By considering the X-composition factors of $L(E_8)$ we see that X is conjugate to $E_8(\#967^{\{0,0\}})$.

9.6.5. $M_1 = G_2G_2$ $(p = 7)$ $(E_8(\#1035))$. From Theorem 6.1, we see that the reductive, maximal connected subgroups of the second G_2 factor are contained in other reductive, maximal connected subgroups of F_4 and have thus already been considered.

This completes the case $M = G_2F_4$.

9.7. $M = F_4$ $(p = 3)$ $(E_8(\#1049))$

By Theorem 3.1, the reductive maximal connected subgroups of F_4 are B_4, A_1C_3, A_1G_2 and A_2A_2. It is proven in [**CST**], that the first three subgroups are

G-irreducible but the fourth is not. Moreover, the subgroup B_4 is conjugate to $E_8(\#45)$, the subgroup A_1C_3 is conjugate to $E_8(\#774)$ and the subgroup A_1G_2 is conjugate to $E_8(\#967^{\{1,0\}})$.

9.8. $M = B_2$ ($p \geq 5$) ($E_8(\#101)$)

By Lemma 3.3, the only reductive, maximal connected subgroups of B_2 are A_1^2 and A_1. The subgroup A_1^2 is the centraliser in B_2 of a semisimple involution t by [**GLS98**, Table 4.3.1]. By Lemma 3.9, the centraliser of t in E_8 is either D_8 or $\bar{A}_1E_7$. We have therefore considered the subgroup A_1^2 before and by considering composition factors we find that it is conjugate to $E_8(\#707^{\{0,0,0,0\}})$.

9.9. $M = A_1A_2$ ($p \geq 5$) ($E_8(\#106)$)

The only reductive, maximal connected subgroup of A_1A_2 is A_1A_1, where the second A_1 factor is irreducibly embedded in A_2. By [**Sei91**, p. 31], there exists an involution t in $N_{E_8}(M)$ such that t centralises the A_1 factor of M and acts as a graph automorphism of the A_2 factor. Thus A_1A_1 is contained in $C_{E_8}(t)^{\circ}$. As before, we calculate the connected centraliser of t is $\bar{A}_1E_7$ and hence A_1A_1 is conjugate to $E_8(\#973^{\{0,0\}})$.

This completes the proof of Theorem 9.1 and thus the proof of Theorem 1.

CHAPTER 10

Corollaries

In this chapter we give the proofs of Corollaries 1–7, as well as giving further corollaries not mentioned in the introduction. Let G be a simple exceptional algebraic group over an algebraically closed field of characteristic p.

Corollaries 4 and 6 are immediate from Theorem 1. Similarly, Corollaries 5 and 7 follow from the lattice structure given in Tables 1–11.5. Further, the proofs of Corollaries 1 and 2 follow from careful inspection of Tables 12.1–12.5, recalling that one can deduce the composition factors of all G-irreducible connected subgroups from the composition factors given for the non-diagonally embedded subgroups. We are left to prove Corollary 3.

Proof of Corollary 3 The strategy for the proof is as follows. For each simple exceptional algebraic group G we find all M-irreducible connected subgroups that are not G-irreducible. Given such a subgroup X we then check whether it satisfies the hypothesis of Corollary 3. That is to say, we check whether X is contained reducibly in another reductive, maximal connected subgroup, or if X is contained in a Levi subgroup of G. To do this we use the composition factors of X on the minimal or adjoint module for G, using restriction from M. Of course, since X is G-reducible there exists some subgroup Z of the same type as X contained in a Levi factor L' having the same composition factors as X. Therefore, we will require the exact module structure of X acting on either the minimal or adjoint module for G to prove that X is not contained in L'.

This has already been done when X is simple in [**Tho16**, Corollary 5] and [**Tho15**, Corollary 2]. Moreover, suppose that Y is a subgroup of M containing such a simple subgroup X and that Y is G-reducible. Then Y also satisfies the hypothesis of the corollary.

By studying the proofs of Theorems 5.1–6.1 we find that there are no non-simple irreducible connected subgroups which are M-irreducible yet G-reducible when G is of type G_2 or F_4. So we need only consider the cases where G has type E_6, E_7 and E_8.

Suppose that G is of type E_6 and consider the M-irreducible non-simple subgroups that are E_6-reducible. These are all found in the proof of Theorem 7.1; let X be such a subgroup. Firstly, consider $M = \bar{A}_1A_5$. Then X is a subgroup of $A_1A_2 \hookrightarrow \bar{A}_1A_1A_2 = E_6(\#28)$ via $(1,1,10)$. It is shown in Section 7.1.2 that A_1A_2 is contained in A_2G_2 and moreover, is A_2G_2-reducible. Thus X does not satisfy the hypothesis of the corollary. Now suppose that $M = F_4$ or C_4 $(p \neq 2)$. Then X is a subgroup of B_4 or B_2^2, respectively, both of which are contained in a Levi subgroup of type D_5. Hence X does not satisfy the hypothesis of Corollary 3.

The remaining case when G is of type E_6 to consider is $X = A_1A_1 < G_2 = M$ when $p = 2$. We prove that X does not satisfy the hypothesis of the corollary by showing that X is contained reducibly in $\bar{A}_1A_5$. Let AB be the subgroup

$A_1A_1 < A_5$ acting as $(W(2),1)$ on $V_{A_5}(\lambda_1)$ and let $Y = A_1A_1 \hookrightarrow \bar{A}_1AB < \bar{A}_1A_5$ via $(1_a, 1_a, 1_b)$. We claim that X is conjugate to Y.

Firstly, note that both X and Y are contained in parabolic subgroups of E_6. Moreover, Y is contained in a parabolic subgroup of $\bar{A}_1A_5$ with Levi factor L' of type $\bar{A}_1\bar{A}_1A_3$ with the projection of Y to L' being L'-irreducible. The only Levi subgroup of E_6 containing $\bar{A}_1^2A_3$ is D_5. Therefore Y is contained in $P = QL$, a D_5-parabolic subgroup of E_6, with irreducible projection to $L' = D_5$. Specifically, the projection of Y to L' is conjugate to $Z = A_1A_1$ acting on $V_{D_5}(\lambda_1)$ as $(1,1) + ((0,0)|((2,0)+(0,4))|(0,0))$. Moreover, using the action of $\bar{A}_1A_5$ on V_{27} we find that Y and Z are non-$\mathrm{GL}_{27}(K)$-conjugate and thus Y is non-E_6-cr.

To prove that X is conjugate to Y we show that there are just two $\mathrm{Aut}(E_6)$-conjugacy classes of subgroups of type A_1A_1 contained in QZ, namely the E_6-cr subgroup Z and the non-E_6-cr subgroup Y. First note that Q is abelian and by [**ABS90**], Q is an L'-module with high weight λ_4 (or λ_5 depending on the choice of D_5-parabolic subgroup). Moreover, the action of Z on either spin module for D_5 is $(1,3) + (2, T(2))$. We need to calculate the Hochschild cohomology group $H^1(Z,Q)$. It is well known that $H^1(A_1, M) = 0$ for the modules $M = 1, T(2), 3$ and $H^1(A_1, 2) \cong K$ so applying Künneth's formula [**Rot09**, 10.85] yields $H^1(Z,Q) \cong K$. Considering the non-trivial action of $Z(L)$ on Q shows that there is just one conjugacy class of subgroups of type A_1A_1 contained in QZ which is not Q-conjugate to Z. This proves the claim.

Now we need to prove that a conjugate of X is contained in P with its projection to L' being conjugate to Z. From this it follows that X does not satisfy the hypothesis of the corollary: X is either contained in a Levi subgroup or contained reducibly in $\bar{A}_1A_5$. In fact, it follows that X is conjugate to Y by considering the action of X on V_{27} (using the action of the maximal subgroup G_2 given in [**LS04**, Table 10.2]). It is shown in the proof of Theorem 7.1 that X is contained in a parabolic subgroup of E_6. The composition factors of the action of X on V_{27} are $(2,2)/(2,0)/(1,3)/(1,1)/(0,4)/(0,2)^2/(0,0)^3$. By Lemma 3.7 there exists an irreducible subgroup of a Levi factor with the same composition factors as X. It is a routine calculation, using the composition factors in Table 13.3, to find that D_5 is the only Levi subgroup with an irreducible subgroup having the same composition factors as X, and that this irreducible subgroup is conjugate to Z. Therefore X is contained in QZ, as required.

We note that if one considers E_6-conjugacy rather than $\mathrm{Aut}(E_6)$-conjugacy then there are two classes of D_5-parabolic subgroups, with representatives $P_1 = Q_1L_1$ and $P_2 = Q_2L_2$, say. Therefore, we have two classes of irreducible subgroups A_1A_1, say Z_1 in P_1 and Z_2 in P_2. There are also two classes of maximal subgroup G_2 and so the class of X splits into two classes, with representatives X_1 and X_2, say. Furthermore, there is no longer an element acting as a graph automorphism on $\bar{A}_1A_5$ and thus the class of Y splits into two classes, with representatives Y_1 and Y_2, say. One then finds that (up to reordering) X_i is conjugate to Y_i, both being non-G-cr and contained in Q_iZ_i.

Next, let G be of type E_7 and consider the M-irreducible non-simple subgroups that are E_7-reducible. These are all found in the proof of Theorem 8.1 and we let X be such a subgroup. First suppose that $M = \bar{A}_1D_6$. Then X is a subgroup of $A_1C_3 \hookrightarrow \bar{A}_1A_1C_3 = E_7(\#42)$ via $(1,1)$ with $p = 2$. In Section 8.1.7, we proved that the subgroup A_1C_3 is contained reducibly in G_2C_3. Therefore X does not satisfy the hypothesis of the corollary.

Now let $M = \bar{A}_2A_5$. Then X is a subgroup of $A_2A_1 \hookrightarrow \bar{A}_2A_2A_1$ via $(10, 01, 1)$ with $p = 3$. It is shown in Section 8.2.1 that A_2A_1 is contained in the maximal subgroup A_1F_4 and is A_1F_4-reducible. Therefore X does not satisfy the hypothesis of Corollary 3.

The last case for G of type E_7 we need to consider is $M = A_7$. Here X is any A_7-irreducible subgroup contained in C_4 when $p = 2$. In particular, either X contains the subgroup $Y = A_1$ acting as $1 \otimes 1^{[r]} \otimes 1^{[s]}$ on $V_{A_7}(\lambda_1)$ or X is of type A_2 acting as 11 on $V_{A_7}(\lambda_1)$. From [**Tho16**, Corollary 5] we know that Y satisfies the hypothesis of the corollary. It remains to consider $X = A_2$. The composition factors of X on V_{56} follow from $V_{56} \downarrow A_7 = V_{A_7}(\lambda_2)/V_{A_7}(\lambda_6)$ and are thus $30^2/11^2/03^2/00^4$. By using them and considering the composition factors of Levi subgroups from Table 13.4 and reductive, maximal connected subgroups from Table 12.4, we conclude that X satisfies the hypothesis of the corollary or is contained in either $\bar{A}_2A_5$ or a Levi subgroup E_6. Using the finite subgroup $S = A_2(4) < X$ and calculation in Magma [**BCP97**] we find that $V_{56} \downarrow X = (00|(30 + 03)|00)^2 + 11^2$. Since X has two direct summands of dimension 20 and no trivial direct summands, it follows that X is not contained in $\bar{A}_2A_5$ or E_6. Hence X satisfies the hypothesis of the corollary.

Finally, suppose that G is of type E_8. There are many subgroups of D_8 to consider when $p = 2$. Specifically, we need to consider all D_8-irreducible subgroups of $A_1C_4(\ddagger)$ and $B_4(\ddagger)$, as defined in Section 9.1. Using Lemma 3.6 we find that all such subgroups are those given in Table 1 as well as all D_8-irreducible subgroups of $Y = A_1\bar{A}_1C_3 < A_1C_4(\ddagger)$. Since Y is a diagonal subgroup of $\bar{A}_1^2A_1C_3 = E_8(\#122)$, it follows that Y is a subgroup of G_2F_4. Moreover, the projection of Y to G_2 is conjugate to $A_1 \hookrightarrow \bar{A}_1A_1 < G_2$ via $(1, 1)$, as may be seen by noting that the $\bar{A}_1C_3$ factor of Y is contained in F_4. Thus Theorem 5.1 implies that Y is G_2F_4-reducible. Hence Y does not satisfy the hypothesis of the corollary.

We need to prove that the remaining D_8-irreducible subgroups of $A_1C_4(\ddagger)$ and $B_4(\ddagger)$ satisfy the hypothesis of the corollary. In all but the case $X = A_1A_2 < A_1C_4$ such subgroups contain the D_8-irreducible subgroup of type A_1 acting as $1 \otimes 1^{[r]} \otimes 1^{[s]} \otimes 1^{[t]}$ $(0 < r < s < t)$. By [**Tho16**, Corollary 5], this subgroup A_1 satisfies the hypothesis of Corollary 3 and thus so does every subgroup containing it.

Now we let $X = A_1A_2 < A_1C_4(\ddagger)$ acting on $V_{D_8}(\lambda_1)$ as $(1, 11)$. We consider the action of the finite subgroup $S = A_1(8)A_2(8) < X$ on $L(E_8)$. From [**LS12**, Lemma 11.2], we have $L(G) \downarrow D_8 = L(D_8) + V_{D_8}(\lambda_7)$. We use the inbuilt functionality of Magma [**BCP97**] to find $L(D_8) \downarrow S$ and $V_{D_8}(\lambda_7) \downarrow S$. It follows that both modules have two direct summands, of dimensions $88, 32$ and $96, 32$, respectively.

We conclude that $L(E_8) \downarrow X$ has a direct summand of dimension at least 96, another of dimension at least 88 and no trivial direct summands. It easily follows that X is not contained in any Levi subgroup or another reductive, maximal connected subgroup of G. Since X is G-reducible it follows that X is non-G-cr and satisfies the hypothesis of Corollary 3.

We now consider the G-reducible subgroups of the other reductive, maximal connected subgroups of G. Let $M = \bar{A}_1E_7$ so X is any M-irreducible subgroup of $A_1F_4 \hookrightarrow \bar{A}_1A_1F_4 = E_8(\#881)$ via $(1, 1, \lambda_1)$ with $p = 2$. Then X does not satisfy the hypothesis of Corollary 3 because A_1F_4 is G_2F_4-reducible. Similarly, no subgroups of $M = \bar{A}_2E_6$ satisfy the hypothesis of the corollary. Indeed, X is any M-irreducible subgroup of $Y = A_2G_2 \hookrightarrow \bar{A}_2A_2G_2 = E_8(\#978)$ via $(10, 10, 10)$ with $p = 3$, and Y is G_2F_4-reducible, as shown in Section 9.3.4. The only M-irreducible subgroups we need to consider of $M = F_4$ $(p = 3)$ are subgroups of $X = A_2A_2$. It is proved in [**CST**] that X is conjugate to a subgroup of $E_8(\#1012)$, which is contained reducibly in the maximal subgroup G_2F_4. Thus X, and all of its subgroups, do not satisfy the hypothesis of the corollary.

The last case to consider is $M = A_8$ and $X = A_2^2$ acting as $(10, 10)$ on $V_{A_8}(\lambda_1)$. Since the diagonal M-irreducible subgroups of X satisfy the hypothesis of the corollary it follows that X itself does. □

10.1. Variations of Steinberg's Tensor Product Theorem

We need some background for the next corollary. Let X be a simple, simply connected algebraic group over an algebraically closed field K of characteristic $p < \infty$. We recall Steinberg's tensor product theorem [**Ste63**]. It states that if $\phi\colon X \to \mathrm{SL}(V)$ is an irreducible rational representation, then we can write $V = V_1^{[r_1]} \otimes \cdots \otimes V_k^{[r_k]}$, where the V_i are restricted X-modules and the r_i are distinct. The main result of [**LS03**] generalises this conclusion to the situation where ϕ is a rational homomorphism from X to an arbitrary simple algebraic group G. To describe this generalisation we need the following definition. Throughout this section we let G be a simple exceptional algebraic group.

DEFINITION 10.1. [**LS03**, p. 263] A simple, simply connected subgroup X of G is *restricted* if all composition factors of $L(G) \downarrow X$ are restricted if X is not of type A_1, and are of high weight at most $2p - 2$ if X is of type A_1.

THEOREM 10.2. [**LS03**, Corollary 1] *Assume p is good for G. If X is a connected simple G-cr subgroup of G, then there is a uniquely determined commuting product $E_1 \dots E_k$ with $X \leq E_1 \dots E_k \leq G$, such that each E_i is a simple restricted subgroup of the same type as X, and each of the projections $X \to E_i/Z(E_i)$ is non-trivial and involves a different field twist.*

Using our classification of G-irreducible connected subgroups, we investigate to what extent Theorem 10.2 is true in bad characteristics for simple G-irreducible connected subgroups. To save repeating ourselves, we say a subgroup X satisfies the conclusion of Theorem 10.2 if there is a uniquely determined commuting product $E_1 \dots E_k$ with $X \leq E_1 \dots E_k \leq G$, such that each E_i is a simple restricted subgroup of the same type as X, and each of the projections $X \to E_i/Z(E_i)$ is non-trivial and involves a different field twist. We have already considered the simple irreducible subgroups of rank at least 2 in [**Tho15**, Section 9] (apart from the maximal subgroup $F_4 < E_8$ when $p = 3$ but this is itself a restricted subgroup and does satisfy the conclusion of Theorem 10.2) and so we need only prove the following results for subgroups of type A_1. The method is similar for all of them and so we give the proof for G of type E_8 only.

COROLLARY 10.3. *Let G be a simple algebraic group of exceptional type and X be a simple G-irreducible connected subgroup of G. Then either X satisfies the conclusion of Theorem* 10.2 *or $p \leq 5$ and X is conjugate to one of subgroups in the* *following table. We list irreducible subgroups by their identification number n and any conditions given with n refer to the field twists associated with $G(\#n)$.*

Table 1:

G	p	n
G_2	3	5
	2	1
F_4	3	$9^{\{r,r,s\}}$
	2	$2^{\{0,r,r,s\}}$; $2^{\{r,s,0,r\}}$; 14; 15; 17–20; 22
E_6	3	$2^{\{r,r,s\}}$; 11; 12
	2	8; 9; 13; 14; 15; 17; 19
E_7	3	3 if $r = t$; $10^{\{r,r,r,s\}}$; $10^{\{r,r,s,r\}}$; 26
	2	12 if the conditions of lines 4–7, 11, 14, 15 or 16 of Table 11.10 are satisfied; 13 if either s or $t \in \{u, v, w\}$; 28
E_8	5	7; 101
	3	6 if $v \in \{r, s\}$; 24 if $t \in \{r, s\} \cap \{u, v\}$; 90–99
	2	8; 26 if the conditions of lines 3, 5, 8, 12, 15, 16 or 17 of Table 11.26 are satisfied; 28 if one of t, u or $v \in \{0, r, s\}$; 29 if one of t, u, v, w or $x \in \{r, s\}$; 53–57; 66–81

PROOF. Suppose that X is an E_8-irreducible subgroup of type A_1 not satisfying the conclusion of Theorem 10.2. Then $p \leq 5$ and X is given by Theorem 9.1. Moreover, using the lattice structure given in Table 11.5 we find that X is contained in at most one commuting product of restricted groups, of the same type as X, containing X as a diagonal subgroup with distinct field twists. It remains to check whether each subgroup in the commuting product is restricted, using Table 12.5. We illustrate this with two examples; the remaining cases are all similar.

Firstly, suppose that $X = E_8(\#7)$ so $X \hookrightarrow A_1A_1 = Y_1Y_2 = E_8(\#771)$ via $(1^{[r]}, 1^{[s]})$ $(rs = 0; r \neq s)$. From Table 12.5 we have

$$L(E_8) \downarrow Y_1Y_2 = (W(5),3)/(4,W(6))/(4,2)/(3,W(7))/(3,W(5))/(3,1)/(2,W(8))/ \\ (2,4)^2/(2,0)/(1,W(9))/(1,W(5))/(1,3)/(0,W(6))/(0,2).$$

Thus when $p = 5$ we have that Y_1 is p-restricted but Y_2 is not since it has a composition factor of high weight 9 which is greater than $2p - 2 = 8$. Therefore X does not satisfy the conclusion of Theorem 10.2.

Secondly, suppose that $X = E_8(\#28)$ so $p = 2$ and $X \hookrightarrow A_1^6 = E_8(\#641)$ via $(1, 1^{[r]}, 1^{[s]}, 1^{[t]}, 1^{[u]}, 1^{[v]})$ $(0 < r < s; t < u < v$; if $t = 0$ then $r \leq u$; if $t = 0$ and

$r = u$ then $s \leq v$). From Table 12.5 we have

$$\begin{aligned} L(E_8) \downarrow A_1^6 = {} & (2,2,0,0,0,0)/(2,0,2,0,0,0)/(2,0,0,1,1,1)/(2,0,0,0,0,0)^2/ \\ & (1,1,1,2,0,0)/(1,1,1,1,1,1)/(1,1,1,0,2,0)/(1,1,1,0,0,2)/ \\ & (1,1,1,0,0,0)^2/(0,2,2,0,0,0)/(0,2,0,1,1,1)/(0,2,0,0,0,0)^2/ \\ & (0,0,2,1,1,1)/(0,0,2,0,0,0)^2/(0,0,0,2,2,0)/(0,0,0,2,0,2)/ \\ & (0,0,0,2,0,0)^2/(0,0,0,1,1,1)^2/(0,0,0,0,2,2)/(0,0,0,0,2,0)^2/ \\ & (0,0,0,0,0,2)^2/(0,0,0,0,0,0)^8. \end{aligned}$$

If X has distinct field twists then it satisfies the conclusion of Theorem 10.2 since 2 is the highest weight of any composition factor for an A_1 factor of A_1^6. If two of the field twists of the embedding of X are equal then at least one of t, u or $v \in \{0, r, s\}$. Since the composition factors $(1,1,1,2,0,0)$, $(1,1,1,0,2,0)$ and $(1,1,1,0,0,2)$ occur in $L(E_8) \downarrow A_1^6$ it follows that at least one of the subgroups of type A_1 in the commuting product containing X will have a composition factor of high weight 3 and therefore not be restricted. $\square$

CHAPTER 11

Tables for Theorem 1

In this chapter we present the tables referenced in Theorems 5.1–9.1. The notation used is described in Section 2. The tables contain all of the G-irreducible connected subgroups of each simple exceptional algebraic group G. They also contain enough information to find the lattice of connected overgroups of each G-irreducible subgroup. The structure of their presentation reflects the strategy described in Section 4 and the proofs of Theorems 5.1–9.1 in Sections 5–9. Moreover, we give an auxiliary table for each exceptional algebraic group G, giving the immediate connected overgroups of every G-irreducible connected subgroup, except for certain diagonal subgroups. This is intended to help the reader construct the lattice of connected overgroups of each G-irreducible connected subgroup. We give further details regarding this lattice structure below.

We start by explaining how to read Tables 1–11.5. Let G be a simple exceptional algebraic group. Each table is broken into sections, divided by pairs of horizontal lines, and within each section there are parts divided by single horizontal lines. There is one section for each reductive, maximal connected subgroup of G. There is then one part for each G-irreducible connected subgroup X of M, such that X contains a proper G-irreducible connected subgroup. This part gives the G-irreducible maximal connected subgroups of X, as well as all G-irreducible diagonal subgroups. There is a heading for each section and part that gives the type of the subgroup being considered, as well as the identification number and any restrictions on the characteristic p. There is one more piece of information in the heading which is the "$M_i =$" for $i = 0, 1, \dots$ (where M_0 is simply written as M). This is intended to make it easier for the reader to follow the tables, and is explained below.

We need to describe what the entries in each column of a generic row represent. If the first column is empty then this row is a heading of a section or part, as described above. There are two other types of rows, both of which correspond to G-irreducible subgroups X contained in Y, where Y is the subgroup of the current section or part, contained in the maximal connected subgroup M (in this description we are including the case $Y = M$). In both cases, the entry in the first column gives the ID number for the conjugacy class or classes of such irreducible subgroups. The second and third column differ, depending on whether X is a diagonal subgroup of Y or not. If X is a diagonal subgroup of Y then the second and third columns are merged and the information given is of the form "X via ...", denoting the usual notation for an embedding of a diagonal subgroup of Y, as well as any restrictions on the characteristic p. If X is not a diagonal subgroup of Y then the second column gives the isomorphism type of X and any restrictions on the characteristic p. The third column contains the description of $V_M \downarrow X$. Note that for a diagonal subgroup X of Y, it is straightforward to work out $V_M \downarrow X$ from $V_M \downarrow Y$. We note that we do not repeat restrictions on p as we list the subgroups of a G-irreducible connected subgroup X. So if, for example, the maximal connected subgroup M exists only for $p \neq 2$ then we write M $(p \neq 2)$ in the heading and it is assumed that any subgroup X of M inherits this restriction on p without explicit labelling. However, when we consider the subgroups of X we do explicitly repeat any restriction inherited from M in the heading, for clarity.

In Tables 11.4 and 11.5 there is one more possibility for a generic row. To make the tables easier to read we have moved large collections of diagonal subgroups to supplementary tables in Section 11.1. Thus the row "See Table x" means that all diagonal subgroups of the relevant subgroup Y appear in Table x in Section 11.1.

We now explain how to find the lattice of connected overgroups of each G-irreducible connected subgroup given in Tables 1–11.5. Each table starts with the G-irreducible subgroups that are maximal amongst reductive connected subgroups, with their identification number listed in the ID column. A pair of horizontal lines then indicates the end of that list and the beginning of the first section. We then write "In $M = H_1H_2\dots$ $(G(\#n_1))$" where n_1 is the identification number for the first reductive, maximal connected subgroup of type $H_1H_2\dots$, and will include any restrictions on the characteristic p. We then list the G-irreducible maximal connected subgroups of M as well as all diagonal connected subgroups of M, if there are any, not just the maximal ones. Recall that we will not explicitly consider the proper subgroups of any diagonal connected subgroups, as discussed in Section 4. A horizontal line then indicates the end of this list. The next row will be a heading "In $M_1 = X_1X_2\dots$ $(G(\#n_2))$", where $X_1X_2\dots$ is the first G-irreducible maximal connected subgroup of M. The "$M_1 =$" tells the reader that we are now listing the subgroups of a maximal subgroup of a maximal connected subgroup. We then repeat the process, listing the G-irreducible maximal connected subgroups of M_1 and all diagonal subgroups of M_1. The next heading could be "In $M_2 = Y_1Y_2\dots$ $(G(\#n_2))$", where $Y_1Y_2\dots$ is a maximal connected subgroup of M_1 or it could be "In $M_1 = Z_1Z_2\dots$ $(G(\#n_3))$", where $Z_1Z_2\dots$ is the second maximal connected subgroup of M. This will depend on whether M_1 has any proper G-irreducible connected subgroups that need considering or not. Once all G-irreducible connected subgroups of M have been listed in this way, a pair of horizontal lines indicates the end of the subgroups contained in the first reductive, maximal connected subgroup of G. The next heading will be "In $M = K_1K_2\dots$ $(G(\#n_4))$" and we repeat the process again for the second reductive, maximal connected subgroup $K_1K_2\dots$ of G. We iterate this process until we have considered all of the G-irreducible subgroups contained in the final reductive, maximal connected subgroup.

There is an important deviation from the process described above. Suppose that X is a representative for a conjugacy class of G-irreducible connected subgroups with more than one conjugacy class of proper immediate overgroups. Then after one occurrence of X in the table we list all repetitions of X with the ID number in italics and do not reconsider its subgroups at those points in the table. Moreover, if X occurs more than once in the same reductive, maximal connected subgroup M then we only list $V_M \downarrow X$ in the third column once. We note that the first occurrence of a subgroup in the table can sometimes be in italics because it is clearer to consider its subgroups later in the table. For example, the subgroup $X = \bar{A}_1^2B_2\bar{A}_3 = E_8(\#351)$ is listed in italics the first time we reach it in Table 11.5, as a subgroup of $Y = \bar{A}_1^2B_5 = E_8(\#118)$. This is because the subgroup X is defined for all p, whereas it only occurs as a maximal connected subgroup of Y when $p \neq 2$.

The second time we reach X is as a subgroup of $Z = \bar{A}_1^2\bar{A}_3^2 = E_8(\#119)$. This time the ID number is not in italics, $V_M \downarrow X$ is given in the third column and we consider the subgroups of X for all p in the next part.

There is another important example where we do not immediately consider the subgroups contained in X, listed as a subgroup of Y_1, say, even if the ID number for X is not in italics nor is X a diagonal subgroup of Y_1. In this case the ID number will be nb and the subgroup X is defined for all $p \geq k$ but only a maximal connected subgroup of Y_1 for some prime $l \geq k$. At some point later in the table X will be defined for all $p \geq k$ except l and given ID number na. At this point we will consider the subgroups of X for all $p \geq k$ together. There are instances where the subgroup X occurs again in the table. If X is listed for $p = l$, we write nb in the ID column. Similarly, if X is listed for $p \geq k$ except l, we write na. It may be that X is listed for all $p \geq k$ in a later reductive, maximal connected subgroup and in this case we simply write n in the ID column. This could cause some confusion. However, upon finding any subgroup $X = G(\#n)$ we can look up X in the auxiliary table for G. There are two possibilities. Either n occurs in the first column of a row, in which case the third column of that row gives any characteristic restriction on X in full generality; or n does not occur in the first column of the table, in which case X occurs only in the place one has found it, leaving no confusion about the characteristic restrictions.

There are many G-irreducible connected subgroups occurring multiple times, especially when $G = E_7$ or E_8. For this reason we have provided an auxiliary table for each G; these are Tables 11.1A–11.5A. Each row of the table gives a conjugacy class of G-irreducible connected subgroups, with the first column giving the ID number, the second column the isomorphism type of a representative X and the third column gives any restrictions on the characteristic p. The fourth column gives all conjugacy classes of immediate connected overgroups of X. Note that we use the notation $X[\#n]$ to denote the subgroup $X = G(\#n)$, as a shorthand only in these five tables.

If X is a diagonal subgroup of Y and all immediate connected overgroups of X are also subgroups of Y (and hence diagonal subgroups of Y or just Y itself) then X does not appear in the auxiliary table. In this case there will be only one appearance of X in Tables 1–11.5 (or the supplementary tables in Section 11.1) and all of its immediate overgroups can be straightforwardly computed from the other diagonal subgroups that will appear just above X in the table. For this reason, the phrase "immediate overgroups in $Y[\#n]$" sometimes appears in the fourth column of a row corresponding to the irreducible subgroup X. This means that the subgroup X is a diagonal subgroup of Y and the immediate connected overgroups of X in Y are included in the list but not explicitly calculated.

We give some examples of how to use the information in the two sets of tables to recover the lattice of overgroups for a given G-irreducible connected subgroup X. They are chosen to highlight as many different scenarios as possible. Firstly, let $G = G_2$. In this case we can easily recover the lattice of overgroups for all G-irreducible connected subgroups and even present this in a small diagram, see Figure 1. Table 1 contains precisely the same information as this diagram and in this case Table 11.1A is unnecessary, but we include it for completeness.

Now suppose $G = E_8$. For the first examples, we highlight the fact that a condition on the characteristic p of a subgroup $X = G(\#n)$ contained in Y_1 is

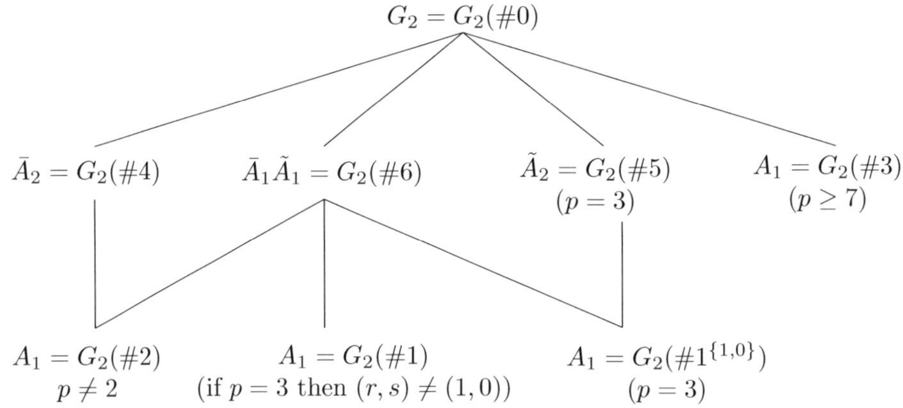

FIGURE 1. The lattice of G_2-irreducible connected subgroups.

based on that specific embedding of X into Y_1 and thus the subgroup X may be considered for other excluded characteristics in a different overgroup Y_2. For example, the subgroup $X = A_1B_6 = E_8(\#110)$ is defined for all p. However, we first arrive at it in Table 11.5 as a maximal connected subgroup of $D_8 = E_8(\#43)$, with ID number 110a and a restriction "$(p \neq 2)$". In this case, we know that the "a" implies that A_1B_6 must be a maximal connected subgroup of at least one other irreducible connected subgroup of G when $p = 2$. To find all of the extra immediate overgroups we go to the row with 110 in the first column in Table 11.5A. We find that A_1B_6 is also a maximal connected subgroup of $B_7 = E_8(\#44)$ when $p = 2$. Indeed, when we come to the $M_1 = B_7$ part of the table we see that X appears with ID number 110b.

For another example, suppose $X = \bar{A}_1^6A_1 = E_8(\#132^{\{0\}})$. Then X is a maximal connected subgroup of $\bar{A}_1^4B_3 = E_8(\#125)$ and $\bar{A}_1^4A_1B_2 = E_8(\#126)$, and in both cases the ID number is listed in italics. The subgroup is not listed again in Table 11.5. Therefore X must be a diagonal subgroup listed in one of the supplementary tables in Section 11.1.2. Using Table 11.5A we immediately find that X is a maximal connected subgroup of $\bar{A}_1^8 = E_8(\#124)$.

Finally, let us suppose that we are interested in subgroups of the reductive, maximal connected subgroup $A_8 = E_8(\#62)$. We see from the $M = A_8$ section of Table 11.5 that there are two irreducible maximal connected subgroups, namely $X_1 = B_4 = E_8(\#46)$ when $p \neq 2$, and $X_2 = A_2^2 = E_8(\#669)$ when $p \neq 3$. Both ID numbers are in italics and therefore both X_1 and X_2 are contained in other irreducible connected subgroups of G. Looking at Table 11.5A we find that X_1 is a maximal connected subgroup of $D_8 = E_8(\#43)$ also when $p \neq 2$, and indeed X_1 is only to be considered when $p \neq 2$, as per the information in the table. Similarly, we find that X_2 is a maximal connected subgroup of $A_2\bar{D}_4 = E_8(\#621)$ when $p \neq 3$ and again X_2 is only considered when $p \neq 3$. We can then continue looking up subgroups in Table 11.5A to find that when $p \neq 3$ the subgroup $A_2\bar{D}_4$ is a maximal connected subgroup of $\bar{D}_4^2 = E_8(\#108)$, which in turn is a maximal connected subgroup of $D_8 = E_8(\#43)$.

Table 1: The lattice structure of irreducible connected subgroups of G_2.

ID	Irreducible subgroup X	$V_M \downarrow X$
4	$\bar{A}_2$	
5	$\tilde{A}_2$ $(p=3)$	
6	$\bar{A}_1\tilde{A}_1$	
3	A_1 $(p \geq 7)$	
	In $M = \bar{A}_2$ $(G_2(\#4))$	
2	A_1 $(p \neq 2)$	2
	In $M = \tilde{A}_2$ $(p=3)$ $(G_2(\#5))$	
$1^{\{\delta_1\}}$	A_1	2
	In $M = \bar{A}_1\tilde{A}_1$ $(G_2(\#6))$	
1	A_1 via $(1^{[r]}, 1^{[s]})$ $(rs = 0; r \neq s)$	
2	A_1 via $(1,1)$ $(p \neq 2)$	

Table 11.1A: The connected overgroups for G-irreducible subgroups of G_2.

ID	X	p	Immediate connected overgroups
$1^{\{\delta_1\}}$	A_1		$\tilde{A}_2[\#5]$ $(p=3)$, $\bar{A}_1\tilde{A}_1[\#6]$
2	A_1	$\neq 2$	$\bar{A}_2[\#4]$, $\bar{A}_1\tilde{A}_1[\#6]$
3	A_1	≥ 7	$G_2[\#0]$
4	$\bar{A}_2$		$G_2[\#0]$
5	$\tilde{A}_2$	3	$G_2[\#0]$
6	$\bar{A}_1\tilde{A}_1$		$G_2[\#0]$

Table 11.2: The lattice structure of irreducible connected subgroups of F_4.

ID	Irred. subgroup X	$V_M \downarrow X$
12	B_4	
14	C_4 $(p=2)$	
24a	$\bar{A}_1C_3$ $(p \neq 2)$	
25	A_1G_2 $(p \neq 2)$	
26	$\bar{A}_2\tilde{A}_2$	
16	G_2 $(p=7)$	
10	A_1 $(p \geq 13)$	
	In $M = B_4$ $(F_4(\#12))$	
13	$\bar{D}_4$	$\lambda_1/0$
27a	$\tilde{A}_1\bar{A}_3$ $(p \neq 2)$	$(2,000)/(0,010)$
28a	$\bar{A}_1^2B_2$ $(p \neq 2)$	$(1,1,00)/(0,0,10)$

29	A_1^2 $(p \neq 2)$	$(2,2)$
30	B_2^2 $(p = 2)$	$(10,00)/(00,10)/(00,00)$
31	$\tilde{A}_1 B_3$ $(p = 2)$	$(2,000)/(0,100)/(0,000)$
7	A_1 $(p \geq 11)$	8
	In $M_1 = \bar{D}_4$ $(F_4(\#13))$	
32	$\bar{A}_1^4$	$(1,1,0,0)/(0,0,1,1)/(0,0,0,0)$
40$^{\{0\}}$	$\tilde{A}_1 B_2$ $(p \neq 2)$	
19	A_2 $(p \neq 3)$	$11/00$
	In $M_2 = \bar{A}_1^4$ $(F_4(\#32))$	
33	$A_1 \bar{A}_1^2$ via $(1_a, 1_a^{[r]}, 1_b, 1_c)$ $(r \neq 0)$	
34	$A_1 \bar{A}_1$ via $(1_a, 1_a^{[r]}, 1_a^{[s]}, 1_b)$ $(0 < r < s)$	
35	$A_1 A_1$ via $(1_a, 1_a^{[r]}, 1_b, 1_b^{[s]})$ $(0 < r \leq s)$	
1	A_1 via $(1, 1^{[r]}, 1^{[s]}, 1^{[t]})$ $(0 < r < s < t)$	
	In $M_1 = \tilde{A}_1 \bar{A}_3$ $(F_4(\#27))$	
40$^{\{0\}}$	$\tilde{A}_1 B_2$ $(p \neq 2)$	
36	$\tilde{A}_1^3$ $(p \neq 2)$	$(2,0,0)/(0,2,0)/(0,0,2)$
	In $M_2 = \tilde{A}_1^3$ $(p \neq 2)$ $(F_4(\#36))$	
37	$A_1 \tilde{A}_1$ via $(1_a, 1_a^{[r]}, 1_b)$ $(r \neq 0)$	
4	A_1 via $(1, 1^{[r]}, 1^{[s]})$ $(0 < r < s)$	
	In $M_1 = \bar{A}_1^2 B_2$ $(F_4(\#28))$	
32	$\bar{A}_1^4$	
38	$\bar{A}_1^2 A_1$ $(p \geq 5)$	$(1,1,0)/(0,0,4)$
39	$\bar{A}_1^2 \tilde{A}_1^2$ $(p = 2)$	$(1,1,0,0)/(0,0,2,0)/(0,0,0,2)/(0,0,0,0)$
40	$A_1 B_2$ via $(1, 1^{[r]}, 10)$ (if $p = 2$ then $r \neq 0$)	
	In $M_2 = \bar{A}_1^2 A_1$ $(p \geq 5)$ $(F_4(\#38))$	
41	$A_1 A_1$ via $(1_a, 1_a^{[r]}, 1_b)$	
42	$A_1 \bar{A}_1$ via $(1_a^{[r]}, 1_b, 1_a^{[s]})$ $(rs = 0)$	
6	A_1 via $(1^{[r]}, 1^{[s]}, 1^{[t]})$ $(rt = 0; r \leq s)$	
	In $M_2 = \bar{A}_1^2 \tilde{A}_1^2$ $(p = 2)$ $(F_4(\#39))$	
43	$A_1 \tilde{A}_1^2$ via $(1_a, 1_a^{[r]}, 1_b, 1_c)$ $(r \neq 0)$	
44	$A_1 \bar{A}_1 \tilde{A}_1$ via $(1_a, 1_b^{[r]}, 1_b^{[s]}, 1_c)$ $(rs = 0)$	
45	$\bar{A}_1^2 A_1$ via $(1_a, 1_b, 1_c, 1_c^{[r]})$ $(r \neq 0)$	
46	$A_1 \tilde{A}_1$ via $(1_a^{[r]}, 1_a^{[s]}, 1_a^{[t]}, 1_b)$ $(rt = 0; r < s)$	
47	$A_1 \bar{A}_1$ via $(1_a^{[r]}, 1_b, 1_a^{[s]}, 1_a^{[t]})$ $(rs = 0; s < t)$	
48	$A_1 A_1$ via $(1_a, 1_a^{[r]}, 1_b, 1_b^{[s]})$ $(rs \neq 0)$	
49	$A_1 A_1$ via $(1_a^{[r]}, 1_b^{[t]}, 1_b^{[u]}, 1_a^{[s]})$ $(rs = tu = 0; r + u \leq t + s)$	
2	A_1 via $(1^{[r]}, 1^{[s]}, 1^{[t]}, 1^{[u]})$ $(rt = 0; r < s; t < u)$	
	In $M_1 = A_1^2$ $(p \neq 2)$ $(F_4(\#29))$	
5	A_1 via $(1, 1^{[r]})$ $(r \neq 0)$	
6$^{\{\underline{0}\}}$	A_1 via $(1,1)$ $(p \geq 5)$	

	In $M_1 = B_2^2$ $(p=2)$ $(F_4(\#30))$	
50	$\tilde{A}_1^2 B_2$	$(2,0,00)/(0,2,00)/(0,0,10)/(0,0,00)$
28b	$\bar{A}_1^2 B_2$	$(1,1,00)/(0,0,10)/(0,0,00)$
21	B_2 via $(10,10^{[r]})$ $(r \neq 0)$	
22	B_2 via $(10,02)$	
23	B_2 via $(10,02^{[r]})$ $(r \neq 0)$	
	In $M_2 = \tilde{A}_1^2 B_2$ $(p=2)$ $(F_4(\#50))$	
51	$\tilde{A}_1^4$	$(2,0,0,0)/(0,2,0,0)/(0,0,2,0)/(0,0,0,2)/(0,0,0,0)$
39	$\tilde{A}_1^2 \bar{A}_1^2$	
52	$A_1 B_2$ via $(1,1^{[r]},10)$ $(r \neq 0)$	
	In $M_3 = \tilde{A}_1^4$ $(p=2)$ $(F_4(\#51))$	
53	$A_1 \tilde{A}_1^2$ via $(1_a, 1_a^{[r]}, 1_b, 1_c)$ $(r \neq 0)$	
54	$A_1 \tilde{A}_1$ via $(1_a, 1_a^{[r]}, 1_a^{[s]}, 1_b)$ $(0 < r < s)$	
55	$A_1 A_1$ via $(1_a, 1_a^{[r]}, 1_b, 1_b^{[s]})$ $(0 < r \leq s)$	
3	A_1 via $(1, 1^{[r]}, 1^{[s]}, 1^{[t]})$ $(0 < r < s < t)$	
	In $M_1 = \tilde{A}_1 B_3$ $(p=2)$ $(F_4(\#31))$	
50	$\tilde{A}_1^2 B_2$	
27b	$\tilde{A}_1 \bar{A}_3$	$(2,000)/(0,010)/(0,000)$
56	$\tilde{A}_1 G_2$	$(2,00)/(0,10)/(0,00)$
	In $M_2 = \tilde{A}_1 G_2$ $(p=2)$ $(F_4(\#56))$	
44$^{\{0\}}$	$\tilde{A}_1 \bar{A}_1 A_1$	
	In $M = C_4$ $(p=2)$ $(F_4(\#14))$	
30	B_2^2	$(01,00)/(00,01)$
15	$\tilde{D}_4$	λ_1
24b	$\bar{A}_1 C_3$	$(1,000)/(0,100)$
	In $M_1 = \tilde{D}_4$ $(p=2)$ $(F_4(\#15))$	
51	$\tilde{A}_1^4$	$(1,1,0,0)/(0,0,1,1)$
18$^{\{\delta_1\}}$	A_2	11
	In $M = \bar{A}_1 C_3$ (max. if $p \neq 2$; contained in C_4 if $p=2$) $(F_4(\#24))$	
28	$\bar{A}_1^2 B_2$	$(1,1,00)/(1,0,01)$
57	$\bar{A}_1 A_1 A_1$ $(p \neq 2)$	$(1,2,1)$
58	$\bar{A}_1 A_1$ $(p \geq 7)$	$(1,5)$
59	$\bar{A}_1 \tilde{A}_3$ $(p=2)$	$(1,010)$
60	$\bar{A}_1 G_2$ $(p=2)$	$(1,10)$
	In $M_1 = \bar{A}_1 A_1 A_1$ $(p \neq 2)$ $(F_4(\#57))$	
61	$A_1 A_1$ via $(1_a^{[r]}, 1_a^{[s]}, 1_b)$ $(rs=0)$	
62	$A_1 A_1$ via $(1_a^{[r]}, 1_b, 1_a^{[s]})$ $(rs=0)$	
63	$\bar{A}_1 A_1$ via $(1_a, 1_b^{[r]}, 1_b^{[s]})$ $(rs=0; r \neq s)$	
42$^{\{0\}}$	$\bar{A}_1 A_1$ via $(1_a, 1_b, 1_b)$ $(p \geq 5)$	
9	A_1 via $(1^{[r]}, 1^{[s]}, 1^{[t]})$ $(rst=0; s \neq t)$	

	In $M_1 = \bar{A}_1 A_1$ $(p \geq 7)$ $(F_4(\#58))$	
8	A_1 via $(1^{[r]}, 1^{[s]})$ $(rs = 0)$	
	In $M_1 = \bar{A}_1 G_2$ $(p = 2)$ $(F_4(\#60))$	
$\mathit{44}^{\{\delta_2\}}$	$\bar{A}_1 \tilde{A}_1 A_1$	$(1,1,1)/(1,0,2)$
	In $M = A_1 G_2$ $(p \neq 2)$ $(F_4(\#25))$	
64	$A_1 \bar{A}_2$	$(1, G_2(\#4))$
65	$A_1 A_2$ $(p = 3)$	$(1, G_2(\#5))$
$\mathit{57}$	$A_1 \bar{A}_1 A_1$	$(1, G_2(\#6))$
66	$A_1 A_1$ $(p \geq 7)$	$(1, G_2(\#3))$
	In $M_1 = A_1 \bar{A}_2$ $(p \neq 2)$ $(F_4(\#64))$	
$\mathit{62}^{\{\underline{0}\}}$	$A_1 A_1$	$(1, G_2(\#2))$
	In $M_1 = A_1 A_2$ $(p = 3)$ $(F_4(\#65))$	
$\mathit{62}^{\{\delta_1\}}$	$A_1 A_1$	$(1, G_2(\#1^{\{\delta_1\}}))$
	In $M_1 = A_1 A_1$ $(p \geq 7)$ $(F_4(\#66))$	
11	A_1 via $(1^{[r]}, 1^{[s]})$ $(rs = 0; r \neq s)$	
$\mathit{8}^{\{\underline{0}\}}$	A_1 via $(1,1)$	
	In $M = \bar{A}_2 \tilde{A}_2$ $(F_4(\#26))$	
67	$A_1 \tilde{A}_2$ $(p \neq 2)$	$(2, 10)$
$\mathit{64}$	$\bar{A}_2 A_1$ $(p \neq 2)$	$(10, 2)$
17	A_2 via $(10, 10)$	
18	A_2 via $(10^{[r]}, 10^{[s]})$ $(rs = 0; r \neq s)$	
19	A_2 via $(10, 01)$ $(p \neq 3)$	
20	A_2 via $(10^{[r]}, 01^{[s]})$ $(rs = 0; r \neq s)$	
	In $M_1 = A_1 \tilde{A}_2$ $(p \neq 2)$ $(F_4(\#67))$	
$\mathit{62}^{\{\underline{0}\}}$	$A_1 A_1$	$(2, 2)$
	In $M = G_2$ $(p = 7)$ $(F_4(\#16))$	
$\mathit{17}$	A_2	$G_2(\#4)$
$\mathit{61}^{\{\underline{0}\}}$	$A_1 A_1$	$G_2(\#6)$
$\mathit{8}^{\{\delta_1\}}$	A_1	$G_2(\#3)$

Table 11.2A: The connected overgroups for irreducible subgroups of F_4.

ID	X	p	Immediate connected overgroups
$6^{\{0\}}$	A_1	≥ 5	$A_1^2[\#29]$, $A_1 A_1[\#41^{\{0\}}]$, $A_1 \bar{A}_1[\#42^{\{\underline{0}\}}]$
7	A_1	≥ 11	$B_4[\#12]$
$8^{\{\underline{0}\}}$	A_1	≥ 7	$\bar{A}_1 A_1[\#58]$, $A_1 A_1[\#66]$
$8^{\{\delta_1\}}$	A_1	≥ 7	$G_2[\#16]$ $(p = 7)$, $\bar{A}_1 A_1[\#58]$
10	A_1	≥ 13	$F_4[\#0]$
12	B_4		$F_4[\#0]$
13	$\bar{D}_4$		$B_4[\#12]$

14	C_4	2	$F_4[\#0]$
15	$\tilde{D}_4$	2	$C_4[\#14]$
16	G_2	7	$F_4[\#0]$
17	A_2		$G_2[\#16]$ $(p = 7)$, $\bar{A}_2\tilde{A}_2[\#26]$
$18^{\{\delta_1\}}$	A_2		$\tilde{D}_4[\#15]$ $(p = 2)$, $\bar{A}_2\tilde{A}_2[\#26]$
19	A_2	$\neq 3$	$\bar{D}_4[\#13]$, $\bar{A}_2\tilde{A}_2[\#26]$
24	$\bar{A}_1C_3$		$F_4[\#0]$ $(p \neq 2)$, $C_4[\#14]$ $(p = 2)$
25	A_1G_2	$\neq 2$	$F_4[\#0]$
26	$\bar{A}_2\tilde{A}_2$		$F_4[\#0]$
27	$\tilde{A}_1\bar{A}_3$		$B_4[\#12]$ $(p \neq 2)$, $\tilde{A}_1B_3[\#31]$ $(p = 2)$
28	$\bar{A}_1^2B_2$		$B_4[\#12]$ $(p \neq 2)$, $\bar{A}_1C_3[\#24]$, $B_2^2[\#30]$ $(p = 2)$
29	A_1^2	$\neq 2$	$B_4[\#12]$
30	B_2^2	2	$B_4[\#12]$, $C_4[\#14]$
31	$\tilde{A}_1B_3$	2	$B_4[\#12]$
32	$\bar{A}_1^4$		$\bar{D}_4[\#13]$, $\bar{A}_1^2B_2[\#28]$
36	$\tilde{A}_1^3$	$\neq 2$	$\tilde{A}_1\bar{A}_3[\#27]$
38	$\bar{A}_1^2A_1$	≥ 5	$\bar{A}_1^2B_2[\#28]$
39	$\bar{A}_1^2\tilde{A}_1^2$	2	$\bar{A}_1^2B_2[\#28]$, $\tilde{A}_1^2B_2[\#50]$
$40^{\{0\}}$	$\tilde{A}_1B_2$	$\neq 2$	$\bar{D}_4[\#13]$, $\tilde{A}_1\bar{A}_3[\#27]$, $\bar{A}_1^2B_2[\#28]$
$42^{\{\underline{0}\}}$	$A_1\bar{A}_1$	≥ 5	$\bar{A}_1^2A_1[\#38]$, $\bar{A}_1A_1A_1[\#57]$
$44^{\{\underline{0}\}}$	$A_1\bar{A}_1\tilde{A}_1$	2	$\bar{A}_1^2\tilde{A}_1^2[\#39]$, $\tilde{A}_1G_2[\#56]$
$44^{\{\delta_2\}}$	$A_1\bar{A}_1\tilde{A}_1$	2	$\bar{A}_1^2\tilde{A}_1^2[\#39]$, $\bar{A}_1G_2[\#60]$
50	$\tilde{A}_1^2B_2$	2	$B_2^2[\#30]$, $\tilde{A}_1B_3[\#31]$
51	$\tilde{A}_1^4$	2	$\tilde{D}_4[\#15]$, $\tilde{A}_1^2B_2[\#50]$
56	$\tilde{A}_1G_2$	2	$\tilde{A}_1B_3[\#31]$
57	$\bar{A}_1A_1A_1$	$\neq 2$	$\bar{A}_1C_3[\#24]$, $A_1G_2[\#25]$
58	$\bar{A}_1A_1$	≥ 7	$\bar{A}_1C_3[\#24]$
59	$\bar{A}_1\tilde{A}_3$	2	$\bar{A}_1C_3[\#24]$
60	$\bar{A}_1G_2$	2	$\bar{A}_1C_3[\#24]$
$61^{\{\underline{0}\}}$	A_1A_1	$\neq 2$	$G_2[\#16]$ $(p = 7)$, $\bar{A}_1A_1A_1[\#57]$
$62^{\{\underline{0}\}}$	A_1A_1	$\neq 2$	$\bar{A}_1A_1A_1[\#57]$, $A_1\bar{A}_2[\#64]$, $A_1\tilde{A}_2[\#67]$
$62^{\{\delta_1\}}$	A_1A_1	$\neq 2$	$\bar{A}_1A_1A_1[\#57]$, $A_1A_2[\#65]$ $(p = 3)$
64	$A_1\bar{A}_2$	$\neq 2$	$A_1G_2[\#25]$, $\bar{A}_2\tilde{A}_2[\#26]$
65	A_1A_2	3	$A_1G_2[\#25]$
66	A_1A_1	≥ 7	$A_1G_2[\#25]$
67	$A_1\tilde{A}_2$	$\neq 2$	$\bar{A}_2\tilde{A}_2[\#26]$

Table 11.3: The lattice structure of irreducible connected subgroups of E_6.

ID	Irred. subgroup X	$V_M \downarrow X$
24	$\bar{A}_1 A_5$	
25	$\bar{A}_2^3$	
26	$A_2 G_2$	
7	F_4	
8a	C_4 $(p \neq 2)$	
10a	G_2 $(p \neq 7)$	
11a	A_2 $(p \geq 5)$	
	In $M = \bar{A}_1 A_5$ $(E_6(\#24))$	
27	$\bar{A}_1 C_3$	$(1, 100)$
28	$\bar{A}_1 A_1 A_2$	$(1, 1, 10)$
29a	$\bar{A}_1 A_3$ $(p \neq 2)$	$(1, 010)$
30	$\bar{A}_1 A_2$ $(p \neq 2)$	$(1, 20)$
	In $M_1 = \bar{A}_1 C_3$ $(E_6(\#27))$	
31	$\bar{A}_1 A_1 A_1$ $(p \neq 2)$	$(1, 2, 1)$
32	$\bar{A}_1 A_1$ $(p \geq 7)$	$(1, 5)$
29b	$\bar{A}_1 A_3$ $(p = 2)$	$(1, 010)$
33	$\bar{A}_1 G_2$ $(p = 2)$	$(1, 10)$
	In $M_2 = \bar{A}_1 A_1 A_1$ $(p \neq 2)$ $(E_6(\#31))$	
34	$A_1 A_1$ via $(1_a^{[r]}, 1_a^{[s]}, 1_b)$ $(rs = 0)$	
35	$A_1 A_1$ via $(1_a^{[r]}, 1_b, 1_a^{[s]})$ $(rs = 0)$	
36	$\bar{A}_1 A_1$ via $(1_a, 1_b^{[r]}, 1_b^{[s]})$ $(rs = 0; r \neq s)$	
2	A_1 via $(1^{[r]}, 1^{[s]}, 1^{[t]})$ $(rst = 0; s \neq t)$	
	In $M_2 = \bar{A}_1 A_1$ $(p \geq 7)$ $(E_6(\#32))$	
1	A_1 via $(1^{[r]}, 1^{[s]})$ $(rs = 0)$	
	In $M_1 = \bar{A}_1 A_1 A_2$ $(E_6(\#28))$	
31	$\bar{A}_1 A_1 A_1$ $(p \neq 2)$	
37	$A_1 A_2$ via $(1^{[r]}, 1^{[s]}, 10)$ $(rs = 0;\ r \neq s$ if $p = 2)$	
	In $M = \bar{A}_2^3$ $(E_6(\#25))$	
38	$A_1 \bar{A}_2^2$ $(p \neq 2)$	$(2, 10, 10)$
39	$\bar{A}_2 A_2$ via $(10_a, 10_b, 10_b^{[r]})$	
40	$\bar{A}_2 A_2$ via $(10_a, 10_b, 01_b^{[r]})$	
13	A_2 via $(10, 10, 01)$	
14	A_2 via $(10, 10, 10^{[r]})$ $(r \neq 0)$	
15	A_2 via $(10, 10, 01^{[r]})$ $(r \neq 0)$	
16	A_2 via $(10, 10^{[r]}, 01)$ $(r \neq 0)$	
17	A_2 via $(10, 10^{[r]}, 10^{[r]})$ $(r \neq 0)$	
18	A_2 via $(10, 10^{[r]}, 01^{[r]})$ $(r \neq 0)$	
19	A_2 via $(10, 01^{[r]}, 01^{[r]})$ $(r \neq 0)$	

20	A_2 via $(10, 10^{[r]}, 10^{[s]})$ $(0 < r < s)$	
21	A_2 via $(10, 10^{[r]}, 01^{[s]})$ $(0 < r < s)$	
22	A_2 via $(10, 01^{[r]}, 10^{[s]})$ $(0 < r < s)$	
23	A_2 via $(10, 01^{[r]}, 01^{[s]})$ $(0 < r < s)$	
	In $M_1 = A_1\bar{A}_2^2$ $(p \neq 2)$ $(E_6(\#38))$	
41	$A_1^2\bar{A}_2$	$(2, 2, 10)$
42	A_1A_2 via $(1, 10, 10^{[r]})$	
43	A_1A_2 via $(1, 10, 01^{[r]})$ $(r \neq 0)$	
37$^{\{\underline{0}\}}$	A_1A_2 via $(1, 10, 01)$	
	In $M_2 = A_1^2\bar{A}_2$ $(p \neq 2)$ $(E_6(\#41))$	
44	A_1^3	$(2, 2, 2)$
45	$A_1\bar{A}_2$ via $(1, 1^{[r]}, 10)$	
	In $M_3 = A_1^3$ $(p \neq 2)$ $(E_6(\#44))$	
46	A_1A_1 via $(1_a, 1_a^{[r]}, 1_b)$ $(r \neq 0)$	
35$^{\{\underline{0}\}}$	A_1A_1 via $(1_a, 1_a, 1_b)$	
3	A_1 via $(1, 1^{[r]}, 1^{[s]})$ $(0 < r < s)$	
	In $M = A_2G_2$ $(E_6(\#26))$	
47	A_1G_2 $(p \neq 2)$	$(2, G_2(\#0))$
40$^{\{0\}}$	$A_2\bar{A}_2$	$(10, G_2(\#4))$
48	A_2A_2 $(p = 3)$	$(10, G_2(\#5))$
28	$A_2\bar{A}_1A_1$	$(10, G_2(\#6))$
49	A_2A_1 $(p \geq 7)$	$(10, G_2(\#3))$
	In $M_1 = A_1G_2$ $(p \neq 2)$ $(E_6(\#47))$	
45$^{\{0\}}$	$A_1\bar{A}_2$	$(2, G_2(\#4))$
50	A_1A_2 $(p = 3)$	$(2, G_2(\#5))$
31	$A_1\bar{A}_1A_1$	$(2, G_2(\#6))$
51	A_1A_1 $(p \geq 7)$	$(2, G_2(\#3))$
	In $M_2 = A_1A_2$ $(p = 3)$ $(E_6(\#50))$	
35$^{\{\delta_1\}}$	A_1A_1	$(2, G_2(\#1^{\{\delta_1\}}))$
	In $M_2 = A_1A_1$ $(p \geq 7)$ $(E_6(\#51))$	
4	A_1 via $(1^{[r]}, 1^{[s]})$ $(rs = 0; r \neq s)$	
1$^{\{\underline{0}\}}$	A_1 via $(1, 1)$	
	In $M_1 = A_2A_2$ $(p = 3)$ $(E_6(\#48))$	
50	A_1A_2	$(2, G_2(\#5))$
37$^{\{\delta_1\}}$	A_2A_1	$(10, G_2(\#1^{\{\delta_1\}}))$
12	A_2 via $(10^{[r]}, 10^{[s]})$ $(rs = 0; r \neq s)$	
	In $M_1 = A_2A_1$ $(p \geq 7)$ $(E_6(\#49))$	
51	A_1A_1	
	In $M = F_4$ $(E_6(\#7))$	
8b	C_4 $(p = 2)$	$F_4(\#14)$
27	$\bar{A}_1C_3$ $(p \neq 2)$	$F_4(\#24a)$

47	A_1G_2 ($p \neq 2$)	$F_4(\#25)$
39$^{\{0\}}$	$\bar{A}_2A_2$	$F_4(\#26)$
10b	G_2 ($p = 7$)	$F_4(\#16)$
5	A_1 ($p \geq 13$)	$F_4(\#10)$
	In $M = C_4$ (max. if $p \neq 2$; contained in F_4 if $p = 2$) ($E_6(\#8)$)	
27	$\bar{A}_1C_3$	$(1,000)/(0,100)$
44	A_1^3 ($p \neq 2$)	$(1,1,1)$
6	A_1 ($p \geq 11$)	7
9	D_4 ($p = 2$)	λ_1
	In $M_1 = D_4$ ($p = 2$) ($E_6(\#9)$)	
15$^{\{1\}}$	A_2	11
	In $M = G_2$ (max. if $p \neq 7$; contained in F_4 if $p = 7$) ($E_6(\#10)$)	
13	A_2	$G_2(\#4)$
11b	A_2 ($p = 3$)	$G_2(\#5)$
34$^{\{\underline{0}\}}$	A_1A_1 ($p \neq 2$)	$G_2(\#6)$
6	A_1 ($p \geq 11$)	$G_2(\#3)$
1$^{\{\delta_1\}}$	A_1 ($p = 7$)	$G_2(\#3)$
	In $M = A_2$ ($p \neq 2$) (max. if $p \geq 5$; contained in G_2 if $p = 3$) ($E_6(\#11)$)	
1$^{\{\underline{0}\}}$	A_1 ($p \geq 7$)	2
2$^{\{\delta_3\}}$	A_1 ($p = 3$)	2

Table 11.3A: The connected overgroups for irreducible subgroups of E_6.

ID	X	p	Immediate connected overgroups
$1^{\{\underline{0}\}}$	A_1	≥ 7	$A_2[\#11]$, $\bar{A}_1A_1[\#32]$, $A_1A_1[\#51]$
$1^{\{\delta_1\}}$	A_1	≥ 7	$G_2[\#10]$ ($p = 7$), $\bar{A}_1A_1[\#32]$
$2^{\{\delta_3\}}$	A_1	$\neq 2$	$A_2[\#11]$ ($p = 3$), $A_1A_1[\#34^{\{\underline{0}\}}]$, $A_1A_1[\#35^{\{\delta_2\}}]$, $\bar{A}_1A_1[\#36^{\{\delta_2\}}]$
5	A_1	≥ 13	$F_4[\#7]$
6	A_1	≥ 11	$C_4[\#8]$, $G_2[\#10]$
7	F_4		$E_6[\#0]$
8	C_4		$E_6[\#0]$ ($p \neq 2$), $F_4[\#7]$ ($p = 2$)
9	D_4	2	$C_4[\#8]$
10	G_2		$E_6[\#0]$ ($p \neq 7$), $F_4[\#7]$ ($p = 7$)
11	A_2	$\neq 2$	$E_6[\#0]$ ($p \geq 5$), $G_2[\#10]$ ($p = 3$)
13	A_2		$G_2[\#10]$, $\bar{A}_2A_2[\#39^{\{0\}}]$, $\bar{A}_2A_2[\#40^{\{0\}}]$
$15^{\{1\}}$	A_2		$D_4[\#9]$ ($p = 2$), $\bar{A}_2A_2[\#39^{\{0\}}]$, $\bar{A}_2A_2[\#40^{\{1\}}]$
24	$\bar{A}_1A_5$		$E_6[\#0]$
25	$\bar{A}_2^3$		$E_6[\#0]$
26	A_2G_2		$E_6[\#0]$

27	$\bar{A}_1C_3$		$F_4[\#7]$ $(p \neq 2)$, $C_4[\#8]$, $\bar{A}_1A_5[\#24]$
28	$\bar{A}_1A_1A_2$		$\bar{A}_1A_5[\#24]$, $A_2G_2[\#26]$
29	$\bar{A}_1A_3$		$\bar{A}_1A_5[\#24]$ $(p \neq 2)$, $\bar{A}_1C_3[\#27]$ $(p = 2)$
30	$\bar{A}_1A_2$	$\neq 2$	$\bar{A}_1A_5[\#24]$
31	$\bar{A}_1A_1A_1$	$\neq 2$	$\bar{A}_1C_3[\#27]$, $\bar{A}_1A_1A_2[\#28]$, $A_1G_2[\#47]$
32	$\bar{A}_1A_1$	≥ 7	$\bar{A}_1C_3[\#27]$
33	$\bar{A}_1G_2$	2	$\bar{A}_1C_3[\#27]$
$34^{\{\underline{0}\}}$	A_1A_1	$\neq 2$	$G_2[\#10]$, $\bar{A}_1A_1A_1[\#31]$
$35^{\{\underline{0}\}}$	A_1A_1	$\neq 2$	$\bar{A}_1A_1A_1[\#31]$, $A_1^3[\#44]$
$35^{\{\delta_1\}}$	A_1A_1	$\neq 2$	$\bar{A}_1A_1A_1[\#31]$, $A_1A_2[\#50]$ $(p = 3)$
$37^{\{\underline{0}\}}$	A_1A_2	$\neq 2$	$\bar{A}_1A_1A_2[\#28]$, $A_1\bar{A}_2^2[\#38]$
$37^{\{\delta_1\}}$	A_1A_2		$\bar{A}_1A_1A_2[\#28]$, $A_2A_2[\#48]$ $(p = 3)$
38	$A_1\bar{A}_2^2$	$\neq 2$	$\bar{A}_2^3[\#25]$
$39^{\{0\}}$	$\bar{A}_2A_2$		$F_4[\#7]$, $\bar{A}_2^3[\#25]$
$40^{\{0\}}$	$\bar{A}_2A_2$		$\bar{A}_2^3[\#25]$, $A_2G_2[\#26]$
41	$A_1^2\bar{A}_2$	$\neq 2$	$A_1\bar{A}_2^2[\#38]$
44	A_1^3	$\neq 2$	$C_4[\#8]$, $A_1^2\bar{A}_2[\#41]$
$45^{\{0\}}$	$A_1\bar{A}_2$	$\neq 2$	$A_1^2\bar{A}_2[\#41]$, $A_1G_2[\#47]$
47	A_1G_2	$\neq 2$	$F_4[\#7]$, $A_2G_2[\#26]$
48	A_2A_2	3	$A_2G_2[\#26]$
49	A_2A_1	≥ 7	$A_2G_2[\#26]$
50	A_1A_2	3	$A_1G_2[\#47]$, $A_2A_2[\#48]$
51	A_1A_1	≥ 7	$A_1G_2[\#47]$, $A_2A_1[\#49]$

Table 11.4: The lattice structure of irreducible connected subgroups of E_7.

ID	Irred. subgroup X	$V_M \downarrow X$
30	$\bar{A}_1D_6$	
31	$\bar{A}_2A_5$	
22	A_7	
32	G_2C_3	
33	A_1F_4	
34	A_1G_2 $(p \neq 2)$	
35	A_1A_1 $(p \geq 5)$	
29	A_2 $(p \geq 5)$	
20	A_1 $(p \geq 17)$	
21	A_1 $(p \geq 19)$	
	In $M = \bar{A}_1D_6$ $(E_7(\#30))$	
36	$\bar{A}_1^3\bar{D}_4$	$(1,1,1,0)/(1,0,0,\lambda_1)$
37a	$\bar{A}_1A_1B_4$ $(p \neq 2)$	$(1,2,0)/(1,0,\lambda_1)$
38a	$\bar{A}_1B_2B_3$ $(p \neq 2)$	$(1,10,000)/(1,00,100)$

39	$\bar{A}_1\bar{A}_3^2$	$(1,010,000)/(1,000,010)$
40	$\bar{A}_1B_5$	$(1,W(\lambda_1))/(1,0)$
41	$\bar{A}_1A_1C_3$	$(1,1,100)$
42	$\bar{A}_1A_1C_3$	$(1,1,100)$
	In $M_1=\bar{A}_1^3\bar{D}_4$ ($E_7(\#36)$)	
43	$\bar{A}_1^7$	$(1,1,1,0,0,0,0)/(1,0,0,1,1,0,0)/(1,0,0,0,0,1,1)$
44	$\bar{A}_1^3B_3$	$(1,1,1,000)/(1,0,0,W(100))/(1,0,0,000)$
45a	$\bar{A}_1^3A_1B_2$ $(p\neq 2)$	$(1,1,1,0,00)/(1,0,0,2,00)/(1,0,0,0,10)$
46a	$\bar{A}_1^3A_2$ $(p\neq 3)$	$(1,1,1,00)/(1,0,0,11)$
47	$A_1\bar{A}_1\bar{D}_4$ via $(1_a,1_a^{[r]},1_b,\lambda_1)$	
48	$A_1\bar{D}_4$ via $(1,1^{[r]},1^{[s]},\lambda_1)$ $(r\leq s)$	
	In $M_2=\bar{A}_1^7$ ($E_7(\#43)$)	
	See Table 11.6	
	In $M_2=\bar{A}_1^3B_3$ ($E_7(\#44)$)	
49{0}	$\bar{A}_1^5A_1$ $(p\neq 2)$	$(1,1,1,0,0,0)/(1,0,0,1,1,0)/(1,0,0,0,0,2)/$ $(1,0,0,0,0,0)$
75	$\bar{A}_1^3G_2$	$(1,1,1,00)/(1,0,0,W(10))/(1,0,0,00)$
45b	$\bar{A}_1^3A_1B_2$ $(p=2)$	$(1,1,1,0,00)/(1,0,0,2,00)/(1,0,0,0,10)/$ $(1,0,0,0,00)^2$
76	$\bar{A}_1A_1B_3$ via $(1_a,1_a^{[r]},1_b,100)$ $(r\neq 0)$	
77	$A_1\bar{A}_1B_3$ via $(1_a^{[r]},1_a^{[s]},1_b,100)$ $(rs=0)$	
78	A_1B_3 via $(1^{[r]},1^{[s]},1^{[t]},100)$ $(rs=0; s<t)$	
	In $M_3=\bar{A}_1^3G_2$ ($E_7(\#75)$)	
46b	$\bar{A}_1^3A_2$ $(p=3)$	$(1,1,1,00)/(1,0,0,11)/(1,0,0,00)$
51{0}	$\bar{A}_1^4A_1$	
79	$\bar{A}_1^3A_1$ $(p\geq 7)$	$(1,1,1,0)/(1,0,0,6)/(1,0,0,0)$
80	$A_1\bar{A}_1G_2$ via $(1_a,1_a^{[r]},1_b,10)$ $(r\neq 0)$	
81	A_1G_2 via $(1,1^{[r]},1^{[s]},10)$ $(0<r<s)$	
	In $M_4=\bar{A}_1^3A_1$ $(p\geq 7)$ ($E_7(\#79)$)	
82	$A_1\bar{A}_1A_1$ via $(1_a,1_a^{[r]},1_b,1_c)$ $(r\neq 0)$	
83	$A_1\bar{A}_1^2$ via $(1_a^{[r]},1_b,1_c,1_a^{[s]})$ $(rs=0)$	
84	A_1A_1 via $(1_a,1_a^{[r]},1_a^{[s]},1_b)$ $(0<r<s)$	
85	$A_1\bar{A}_1$ via $(1_a^{[r]},1_a^{[s]},1_b,1_a^{[t]})$ $(rt=0; r<s)$	
86	A_1A_1 via $(1_a,1_a^{[r]},1_b^{[s]},1_b^{[t]})$ $(st=0\neq r)$	
7	A_1 via $(1^{[r]},1^{[s]},1^{[t]},1^{[u]})$ $(ru=0; r<s<t)$	
	In $M_2=\bar{A}_1^3A_1B_2$ ($E_7(\#45)$)	
49{0}	$\bar{A}_1^3A_1\bar{A}_1^2$	
87	$\bar{A}_1^3A_1A_1$ $(p\geq 5)$	$(1,1,1,0,0)/(1,0,0,2,0)/(1,0,0,0,4)$
88	$\bar{A}_1^3A_1^3$ $(p=2)$	$(1,1,1,0,0,0)/(1,0,0,2,0,0)/(1,0,0,0,2,0)/$ $(1,0,0,0,0,2)/(1,0,0,0,0,0)^2$
	See Table 11.11	
	In $M_3=\bar{A}_1^3A_1A_1$ $(p\geq 5)$ ($E_7(\#87)$)	

	See Table 11.12	
	In $M_3 = \bar{A}_1^3 A_1^3$ $(p = 2)$ $(E_7(\#88))$	
	See Table 11.13	
	In $M_2 = \bar{A}_1^3 A_2$ $(E_7(\#46))$	
$105^{\{0\}}$	$\bar{A}_1^3 A_1$ $(p \geq 5)$	
$56^{\{\delta_3\}}$	$\bar{A}_1^3 A_1$ $(p = 3)$	
183	$A_1 \bar{A}_1 A_2$ via $(1_a, 1_a^{[r]}, 1_b, 10)$ (if $p = 3$ then $r \neq 0$)	
184	$A_1 A_2$ via $(1, 1^{[r]}, 1^{[s]}, 10)$ ($r \leq s$; if $p = 3$ then $0 < r < s$)	
	In $M_1 = \bar{A}_1 A_1 B_4$ $(E_7(\#37))$	
$47^{\{0\}}$	$\bar{A}_1 A_1 \bar{D}_4$	
185a	$\bar{A}_1 A_1^2 \bar{A}_3$ $(p \neq 2)$	$(1,2,0,000)/(1,0,2,000)/(1,0,0,010)$
45a	$\bar{A}_1^3 A_1 B_2$ $(p \neq 2)$	
186	$\bar{A}_1 A_1 A_1^2$ $(p \neq 2)$	$(1,2,0,0)/(1,0,2,2)$
187	$\bar{A}_1 A_1 A_1$ $(p \geq 11)$	$(1,2,0)/(1,0,8)$
188	$\bar{A}_1 A_1 B_2^2$ $(p = 2)$	$(1,2,00,00)/(1,0,10,00)/(1,0,00,10)/$ $(1,0,00,00)^2$
189	$\bar{A}_1 A_1^2 B_3$ $(p = 2)$	$(1,2,0,000)/(1,0,2,000)/(1,0,0,100)/$ $(1,0,0,000)^2$
190	$A_1 B_4$ via $(1^{[r]}, 1^{[s]}, \lambda_1)$ $(rs = 0)$	
	In $M_2 = \bar{A}_1 A_1^2 \bar{A}_3$ $(E_7(\#185))$	
191	$\bar{A}_1 A_1^4$ $(p \neq 2)$	$(1,2,0,0,0)/(1,0,2,0,0)/(1,0,0,2,0)/(1,0,0,0,2)$
$91^{\{0\}}$	$\bar{A}_1 A_1^2 B_2$ $(p \neq 2)$	
192	$A_1 A_1 \bar{A}_3$ via $(1_a^{[r]}, 1_a^{[s]}, 1_b, 100)$ $(rs = 0)$	
193	$\bar{A}_1 A_1 \bar{A}_3$ via $(1_a, 1_b, 1_b^{[r]}, 100)$ $(r \neq 0)$	
194	$A_1 \bar{A}_3$ via $(1^{[r]}, 1^{[s]}, 1^{[t]}, 100)$ $(rs = 0; s < t)$	
	In $M_3 = \bar{A}_1 A_1^4$ $(p \neq 2)$ $(E_7(\#191))$	
	See Table 11.14	
	In $M_2 = \bar{A}_1 A_1 A_1^2$ $(p \neq 2)$ $(E_7(\#186))$	
	See Table 11.15	
	In $M_2 = \bar{A}_1 A_1 A_1$ $(p \geq 11)$ $(E_7(\#187))$	
214	$A_1 A_1$ via $(1_a^{[r]}, 1_a^{[s]}, 1_b)$ $(rs = 0)$	
215	$A_1 A_1$ via $(1_a^{[r]}, 1_b, 1_a^{[s]})$ $(rs = 0)$	
216	$\bar{A}_1 A_1$ via $(1_a, 1_b^{[r]}, 1_b^{[s]})$ $(rs = 0)$	
5	A_1 via $(1^{[r]}, 1^{[s]}, 1^{[t]})$ $(rst = 0)$	
	In $M_2 = \bar{A}_1 A_1 B_2^2$ $(p = 2)$ $(E_7(\#188))$	
217	$\bar{A}_1 A_1^3 B_2$	$(1,2,0,0,00)/(1,0,2,0,00)/(1,0,0,2,00)/$ $(1,0,0,0,10)/(1,0,0,0,00)^2$
45b	$\bar{A}_1^3 A_1 B_2$	
218	$A_1 B_2^2$ via $(1^{[r]}, 1^{[s]}, 10, 10)$ $(rs = 0)$	
219	$\bar{A}_1 A_1 B_2$ via $(1, 1, 10, 10^{[r]})$ $(r \neq 0)$	
220	$\bar{A}_1 A_1 B_2$ via $(1, 1, 10, 02^{[r]})$	
221	$A_1 B_2$ via $(1^{[r]}, 1^{[s]}, 10, 10^{[t]})$ $(rs = 0 \neq t)$	

222	A_1B_2 via $(1^{[r]}, 1^{[s]}, 10, 02^{[t]})$ $(rs = 0)$	
	In $M_3 = \bar{A}_1A_1^3B_2$ $(p = 2)$ $(E_7(\#217))$	
88	$\bar{A}_1^3A_1^3$	
223	$\bar{A}_1A_1^5$	$(1,2,0,0,0,0)/(1,0,2,0,0,0)/(1,0,0,2,0,0)/$ $(1,0,0,0,2,0)/(1,0,0,0,0,2)/(1,0,0,0,0,0)^2$
224	$A_1A_1^2B_2$ via $(1_a^{[r]}, 1_a^{[s]}, 1_b, 1_c, 10)$ $(rs = 0)$	
225	$\bar{A}_1A_1A_1B_2$ via $(1_a, 1_b, 1_b^{[r]}, 1_c, 10)$ $(r \neq 0)$	
226	$A_1A_1B_2$ via $(1_a^{[r]}, 1_a^{[s]}, 1_a^{[t]}, 1_b, 10)$ $(rs = 0; s < t)$	
227	$\bar{A}_1A_1B_2$ via $(1_a, 1_b, 1_b^{[r]}, 1_b^{[s]}, 10)$ $(0 < r < s)$	
228	$A_1A_1B_2$ via $(1_a^{[r]}, 1_a^{[s]}, 1_b, 1_b^{[t]}, 10)$ $(rs = 0 \neq t)$	
229	A_1B_2 via $(1^{[r]}, 1^{[s]}, 1^{[t]}, 1^{[u]}, 10)$ $(rs = 0; s < t < u)$	
	In $M_4 = \bar{A}_1A_1^5$ $(p = 2)$ $(E_7(\#223))$	
	See Table 11.16	
	In $M_2 = \bar{A}_1A_1^2B_3$ $(p = 2)$ $(E_7(\#189))$	
217	$\bar{A}_1A_1^3B_2$	
185b	$\bar{A}_1A_1^2\bar{A}_3$	$(1,2,0,000)/(1,0,2,000)/(1,0,0,010)/$ $(1,0,0,000)^2$
244	$\bar{A}_1A_1^2G_2$	$(1,2,0,00)/(1,0,2,00)/(1,0,0,10)/(1,0,0,00)^2$
245	$A_1A_1B_3$ via $(1_a^{[r]}, 1_a^{[s]}, 1_b, 100)$ $(rs = 0)$	
246	$\bar{A}_1A_1B_3$ via $(1_a, 1_b, 1_b^{[r]}, 100)$ $(r \neq 0)$	
247	A_1B_3 via $(1^{[r]}, 1^{[s]}, 1^{[t]}, 100)$ $(rs = 0; s < t)$	
	In $M_3 = \bar{A}_1A_1^2G_2$ $(p = 2)$ $(E_7(\#244))$	
134$^{\{\underline{0}\}}$	$\bar{A}_1A_1^2\bar{A}_1A_1$	
248	$A_1A_1G_2$ via $(1_a^{[r]}, 1_a^{[s]}, 1_b, 10)$ $(rs = 0)$	
249	$\bar{A}_1A_1G_2$ via $(1_a, 1_b, 1_b^{[r]}, 10)$ $(r \neq 0)$	
250	A_1G_2 via $(1^{[r]}, 1^{[s]}, 1^{[t]}, 10)$ $(rs = 0; s < t)$	
	In $M_1 = \bar{A}_1B_2B_3$ $(E_7(\#38))$	
44	$\bar{A}_1^3B_3$	
251	$\bar{A}_1A_1B_3$ $(p \geq 5)$	$(1,4,000)/(1,0,100)$
189	$\bar{A}_1A_1^2B_3$ $(p = 2)$	
252	$\bar{A}_1B_2\bar{A}_3$	$(1,W(10),000)/(1,00,010)/(1,00,000)$
45a	$\bar{A}_1B_2\bar{A}_1^2A_1$ $(p \neq 2)$	
253	$\bar{A}_1B_2G_2$	$(1,W(10),00)/(1,00,W(10))$
188	$\bar{A}_1A_1B_2^2$ $(p = 2)$	
	In $M_2 = \bar{A}_1A_1B_3$ $(p \geq 5)$ $(E_7(\#251))$	
254	$\bar{A}_1A_1\bar{A}_3$	$(1,4,000)/(1,0,010)/(1,0,000)$
87	$\bar{A}_1A_1\bar{A}_1^2A_1$	
255	$\bar{A}_1A_1G_2$	$(1,4,00)/(1,0,10)$
256	A_1B_3 via $(1^{[r]}, 1^{[s]}, 100)$ $(rs = 0)$	
	In $M_3 = \bar{A}_1A_1\bar{A}_3$ $(p \geq 5)$ $(E_7(\#254))$	
102$^{\{0\}}$	$\bar{A}_1A_1A_1^2$	

257	$A_1\bar{A}_3$ via $(1^{[r]}, 1^{[s]}, 100)$ $(rs = 0)$	
	In $M_3 = \bar{A}_1A_1G_2$ $(p \geq 5)$ $(E_7(\#255))$	
103$^{\{\underline{0}\}}$	$\bar{A}_1^2A_1A_1$	
258	$\bar{A}_1A_1A_1$ $(p \geq 7)$	$(1,4,0)/(1,0,6)$
259	A_1G_2 via $(1^{[r]}, 1^{[s]}, 10)$ $(rs = 0)$	
	In $M_4 = \bar{A}_1A_1A_1$ $(p \geq 7)$ $(E_7(\#258))$	
260	A_1A_1 via $(1_a^{[r]}, 1_a^{[s]}, 1_b)$ $(rs = 0)$	
261	A_1A_1 via $(1_a^{[r]}, 1_b, 1_a^{[s]})$ $(rs = 0)$	
262	$\bar{A}_1A_1$ via $(1_a, 1_b^{[r]}, 1_b^{[s]})$ $(rs = 0)$	
6	A_1 via $(1^{[r]}, 1^{[s]}, 1^{[t]})$ $(rst = 0)$	
	In $M_2 = \bar{A}_1B_2\bar{A}_3$ $(E_7(\#252))$	
254	$\bar{A}_1A_1\bar{A}_3$ $(p \geq 5)$	
185b	$\bar{A}_1A_1^2\bar{A}_3$ $(p = 2)$	
91$^{\{0\}}$	$\bar{A}_1B_2A_1^2$ $(p \neq 2)$	
	In $M_2 = \bar{A}_1B_2G_2$ $(E_7(\#253))$	
75	$\bar{A}_1^3G_2$	
255	$\bar{A}_1A_1G_2$ $(p \geq 5)$	
244	$\bar{A}_1A_1^2G_2$ $(p = 2)$	
92$^{\{\underline{0}\}}$	$\bar{A}_1^2B_2A_1$	
263	$\bar{A}_1B_2A_2$ $(p = 3)$	$(1,10,00)/(1,00,11)$
264	$\bar{A}_1B_2A_1$ $(p \geq 7)$	$(1,10,0)/(1,00,6)$
	In $M_3 = \bar{A}_1B_2A_2$ $(p = 3)$ $(E_7(\#263))$	
46b	$\bar{A}_1^3A_2$	
94$^{\{\delta_2\}}$	$\bar{A}_1B_2A_1$	
	In $M_3 = \bar{A}_1B_2A_1$ $(p \geq 7)$ $(E_7(\#264))$	
79	$\bar{A}_1^3A_1$	
258	$\bar{A}_1A_1A_1$	
	In $M_1 = \bar{A}_1\bar{A}_3^2$ $(E_7(\#39))$	
252	$\bar{A}_1B_2\bar{A}_3$	
185a	$\bar{A}_1A_1^2\bar{A}_3$ $(p \neq 2)$	
265	$\bar{A}_1A_3$ via $(1, 100_a, 100_a^{[r]})$ $(r \neq 0)$	
266	$\bar{A}_1A_3$ via $(1, 100_a, 001_a^{[r]})$ $(r \neq 0)$	
	In $M_1 = \bar{A}_1B_5$ $(E_7(\#40))$	
47$^{\{0\}}$	$\bar{A}_1A_1\bar{D}_4$ $(p \neq 2)$	
44	$\bar{A}_1^3B_3$ $(p \neq 2)$	
252	$\bar{A}_1B_2\bar{A}_3$ $(p \neq 2)$	
267	$\bar{A}_1A_1$ $(p \geq 11)$	$(1,10)/(1,0)$
37b	$\bar{A}_1A_1B_4$ $(p = 2)$	$(1,2,0)/(1,0,\lambda_1)/(1,0,0)^2$
38b	$\bar{A}_1B_2B_3$ $(p = 2)$	$(1,10,000)/(1,00,100)/(1,00,000)^2$
	In $M_2 = \bar{A}_1A_1$ $(p \geq 11)$ $(E_7(\#267))$	
4	A_1 via $(1^{[r]}, 1^{[s]})$ $(rs = 0)$	
	In $M_1 = \bar{A}_1A_1C_3$ $(E_7(\#41))$	

$\mathit{90}^{\{\underline{0}\}}$	$\bar{A}_1A_1\bar{A}_1B_2$	
268	$\bar{A}_1A_1A_1A_1$ $(p \neq 2)$	$(1,1,2,1)$
269	$\bar{A}_1A_1A_1$ $(p \geq 7)$	$(1,1,5)$
270	$\bar{A}_1A_1A_3$ $(p = 2)$	$(1,1,010)$
271	$\bar{A}_1A_1G_2$ $(p = 2)$	$(1,1,10)$
272	A_1C_3 via $(1^{[r]},1^{[s]},100)$ $(rs = 0)$	
	In $M_2 = \bar{A}_1A_1A_1A_1$ $(p \neq 2)$ $(E_7(\#268))$	
	See Table 11.17	
	In $M_2 = \bar{A}_1A_1A_1$ $(p \geq 7)$ $(E_7(\#269))$	
286	A_1A_1 via $(1_a^{[r]},1_a^{[s]},1_b)$ $(rs = 0)$	
287	A_1A_1 via $(1_a^{[r]},1_b,1_a^{[s]})$ $(rs = 0)$	
288	$\bar{A}_1A_1$ via $(1_a,1_b^{[r]},1_b^{[s]})$ $(rs = 0; r \neq s)$	
$\mathit{262}^{\{\underline{0}\}}$	$\bar{A}_1A_1$ via $(1_a,1_b,1_b)$	
1	A_1 via $(1^{[r]},1^{[s]},1^{[t]})$ $(rst = 0; s \neq t)$	
	In $M_2 = \bar{A}_1A_1A_3$ $(p = 2)$ $(E_7(\#270))$	
289	A_1A_3 via $(1^{[r]},1^{[s]},100)$ $(rs = 0)$	
	In $M_2 = \bar{A}_1A_1G_2$ $(p = 2)$ $(E_7(\#271))$	
$\mathit{147}^{\{\delta_3\}}$	$\bar{A}_1A_1A_1A_1$	
290	A_1G_2 via $(1^{[r]},1^{[s]},10)$ $(rs = 0)$	
	In $M_1 = \bar{A}_1A_1C_3$ $(E_7(\#42))$	
$\mathit{92}^{\{\underline{0}\}}$	$\bar{A}_1^2A_1B_2$	
268	$\bar{A}_1A_1A_1A_1$ $(p \neq 2)$	
291	$\bar{A}_1A_1A_1$ $(p \geq 7)$	$(1,1,5)$
292	$\bar{A}_1A_1A_3$ $(p = 2)$	$(1,1,010)$
293	$\bar{A}_1A_1G_2$ $(p = 2)$	$(1,1,10)$
294	A_1C_3 via $(1^{[r]},1^{[s]},100)$ $(rs = 0$; if $p = 2$ then $r \neq s)$	
	In $M_2 = \bar{A}_1A_1A_1$ $(p \geq 7)$ $(E_7(\#291))$	
295	A_1A_1 via $(1_a^{[r]},1_a^{[s]},1_b)$ $(rs = 0)$	
296	A_1A_1 via $(1_a^{[r]},1_b,1_a^{[s]})$ $(rs = 0)$	
297	$\bar{A}_1A_1$ via $(1_a,1_b^{[r]},1_b^{[s]})$ $(rs = 0; r \neq s)$	
$\mathit{262}^{\{\underline{0}\}}$	$\bar{A}_1A_1$ via $(1_a,1_b,1_b)$	
2	A_1 via $(1^{[r]},1^{[s]},1^{[t]})$ $(rst = 0; s \neq t)$	
	In $M_2 = \bar{A}_1A_1A_3$ $(p = 2)$ $(E_7(\#292))$	
298	A_1A_3 via $(1^{[r]},1^{[s]},100)$ $(rs = 0; r \neq s)$	
	In $M_2 = \bar{A}_1A_1G_2$ $(p = 2)$ $(E_7(\#293))$	
$\mathit{147}^{\{\delta_1\}}$	$\bar{A}_1A_1A_1A_1$	
299	A_1G_2 via $(1^{[r]},1^{[s]},10)$ $(rs = 0; r \neq s)$	
	In $M = \bar{A}_2A_5$ $(E_7(\#31))$	
300	$\bar{A}_2A_2A_1$	$(10,10,1)$
301	$\bar{A}_2C_3$	$(10,100)$
302a	$\bar{A}_2A_3$ $(p \neq 2)$	$(10,010)$

303	$\bar{A}_2A_2$ $(p \neq 2)$	$(10, 20)$
304	A_1A_5 $(p \neq 2)$	$(2, \lambda_1)$
	In $M_1 = \bar{A}_2A_2A_1$ $(E_7(\#300))$	
305	$A_1A_2A_1$ $(p \neq 2)$	$(2, 10, 1)$
306	$\bar{A}_2A_1A_1$ $(p \neq 2)$	$(10, 2, 1)$
307	A_2A_1 via $(10^{[r]}, 10^{[s]}, 1)$ $(rs = 0)$	
308	A_2A_1 via $(10^{[r]}, 01^{[s]}, 1)$ $(rs = 0; r \neq s)$	
$\mathit{184}^{\{\underline{0}\}}$	A_2A_1 via $(10, 01, 1)$ $(p \neq 3)$	
	In $M_2 = A_1A_2A_1$ $(p \neq 2)$ $(E_7(\#305))$	
$\mathit{275}^{\{\underline{0}\}}$	$A_1A_1A_1$	$(2, 2, 1)$
309	A_1A_2 via $(1^{[r]}, 10, 1^{[s]})$ $(rs = 0)$	
	In $M_2 = \bar{A}_2A_1A_1$ $(p \neq 2)$ $(E_7(\#306))$	
$\mathit{275}^{\{\underline{0}\}}$	$A_1A_1A_1$	
310	$\bar{A}_2A_1$ via $(10, 1^{[r]}, 1^{[s]})$ $(rs = 0; r \neq s)$	
	In $M_1 = \bar{A}_2C_3$ $(E_7(\#301))$	
$\mathit{294}^{\{\underline{0}\}}$	A_1C_3 $(p \neq 2)$	$(2, 100)$
306	$\bar{A}_2A_1A_1$ $(p \neq 2)$	
311	$\bar{A}_2A_1$ $(p \geq 7)$	$(10, 5)$
302b	$\bar{A}_2A_3$ $(p = 2)$	$(10, 010)$
312	$\bar{A}_2G_2$ $(p = 2)$	$(10, 10)$
	In $M_2 = \bar{A}_2A_1$ $(p \geq 7)$ $(E_7(\#311))$	
$\mathit{295}^{\{\underline{0}\}}$	A_1A_1	$(2, 5)$
	In $M_1 = \bar{A}_2A_3$ $(E_7(\#302))$	
313	A_1A_3 $(p \neq 2)$	$(2, 010)$
	In $M_1 = \bar{A}_2A_2$ $(p \neq 2)$ $(E_7(\#303))$	
314	A_1A_2	$(2, 20)$
24	A_2 via $(10, 10)$ $(p \geq 5)$	
25	A_2 via $(10^{[r]}, 10^{[s]})$ $(rs = 0; r \neq s)$	
26	A_2 via $(10, 01)$	
27	A_2 via $(10^{[r]}, 01^{[s]})$ $(rs = 0; r \neq s)$	
	In $M_1 = A_1A_5$ $(p \neq 2)$ $(E_7(\#304))$	
305	$A_1A_2A_1$	
$\mathit{294}^{\{\underline{0}\}}$	A_1C_3	
313	A_1A_3	
314	A_1A_2	
	In $M = A_7$ $(E_7(\#22))$	
23	D_4 $(p \neq 2)$	λ_1
313	A_1A_3 $(p \neq 2)$	$(1, 100)$
	In $M_1 = D_4$ $(p \neq 2)$ $(E_7(\#23))$	
24	A_2 $(p \geq 5)$	11
	In $M = G_2C_3$ $(E_7(\#32))$	

301	$\bar{A}_2C_3$	$(G_2(\#4), 100)$
315	A_2C_3 $(p=3)$	$(G_2(\#5), 100)$
42	$\bar{A}_1A_1C_3$	$(G_2(\#6), 100)$
316	A_1C_3 $(p \geq 7)$	$(G_2(\#3), 100)$
253	$G_2\bar{A}_1B_2$	$(G_2(\#0), (1,00)/(0,01))$
317	$G_2A_1A_1$ $(p \neq 2)$	$(G_2(\#0), (2,1))$
318	G_2A_1 $(p \geq 7)$	$(G_2(\#0), 5)$
319	G_2A_3 $(p=2)$	$(G_2(\#0), 010)$
320	G_2G_2 $(p=2)$	$(G_2(\#0), 10)$
	In $M_1 = A_2C_3$ $(p=3)$ $(E_7(\#315))$	
$\mathit{294}^{\{\delta_1\}}$	A_1C_3	$(G_2(\#1^{\{\delta_1\}}), 100)$
263	$A_2\bar{A}_1B_2$	$(G_2(\#5), (1,00)/(0,01))$
321	$A_2A_1A_1$	$(G_2(\#5), (2,1))$
	In $M_2 = A_2A_1A_1$ $(p=3)$ $(E_7(\#321))$	
$\mathit{275}^{\{\delta_1\}}$	$A_1A_1A_1$	$(G_2(\#1^{\{\delta_1\}}), (2,1))$
322	A_2A_1 via $(10, 1^{[r]}, 1^{[s]})$ $(rs=0; r \neq s)$	
	In $M_1 = A_1C_3$ $(p \geq 7)$ $(E_7(\#316))$	
264	$A_1\bar{A}_1B_2$	$(G_2(\#3), (1,00)/(0,01))$
323	$A_1A_1A_1$	$(G_2(\#3), (2,1))$
324	A_1A_1	$(G_2(\#3), 5)$
	In $M_2 = A_1A_1A_1$ $(p \geq 7)$ $(E_7(\#323))$	
325	A_1A_1 via $(1_a^{[r]}, 1_a^{[s]}, 1_b)$ $(rs=0;\ r \neq s)$	
$\mathit{287}^{\{\underline{\varrho}\}}$	A_1A_1 via $(1_a, 1_a, 1_b)$	
326	A_1A_1 via $(1_a^{[r]}, 1_b, 1_a^{[s]})$ $(rs=0)$	
327	A_1A_1 via $(1_a, 1_b^{[r]}, 1_b^{[s]})$ $(rs=0; r \neq s)$	
$\mathit{260}^{\{\underline{\varrho}\}}$	A_1A_1 via $(1_a, 1_b, 1_b)$	
16	A_1 via $(1^{[r]}, 1^{[s]}, 1^{[t]})$ $(rst=0; r \neq s; s \neq t)$	
	In $M_2 = A_1A_1$ $(p \geq 7)$ $(E_7(\#324))$	
15	A_1 via $(1^{[r]}, 1^{[s]})$ $(rs=0; r \neq s)$	
$\mathit{5}^{\{\underline{\varrho}\}}$	A_1 via $(1,1)$ $(p \geq 11)$	
	In $M_1 = G_2A_1A_1$ $(p \neq 2)$ $(E_7(\#317))$	
306	$\bar{A}_2A_1A_1$	$(G_2(\#4), (2,1))$
321	$A_2A_1A_1$ $(p=3)$	
268	$\bar{A}_1A_1A_1A_1$	$(G_2(\#6), (2,1))$
323	$A_1A_1A_1$ $(p \geq 7)$	
328	G_2A_1 via $(10, 1^{[r]}, 1^{[s]})$ $(rs=0; r \neq s)$	
$\mathit{259}^{\{\underline{\varrho}\}}$	G_2A_1 via $(10,1,1)$ $(p \geq 5)$	
	In $M_1 = G_2A_1$ $(p \geq 7)$ $(E_7(\#318))$	
311	$\bar{A}_2A_1$	$(G_2(\#4), 5)$
291	$\bar{A}_1A_1A_1$	$(G_2(\#6), 5)$
324	A_1A_1	
	In $M_1 = G_2A_3$ $(p=2)$ $(E_7(\#319))$	

302b	$\bar{A}_2A_3$	$(G_2(\#4), 010)$
292	$\bar{A}_1A_1A_3$	$(G_2(\#6), 010)$
	In $M_1 = G_2G_2$ $(p = 2)$ $(E_7(\#320))$	
312	$\bar{A}_2G_2$	$(G_2(\#4), 10)$
293	$\bar{A}_1A_1G_2$	$(G_2(\#6), 10)$
92$^{\{\underline{0}\}}$	$G_2A_1A_1$	$(G_2(\#0), (1,1)/(0,2))$
28	G_2 via $(10^{[r]}, 10^{[s]})$ $(rs = 0; r \neq s)$	
	In $M = A_1F_4$ $(E_7(\#33))$	
190$^{\{\underline{0}\}}$	A_1B_4	$(1, F_4(\#12))$
329	A_1C_4 $(p = 2)$	$(1, F_4(\#14))$
41a	$A_1\bar{A}_1C_3$ $(p \neq 2)$	$(1, F_4(\#24\text{a}))$
317	$A_1A_1G_2$ $(p \neq 2)$	$(1, F_4(\#25))$
300	$A_1\bar{A}_2A_2$	$(1, F_4(\#26))$
330	A_1G_2 $(p = 7)$	$(1, F_4(\#16))$
331	A_1A_1 $(p \geq 13)$	$(1, F_4(\#10))$
	In $M_1 = A_1C_4$ $(p = 2)$ $(E_7(\#329))$	
332	A_1D_4	$(1, F_4(\#15))$
218$^{\{\underline{0}\}}$	$A_1B_2^2$	$(1, F_4(\#30))$
41b	$A_1\bar{A}_1C_3$	$(1, F_4(\#24\text{b}))$
	In $M_2 = A_1D_4$ $(p = 2)$ $(E_7(\#332))$	
230$^{\{\underline{0}\}}$	$A_1A_1^4$	$(1, F_4(\#51))$
307$^{\{\delta_1\}}$	A_1A_2	$(1, F_4(\#18^{\{\delta_1\}}))$
	In $M_1 = A_1G_2$ $(p = 7)$ $(E_7(\#330))$	
307$^{\{\underline{0}\}}$	A_1A_2	$(1, F_4(\#17))$
274$^{\{\underline{0}\}}$	$A_1A_1A_1$	$(1, F_4(\#61^{\{\underline{0}\}}))$
287$^{\{\delta_1\}}$	A_1A_1	$(1, F_4(\#8^{\{\delta_1\}}))$
	In $M_1 = A_1A_1$ $(p \geq 13)$ $(E_7(\#331))$	
18	A_1 via $(1^{[r]}, 1^{[s]})$ $(rs = 0)$	
	In $M = A_1G_2$ $(p \neq 2)$ $(E_7(\#34))$	
309$^{\{\underline{0}\}}$	A_1A_2	$(1, G_2(\#4))$
333	A_1A_2 $(p = 3)$	$(1, G_2(\#5))$
206$^{\{\underline{0}\}}$	$A_1A_1A_1$	$(1, G_2(\#6))$
334	A_1A_1 $(p \geq 7)$	$(1, G_2(\#3))$
	In $M_1 = A_1A_2$ $(p = 3)$ $(E_7(\#333))$	
213$^{\{\delta_3\}}$	A_1A_1 $(p = 3)$	$(1, G_2(\#1^{\{\delta_1\}}))$
	In $M_1 = A_1A_1$ $(p \geq 7)$ $(E_7(\#334))$	
17	A_1 via $(1^{[r]}, 1^{[s]})$ $(rs = 0; r \neq s)$	
5$^{\{\underline{0}\}}$	A_1 via $(1, 1)$ $(p \geq 11)$	
	In $M = A_1A_1$ $(p \geq 5)$ $(E_7(\#35))$	
19	A_1 via $(1^{[r]}, 1^{[s]})$ $(rs = 0; r \neq s)$	

Table 11.4A: The connected overgroups for irreducible subgroups of E_7.

ID	X	p	Immediate connected overgroups
$5^{\{\underline{0}\}}$	A_1	≥ 11	immediate overgroups in $\bar{A}_1A_1A_1[\#187]$, $A_1A_1[\#324]$, $A_1A_1[\#334]$
20	A_1	≥ 17	$E_7[\#0]$
21	A_1	≥ 19	$E_7[\#0]$
22	A_7		$E_7[\#0]$
23	D_4	$\neq 2$	$A_7[\#22]$
24	A_2	≥ 5	$D_4[\#23]$, $\bar{A}_2A_2[\#303]$
29	A_2	≥ 5	$E_7[\#0]$
30	$\bar{A}_1D_6$		$E_7[\#0]$
31	$\bar{A}_2A_5$		$E_7[\#0]$
32	G_2C_3		$E_7[\#0]$
33	A_1F_4		$E_7[\#0]$
34	A_1G_2	$\neq 2$	$E_7[\#0]$
35	A_1A_1	≥ 5	$E_7[\#0]$
36	$\bar{A}_1^3\bar{D}_4$		$\bar{A}_1D_6[\#30]$
37	$\bar{A}_1A_1B_4$		$\bar{A}_1D_6[\#30]$ $(p \neq 2)$, $\bar{A}_1B_5[\#40]$ $(p = 2)$
38	$\bar{A}_1B_2B_3$		$\bar{A}_1D_6[\#30]$ $(p \neq 2)$, $\bar{A}_1B_5[\#40]$ $(p = 2)$
39	$\bar{A}_1\bar{A}_3^2$		$\bar{A}_1D_6[\#30]$
40	$\bar{A}_1B_5$		$\bar{A}_1D_6[\#30]$
41	$\bar{A}_1A_1C_3$		$\bar{A}_1D_6[\#30]$, $A_1F_4[\#33]$ $(p \neq 2)$, $A_1C_4[\#329]$ $(p = 2)$
42	$\bar{A}_1A_1C_3$		$\bar{A}_1D_6[\#30]$, $G_2C_3[\#32]$
43	$\bar{A}_1^7$		$\bar{A}_1^3\bar{D}_4[\#36]$
44	$\bar{A}_1^3B_3$		$\bar{A}_1^3\bar{D}_4[\#36]$, $\bar{A}_1B_2B_3[\#38]$, $\bar{A}_1B_5[\#40]$ $(p \neq 2)$
45	$\bar{A}_1^3A_1B_2$		$\bar{A}_1^3\bar{D}_4[\#36]$ $(p \neq 2)$, $\bar{A}_1A_1B_4[\#37]$ $(p \neq 2)$, $\bar{A}_1B_2B_3[\#38]$ $(p \neq 2)$, $\bar{A}_1^3B_3[\#44]$ $(p = 2)$, $\bar{A}_1A_1B_2^2[\#188]$ $(p = 2)$
46	$\bar{A}_1^3A_2$		$\bar{A}_1^3\bar{D}_4[\#36]$ $(p \neq 3)$, $\bar{A}_1^3G_2[\#75]$ $(p = 3)$, $\bar{A}_1B_2A_2[\#263]$ $(p = 3)$
$47^{\{0\}}$	$A_1\bar{A}_1\bar{D}_4$		$\bar{A}_1^3\bar{D}_4[\#36]$, $\bar{A}_1A_1B_4[\#37]$, $\bar{A}_1B_5[\#40]$ $(p \neq 2)$
$49^{\{0\}}$	$\bar{A}_1^5A_1$		$\bar{A}_1^7[\#43]$, $\bar{A}_1^3B_3[\#44]$, $\bar{A}_1^3A_1B_2[\#45]$
$51^{\{\underline{0}\}}$	$\bar{A}_1^4A_1$		$\bar{A}_1^5A_1[\#49^{\{0\}}]$, $\bar{A}_1^3G_2[\#75]$
$56^{\{\delta_3\}}$	$\bar{A}_1^3A_1$		immediate overgroups in $\bar{A}_1^7[\#43]$, $\bar{A}_1^3A_2[\#46]$ $(p = 3)$
75	$\bar{A}_1^3G_2$		$\bar{A}_1^3B_3[\#44]$, $\bar{A}_1B_2G_2[\#253]$
79	$\bar{A}_1^3A_1$	≥ 7	$\bar{A}_1^3G_2[\#75]$, $\bar{A}_1B_2A_1[\#264]$
87	$\bar{A}_1^3A_1A_1$	≥ 5	$\bar{A}_1^3A_1B_2[\#45]$, $\bar{A}_1A_1B_3[\#251]$
88	$\bar{A}_1^3A_1^3$	2	$\bar{A}_1^3A_1B_2[\#45]$, $\bar{A}_1A_1^3B_2[\#217]$
$90^{\{0\}}$	$\bar{A}_1^2A_1B_2$		$\bar{A}_1A_1C_3[\#41]$, $\bar{A}_1^3A_1B_2[\#45]$
$91^{\{0\}}$	$\bar{A}_1A_1^2B_2$	$\neq 2$	$\bar{A}_1^3A_1B_2[\#45]$, $\bar{A}_1A_1^2\bar{A}_3[\#185]$, $\bar{A}_1B_2\bar{A}_3[\#252]$

$92^{\{\underline{0}\}}$	$\bar{A}_1^2A_1B_2$		$\bar{A}_1A_1C_3[\#42]$, $\bar{A}_1^3A_1B_2[\#45]$, $\bar{A}_1B_2G_2[\#253]$, $G_2G_2[\#320]$ $(p=2)$
$94^{\{\delta_2\}}$	$\bar{A}_1A_1B_2$		immediate overgroups in $\bar{A}_1^3A_1B_2[\#45]$, $\bar{A}_1B_2A_2[\#263]$ $(p=3)$
$102^{\{0\}}$	$\bar{A}_1A_1A_1^2$	≥ 5	$\bar{A}_1^3A_1A_1[\#87]$, $\bar{A}_1A_1\bar{A}_3[\#254]$
$103^{\{\underline{0}\}}$	$\bar{A}_1^2A_1A_1$	≥ 5	$\bar{A}_1^3A_1A_1[\#87]$, $\bar{A}_1A_1G_2[\#255]$
$105^{\{\underline{0}\}}$	$\bar{A}_1^3A_1$	≥ 5	$\bar{A}_1^3A_2[\#46]$, $\bar{A}_1^3A_1A_1[\#87]$
$115^{\{\underline{0}\}}$	$\bar{A}_1A_1A_1$	≥ 5	$\bar{A}_1A_1^2A_1[\#99^{\{\underline{0}\}}]$, $\bar{A}_1^3A_1[\#105^{\{\underline{0}\}}]$, $\bar{A}_1A_1A_1^2[\#186]$
$117^{\{\underline{0}\}}$	$\bar{A}_1A_1A_1$	≥ 5	$\bar{A}_1^2A_1A_1[\#100^{\{\underline{0}\}}]$, $\bar{A}_1^2A_1A_1[\#104^{\{\underline{0}\}}]$, $\bar{A}_1A_1A_1A_1[\#268]$
$119^{\{\underline{0}\}}$	$\bar{A}_1A_1A_1$	≥ 5	$\bar{A}_1^2A_1A_1[\#101^{\{\underline{0}\}}]$, $\bar{A}_1^2A_1A_1[\#103^{\{\underline{0}\}}]$, $\bar{A}_1A_1A_1A_1[\#268]$
$134^{\{\underline{0}\}}$	$\bar{A}_1^2A_1^2A_1$	2	$\bar{A}_1^3A_1^3[\#88]$, $\bar{A}_1A_1^2G_2[\#244]$
$147^{\{\delta_1\}}$	$\bar{A}_1A_1A_1A_1$	2	$\bar{A}_1^2A_1^2A_1[\#134^{\{\delta_1\}}]$, $\bar{A}_1^2A_1^2A_1[\#136^{\{\underline{0}\}}]$, $\bar{A}_1A_1G_2[\#293]$
$147^{\{\delta_3\}}$	$\bar{A}_1A_1A_1A_1$	2	$\bar{A}_1^2A_1^2A_1[\#134^{\{\underline{0}\}}]$, $\bar{A}_1^2A_1^2A_1[\#136^{\{\delta_1\}}]$, $\bar{A}_1A_1G_2[\#271]$
$184^{\{\underline{0}\}}$	A_1A_2	$\neq 3$	$\bar{A}_1^3A_2[\#46]$, $\bar{A}_2A_2A_1[\#300]$
185	$\bar{A}_1A_1^2\bar{A}_3$		$\bar{A}_1A_1B_4[\#37]$ $(p\neq 2)$, $\bar{A}_1\bar{A}_3^2[\#39]$ $(p\neq 2)$, $\bar{A}_1A_1^2B_3[\#189]$ $(p=2)$, $\bar{A}_1B_2\bar{A}_3[\#252]$ $(p=2)$
186	$\bar{A}_1A_1A_1^2$	$\neq 2$	$\bar{A}_1A_1B_4[\#37]$
187	$\bar{A}_1A_1A_1$	≥ 11	$\bar{A}_1A_1B_4[\#37]$
188	$\bar{A}_1A_1B_2^2$	2	$\bar{A}_1A_1B_4[\#37]$, $\bar{A}_1B_2B_3[\#38]$
189	$\bar{A}_1A_1^2B_3$	2	$\bar{A}_1A_1B_4[\#37]$, $\bar{A}_1B_2B_3[\#38]$
$190^{\{\underline{0}\}}$	A_1B_4		$A_1F_4[\#33]$, $\bar{A}_1A_1B_4[\#37]$
191	$\bar{A}_1A_1^4$	$\neq 2$	$\bar{A}_1A_1^2\bar{A}_3[\#185]$
$206^{\{\underline{0}\}}$	$A_1A_1A_1$	$\neq 2$	$A_1G_2[\#34]$, $\bar{A}_1A_1A_1^2[\#186]$
$207^{\{\underline{0}\}}$	$\bar{A}_1A_1A_1$	$\neq 2$	$\bar{A}_1A_1A_1^2[\#186]$, $\bar{A}_1A_1A_1A_1[\#268]$
$213^{\{\delta_3\}}$	A_1A_1	$\neq 2$	$A_1A_1A_1[\#206^{\{\underline{0}\}}]$, $\bar{A}_1A_1A_1[\#207^{\{\delta_1\}}]$, $A_1A_2[\#333]$ $(p=3)$
217	$\bar{A}_1A_1^3B_2$	2	$\bar{A}_1A_1B_2^2[\#188]$, $\bar{A}_1A_1^2B_3[\#189]$
$218^{\{\underline{0}\}}$	$A_1B_2^2$	2	$\bar{A}_1A_1B_2^2[\#188]$, $A_1C_4[\#329]$
223	$\bar{A}_1A_1^5$	2	$\bar{A}_1A_1^3B_2[\#217]$
$230^{\{\underline{0}\}}$	$A_1A_1^4$	2	$\bar{A}_1A_1^5[\#223]$, $A_1D_4[\#332]$
244	$\bar{A}_1A_1^2G_2$	2	$\bar{A}_1A_1^2B_3[\#189]$, $\bar{A}_1B_2G_2[\#253]$
251	$\bar{A}_1A_1B_3$	≥ 5	$\bar{A}_1B_2B_3[\#38]$
252	$\bar{A}_1B_2\bar{A}_3$		$\bar{A}_1B_2B_3[\#38]$, $\bar{A}_1\bar{A}_3^2[\#39]$, $\bar{A}_1B_5[\#40]$ $(p\neq 2)$
253	$\bar{A}_1B_2G_2$		$G_2C_3[\#32]$, $\bar{A}_1B_2B_3[\#38]$
254	$\bar{A}_1A_1\bar{A}_3$	≥ 5	$\bar{A}_1A_1B_3[\#251]$, $\bar{A}_1B_2\bar{A}_3[\#252]$
255	$\bar{A}_1A_1G_2$	≥ 5	$\bar{A}_1A_1B_3[\#251]$, $\bar{A}_1B_2G_2[\#253]$
258	$\bar{A}_1A_1A_1$	≥ 7	$\bar{A}_1A_1G_2[\#255]$, $\bar{A}_1B_2A_1[\#264]$
$259^{\{\underline{0}\}}$	A_1G_2	≥ 5	$\bar{A}_1A_1G_2[\#255]$, $G_2A_1A_1[\#317]$
$260^{\{\underline{0}\}}$	A_1A_1	≥ 7	$\bar{A}_1A_1A_1[\#258]$, $A_1A_1A_1[\#323]$
$262^{\{\underline{0}\}}$	$\bar{A}_1A_1$	≥ 7	$\bar{A}_1A_1A_1[\#258]$, $\bar{A}_1A_1A_1[\#269]$, $\bar{A}_1A_1A_1[\#291]$
263	$\bar{A}_1B_2A_2$	3	$\bar{A}_1B_2G_2[\#253]$, $A_2C_3[\#315]$

264	$\bar{A}_1B_2A_1$	≥ 7	$\bar{A}_1B_2G_2$[#253], A_1C_3[#316]
267	$\bar{A}_1A_1$	≥ 11	$\bar{A}_1B_5$[#40]
268	$\bar{A}_1A_1A_1A_1$	$\neq 2$	$\bar{A}_1A_1C_3$[#41], $\bar{A}_1A_1C_3$[#42], $G_2A_1A_1$[#317]
269	$\bar{A}_1A_1A_1$	≥ 7	$\bar{A}_1A_1C_3$[#41]
270	$\bar{A}_1A_1A_3$	2	$\bar{A}_1A_1C_3$[#41]
271	$\bar{A}_1A_1G_2$	2	$\bar{A}_1A_1C_3$[#41]
$274^{\{\underline{0}\}}$	$A_1A_1A_1$	$\neq 2$	$\bar{A}_1A_1A_1A_1$[#268], A_1G_2[#330] $(p = 7)$
$275^{\{\underline{0}\}}$	$A_1A_1A_1$	$\neq 2$	$\bar{A}_1A_1A_1A_1$[#268], $A_1A_2A_1$[#305], $\bar{A}_2A_1A_1$[#306]
$275^{\{\delta_1\}}$	$A_1A_1A_1$	$\neq 2$	$\bar{A}_1A_1A_1A_1$[#268], $A_2A_1A_1$[#321] $(p = 3)$
$287^{\{\underline{0}\}}$	A_1A_1	≥ 7	$\bar{A}_1A_1A_1$[#269], $A_1A_1A_1$[#323]
$287^{\{\delta_1\}}$	A_1A_1	≥ 7	$\bar{A}_1A_1A_1$[#269], A_1G_2[#330] $(p = 7)$
291	$\bar{A}_1A_1A_1$	≥ 7	$\bar{A}_1A_1C_3$[#42], G_2A_1[#318]
292	$\bar{A}_1A_1A_3$	2	$\bar{A}_1A_1C_3$[#42], G_2A_3[#319]
293	$\bar{A}_1A_1G_2$	2	$\bar{A}_1A_1C_3$[#42], G_2G_2[#320]
$294^{\{\underline{0}\}}$	A_1C_3	$\neq 2$	$\bar{A}_1A_1C_3$[#42], $\bar{A}_2C_3$[#301], A_1A_5[#304]
$294^{\{\delta_1\}}$	A_1C_3		$\bar{A}_1A_1C_3$[#42], A_2C_3[#315] $(p = 3)$
$295^{\{\underline{0}\}}$	A_1A_1	≥ 7	$\bar{A}_1A_1A_1$[#291], $\bar{A}_2A_1$[#311]
300	$\bar{A}_2A_2A_1$		$\bar{A}_2A_5$[#31], A_1F_4[#33]
301	$\bar{A}_2C_3$		$\bar{A}_2A_5$[#31], G_2C_3[#32]
302	$\bar{A}_2A_3$		$\bar{A}_2A_5$[#31] $(p \neq 2)$, $\bar{A}_2C_3$[#301] $(p = 2)$, G_2A_3[#319] $(p = 2)$
303	$\bar{A}_2A_2$	$\neq 2$	$\bar{A}_2A_5$[#31]
304	A_1A_5	$\neq 2$	$\bar{A}_2A_5$[#31]
305	$A_1A_2A_1$	$\neq 2$	$\bar{A}_2A_2A_1$[#300], A_1A_5[#304]
306	$\bar{A}_2A_1A_1$	$\neq 2$	$\bar{A}_2A_2A_1$[#300], $\bar{A}_2C_3$[#301], $G_2A_1A_1$[#317]
$307^{\{\underline{0}\}}$	A_1A_2		$\bar{A}_2A_2A_1$[#300], A_1G_2[#330] $(p = 7)$
$307^{\{\delta_1\}}$	A_1A_2		$\bar{A}_2A_2A_1$[#300], A_1D_4[#332] $(p = 2)$
$309^{\{\underline{0}\}}$	A_1A_2	$\neq 2$	A_1G_2[#34], $A_1A_2A_1$[#305]
311	$\bar{A}_2A_1$	≥ 7	$\bar{A}_2C_3$[#301], G_2A_1[#318]
312	$\bar{A}_2G_2$	2	$\bar{A}_2C_3$[#301], G_2G_2[#320]
313	A_1A_3	$\neq 2$	A_7[#22], $\bar{A}_2A_3$[#302], A_1A_5[#304]
314	A_1A_2	$\neq 2$	$\bar{A}_2A_2$[#303], A_1A_5[#304]
315	A_2C_3	3	G_2C_3[#32]
316	A_1C_3	≥ 7	G_2C_3[#32]
317	$G_2A_1A_1$	$\neq 2$	G_2C_3[#32], A_1F_4[#33]
318	G_2A_1	≥ 7	G_2C_3[#32]
319	G_2A_3	2	G_2C_3[#32]
320	G_2G_2	2	G_2C_3[#32]
321	$A_2A_1A_1$	3	A_2C_3[#315], $G_2A_1A_1$[#317]
323	$A_1A_1A_1$	≥ 7	A_1C_3[#316], $G_2A_1A_1$[#317]
324	A_1A_1	≥ 7	A_1C_3[#316], G_2A_1[#318]
329	A_1C_4	2	A_1F_4[#33]

330	A_1G_2	7	A_1F_4[#33]
331	A_1A_1	≥ 13	A_1F_4[#33]
332	A_1D_4	2	A_1C_4[#329]
333	A_1A_2	3	A_1G_2[#34]
334	A_1A_1	≥ 7	A_1G_2[#34]

Table 11.5: The lattice structure of irreducible connected subgroups of E_8.

ID	Irred. subgroup X	$V_M \downarrow X$
43	D_8	
102	$\bar{A}_1E_7$	
103	$\bar{A}_2E_6$	
62	A_8	
104	$\bar{A}_4^2$	
105	G_2F_4	
1049	F_4 $(p = 3)$	
101	B_2 $(p \geq 5)$	
106	A_1A_2 $(p \geq 5)$	
40	A_1 $(p \geq 23)$	
41	A_1 $(p \geq 29)$	
42	A_1 $(p \geq 31)$	
	In $M = D_8$ $(E_8(\#43))$	
107	$\bar{A}_1^2D_6$	$(1,1,0)/(0,0,\lambda_1)$
108	$\bar{D}_4^2$	$(\lambda_1,0)/(0,\lambda_1)$
109	$\bar{A}_3D_5$	$(010,0)/(000,\lambda_1)$
44	B_7	$\lambda_1/0$
110a	A_1B_6 $(p \neq 2)$	$(2,0)/(0,\lambda_1)$
111a	B_2B_5 $(p \neq 2)$	$(10,0)/(00,\lambda_1)$
112a	B_3B_4 $(p \neq 2)$	$(100,0)/(000,\lambda_1)$
113a	B_2^2 $(p \neq 2)$	$(01,01)$
114	B_2^2 $(p \neq 2)$	$(01,01)$
115	A_1C_4	$(1,\lambda_1)$
116	A_1C_4 $(p \neq 2)$	$(1,\lambda_1)$
45	B_4	λ_4
46	B_4 $(p \neq 2)$	λ_4
	In $M_1 = \bar{A}_1^2D_6$ $(E_8(\#107))$	
117	$\bar{A}_1^4\bar{D}_4$	$(1,1,0,0,0)/(0,0,1,1,0)/(0,0,0,0,\lambda_1)$
118	$\bar{A}_1^2B_5$	$(1,1,0)/(0,0,W(\lambda_1))/(0,0,0)$
119	$\bar{A}_1^2\bar{A}_3^2$	$(1,1,000,000)/(0,0,010,000)/(0,0,000,010)$
120a	$\bar{A}_1^2A_1B_4$ $(p \neq 2)$	$(1,1,0,0)/(0,0,2,0)/(0,0,0,\lambda_1)$

121a	$\bar{A}_1^2 B_2 B_3$ $(p \neq 2)$	$(1,1,00,000)/(0,0,10,000)/(0,0,00,100)$
122	$\bar{A}_1^2 A_1 C_3$	$(1,1,0,000)/(0,0,1,100)$
123	$A_1 D_6$ via $(1, 1^{[r]}, \lambda_1)$	
	In $M_2 = \bar{A}_1^4 \bar{D}_4$ $(E_8(\#117))$	
124	$\bar{A}_1^8$	$(1,1,0,0,0,0,0,0)/(0,0,1,1,0,0,0,0)/$ $(0,0,0,0,1,1,0,0)/(0,0,0,0,0,0,1,1)$
125	$\bar{A}_1^4 B_3$	$(1,1,0,0,000)/(0,0,1,1,000)/(0,0,0,0,W(100))/$ $(0,0,0,0,000)$
126a	$\bar{A}_1^4 A_1 B_2$ $(p \neq 2)$	$(1,1,0,0,0,00)/(0,0,1,1,0,00)/(0,0,0,0,2,00)/$ $(0,0,0,0,0,10)$
127a	$\bar{A}_1^4 A_2$ $(p \neq 3)$	$(1,1,0,0,00)/(0,0,1,1,00)/(0,0,0,0,11)$
128	$A_1 \bar{A}_1^2 \bar{D}_4$ via $(1_a, 1_a^{[r]}, 1_b, 1_c, \lambda_1)$	
129	$A_1 \bar{A}_1 \bar{D}_4$ via $(1_a, 1_a^{[r]}, 1_a^{[s]}, 1_b, \lambda_1)$ $(r \leq s)$	
130	$A_1 A_1 \bar{D}_4$ via $(1_a, 1_a^{[r]}, 1_b, 1_b^{[s]}, \lambda_1)$ $(r \leq s)$	
131	$A_1 \bar{D}_4$ via $(1, 1^{[r]}, 1^{[s]}, 1^{[t]}, \lambda_1)$ ($r \leq s \leq t$; if two of $0, r, s, t$ are equal then the other two are not equal)	
	In $M_3 = \bar{A}_1^8$ $(E_8(\#124))$	
	See Table 11.18	
	In $M_3 = \bar{A}_1^4 B_3$ $(E_8(\#125))$	
167	$\bar{A}_1^4 G_2$	$(1,1,0,0,00)/(0,0,1,1,00)/(0,0,0,0,W(10))/$ $(0,0,0,0,00)$
132$^{\{0\}}$	$\bar{A}_1^6 A_1$ $(p \neq 2)$	
126b	$\bar{A}_1^4 A_1 B_2$ $(p = 2)$	$(1,1,0,0,0,00)/(0,0,1,1,0,00)/(0,0,0,0,2,00)/$ $(0,0,0,0,0,10)/(0,0,0,0,0,00)^2$
168	$A_1 \bar{A}_1^2 B_3$ via $(1_a, 1_a^{[r]}, 1_b, 1_c, 100)$ $(r \neq 0)$	
169	$A_1 \bar{A}_1^2 B_3$ via $(1_a^{[r]}, 1_b, 1_a^{[s]}, 1_c, 100)$ $(rs = 0)$	
170	$A_1 \bar{A}_1 B_3$ via $(1_a^{[r]}, 1_a^{[s]}, 1_a^{[t]}, 1_b, 100)$ $(rt = 0; r < s)$	
171	$A_1 A_1 B_3$ via $(1_a, 1_a^{[r]}, 1_b, 1_b^{[s]}, 100)$ $(0 < r \leq s)$	
172	$A_1 A_1 B_3$ via $(1_a, 1_b^{[s]}, 1_a^{[r]}, 1_b^{[t]}, 100)$ ($st = 0$; $r \leq s + t$; if $r = 0$ then $t \neq 0$)	
173	$A_1 B_3$ via $(1, 1^{[r]}, 1^{[s]}, 1^{[t]}, 100)$ ($r \neq 0; s < t$; if $s = 0$ then $r < t$)	
	In $M_4 = \bar{A}_1^4 G_2$ $(E_8(\#167))$	
133$^{\{\underline{0}\}}$	$\bar{A}_1^5 A_1$	
174	$\bar{A}_1^4 A_1$ $(p \geq 7)$	$(1,1,0,0,0)/(0,0,1,1,0)/(0,0,0,0,6)/(0,0,0,0,0)$
127b	$\bar{A}_1^4 A_2$ $(p = 3)$	$(1,1,0,0,00)/(0,0,1,1,00)/(0,0,0,0,11)/$ $(0,0,0,0,00)$
175	$\bar{A}_1^2 A_1 G_2$ via $(1_a, 1_a^{[r]}, 1_b, 1_c, 10)$ $(r \neq 0)$	
176	$\bar{A}_1 A_1 G_2$ via $(1_a, 1_a^{[r]}, 1_a^{[s]}, 1_b, 10)$ $(0 < r < s)$	
177	$A_1 A_1 G_2$ via $(1_a, 1_a^{[r]}, 1_b, 1_b^{[s]}, 10)$ $(0 < r \leq s)$	
178	$A_1 G_2$ via $(1, 1^{[r]}, 1^{[s]}, 1_b^{[t]}, 10)$ $(0 < r < s < t)$	
	In $M_5 = \bar{A}_1^4 A_1$ $(p \geq 7)$ $(E_8(\#174))$	
	See Table 11.27	
	In $M_3 = \bar{A}_1^4 A_1 B_2$ $(E_8(\#126))$	

132$^{\{0\}}$	$\bar{A}_1^6 A_1$	
189	$\bar{A}_1^4 A_1 A_1$ $(p \geq 5)$	$(1,1,0,0,0,0)/(0,0,1,1,0,0)/(0,0,0,0,2,0)/$ $(0,0,0,0,0,4)$
190	$\bar{A}_1^4 A_1^3$ $(p = 2)$	$(1,1,0,0,0,0,0)/(0,0,1,1,0,0,0)/(0,0,0,0,2,0,0)/$ $(0,0,0,0,0,2,0)/(0,0,0,0,0,0,2)/(0,0,0,0,0,0,0)^2$
	See Table 11.28	
	In $M_4 = \bar{A}_1^4 A_1 A_1$ $(p \geq 5)$ $(E_8(\#189))$	
	See Table 11.29	
	In $M_4 = \bar{A}_1^4 A_1^3$ $(p = 2)$ $(E_8(\#190))$	
	See Table 11.31	
	In $M_3 = \bar{A}_1^4 A_2$ $(E_8(\#127))$	
211$^{\{0\}}$	$\bar{A}_1^4 A_1$ $(p \geq 5)$	
136$^{\{\delta_3\}}$	$\bar{A}_1^4 A_1$ $(p = 3)$	
346	$\bar{A}_1^2 A_1 A_2$ via $(1_a, 1_a^{[r]}, 1_b, 1_c, 10)$ (if $p = 3$ then $r \neq 0$)	
347	$A_1 \bar{A}_1 A_2$ via $(1_a, 1_a^{[r]}, 1_a^{[s]}, 1_b, 10)$ ($r \leq s$; if $p = 3$ then $0 < r < s$)	
348	$A_1 A_1 A_2$ via $(1_a, 1_a^{[r]}, 1_b, 1_b^{[s]}, 10)$ ($r \leq s$; $s \neq 0$; if $p = 3$ then $r \neq 0$)	
349	$A_1 A_2$ via $(1, 1^{[r]}, 1^{[s]}, 1^{[t]}, 10)$ ($r \leq s \leq t$; if two of $0, r, s, t$ are equal then the other two are not equal; if $p = 3$ then $0 < r < s < t$)	
	In $M_2 = \bar{A}_1^2 B_5$ $(E_8(\#118))$	
128$^{\{0\}}$	$\bar{A}_1^2 A_1 \bar{D}_4$ $(p \neq 2)$	
125	$\bar{A}_1^4 B_3$ $(p \neq 2)$	
350	$\bar{A}_1^2 B_2 \bar{A}_3$ $(p \neq 2)$	
351	$\bar{A}_1^2 A_1$ $(p \geq 11)$	$(1,1,0)/(0,0,10)/(0,0,0)$
120b	$\bar{A}_1^2 A_1 B_4$ $(p = 2)$	$(1,1,0,0)/(0,0,2,0)/(0,0,0,\lambda_1)/(0,0,0,0)^2$
121b	$\bar{A}_1^2 B_2 B_3$ $(p = 2)$	$(1,1,00,000)/(0,0,10,000)/(0,0,00,100)/$ $(0,0,00,000)^2$
352	$A_1 B_5$ via $(1^{[r]}, 1^{[s]}, \lambda_1)$ ($rs = 0$; if $p = 2$ then $r \neq s$)	
	In $M_3 = \bar{A}_1^2 A_1$ $(p \geq 11)$ $(E_8(\#351))$	
353	$A_1 A_1$ via $(1_a, 1_a^{[r]}, 1_b)$ $(r \neq 0)$	
354	$\bar{A}_1 A_1$ via $(1_a, 1_b^{[r]}, 1_b^{[s]})$ $(rs = 0)$	
13	A_1 via $(1^{[r]}, 1^{[s]}, 1^{[t]})$ $(rt = 0; r < s)$	
	In $M_2 = \bar{A}_1^2 \bar{A}_3^2$ $(E_8(\#119))$	
350	$\bar{A}_1^2 B_2 \bar{A}_3$	$(1,1,00,000)/(0,0,10,000)/(0,0,00,010)/$ $(0,0,00,000)$
355a	$\bar{A}_1^2 A_1^2 \bar{A}_3$ $(p \neq 2)$	$(1,1,0,0,000)/(0,0,2,0,000)/(0,0,0,2,000)/$ $(0,0,0,0,010)$
356	$A_1 \bar{A}_3^2$ via $(1_a, 1_a^{[r]}, 100, 100)$	
357	$\bar{A}_1^2 A_3$ via $(1, 1, 100_c, 100_c^{[r]})$ $(r \neq 0)$	
358	$A_1 A_3$ via $(1_a, 1_a^{[r]}, 100_b, 100_b^{[s]})$ $(s \neq 0)$	
359	$A_1 A_3$ via $(1_a, 1_a^{[r]}, 100_b, 001_b^{[s]})$	
	In $M_3 = \bar{A}_1^2 B_2 \bar{A}_3$ $(E_8(\#350))$	
360	$\bar{A}_1^2 A_1 \bar{A}_3$ $(p \geq 5)$	$(1,1,0,000)/(0,0,4,000)/(0,0,0,010)/(0,0,0,000)$

355b	$\bar{A}_1^2 A_1^2 \bar{A}_3$ $(p=2)$	$(1,1,0,0,000)/(0,0,2,0,000)/(0,0,0,2,000)/(0,0,0,0,010)/(0,0,0,0,000)^2$
191$^{\{0\}}$	$\bar{A}_1^2 B_2 A_1^2$ $(p \neq 2)$	
361	$A_1 B_2 \bar{A}_3$ via $(1, 1^{[r]}, 10, 100)$ $(r \neq 0)$	
	In $M_4 = \bar{A}_1^2 A_1 \bar{A}_3$ $(p \geq 5)$ $(E_8(\#360))$	
207$^{\{0\}}$	$\bar{A}_1^2 A_1 A_1^2$	
362	$A_1 A_1 \bar{A}_3$ via $(1_a, 1_a^{[r]}, 1_b, 100)$ $(r \neq 0)$	
363	$A_1 A_1 \bar{A}_3$ via $(1_a, 1_b^{[r]}, 1_b^{[s]}, 100)$ $(rs = 0)$	
364	$A_1 \bar{A}_3$ via $(1^{[r]}, 1^{[s]}, 1^{[t]}, 100)$ $(rt = 0; r < s)$	
	In $M_3 = \bar{A}_1^2 A_1^2 \bar{A}_3$ $(E_8(\#355))$	
191$^{\{0\}}$	$\bar{A}_1^2 A_1^2 B_2$ $(p \neq 2)$	
365	$\bar{A}_1^2 A_1^4$ $(p \neq 2)$	$(1,1,0,0,0,0)/(0,0,2,0,0,0)/(0,0,0,2,0,0)/(0,0,0,0,2,0)/(0,0,0,0,0,2)$
366	$A_1 A_1^2 \bar{A}_3$ via $(1_a, 1_a^{[r]}, 1_b, 1_c, 100)$ (if $p=2$ then $r \neq 0$)	
367	$A_1 \bar{A}_1 A_1 \bar{A}_3$ via $(1_a^{[r]}, 1_b, 1_a^{[s]}, 1_c, 100)$	
368	$\bar{A}_1^2 A_1 \bar{A}_3$ via $(1_a, 1_b, 1_c, 1_c^{[r]}, 100)$ $(r \neq 0)$	
369	$A_1 A_1 \bar{A}_3$ via $(1_a^{[r]}, 1_a^{[s]}, 1_a^{[t]}, 1_b, 100)$ $(rt = 0;\ r \leq s;$ if $r = s$ then $r < t$; if $p = 2$ then $r < s)$	
370	$A_1 \bar{A}_1 \bar{A}_3$ via $(1_a^{[r]}, 1_b, 1_a^{[s]}, 1_a^{[t]}, 100)$ $(rs = 0;\ s < t)$	
371	$A_1 A_1 \bar{A}_3$ via $(1_a, 1_a^{[r]}, 1_b, 1_b^{[s]}, 100)$ $(s \neq 0$; if $p = 2$ then $r \neq 0)$	
372	$A_1 A_1 \bar{A}_3$ via $(1_a^{[r]}, 1_b^{[t]}, 1_a^{[s]}, 1_b^{[u]}, 100)$ $(rs = tu = 0;\ r \leq t$ and if $r = t$ then $s \leq u)$	
373	$A_1 \bar{A}_3$ via $(1^{[r]}, 1^{[s]}, 1^{[t]}, 1^{[u]}, 100)$ $(rt = 0;\ r \leq s;\ t < u$; if $r = s$ then $r < t$; if $p = 2$ then $r < s)$	
	In $M_4 = \bar{A}_1^2 A_1^4$ $(p \neq 2)$ $(E_8(\#365))$ See Table 11.32	
	In $M_2 = \bar{A}_1^2 A_1 B_4$ $(E_8(\#120))$	
128$^{\{0\}}$	$\bar{A}_1^2 A_1 \bar{D}_4$	
126a	$\bar{A}_1^4 A_1 B_2$ $(p \neq 2)$	
355a	$\bar{A}_1^2 A_1^2 \bar{A}_3$ $(p \neq 2)$	
401	$\bar{A}_1^2 A_1 A_1^2$ $(p \neq 2)$	$(1,1,0,0,0)/(0,0,2,0,0)/(0,0,0,2,2)$
402	$\bar{A}_1^2 A_1 A_1$ $(p \geq 11)$	$(1,1,0,0)/(0,0,2,0)/(0,0,0,8)$
403	$\bar{A}_1^2 A_1^2 B_3$ $(p = 2)$	$(1,1,0,0,000)/(0,0,2,0,000)/(0,0,0,2,000)/(0,0,0,0,100)/(0,0,0,0,000)^2$
404	$\bar{A}_1^2 A_1 B_2^2$ $(p = 2)$	$(1,1,0,00,00)/(0,0,2,00,00)/(0,0,0,10,00)/(0,0,0,00,10)/(0,0,0,00,00)^2$
405	$A_1 A_1 B_4$ via $(1_a, 1_a^{[r]}, 1_b, \lambda_1)$ (if $p = 2$ then $r \neq 0$)	
406	$A_1 \bar{A}_1 B_4$ via $(1_a^{[r]}, 1_b, 1_a^{[s]}, \lambda_1)$ $(rs = 0)$	
407	$A_1 B_4$ via $(1^{[r]}, 1^{[s]}, 1^{[t]}, \lambda_1)$ $(rt = 0;\ r \leq s$; if $p = 2$ then $r < s$; if $r = s$ then $r < t)$	
	In $M_3 = \bar{A}_1^2 A_1 A_1^2$ $(p \neq 2)$ $(E_8(\#401))$ See Table 11.33	

	In $M_3 = \bar{A}_1^2 A_1 A_1$ $(p \geq 11)$ $(E_8(\#402))$	
	See Table 11.34	
	In $M_3 = \bar{A}_1^2 A_1^2 B_3$ $(p = 2)$ $(E_8(\#403))$	
441	$\bar{A}_1^2 A_1^3 B_2$	$(1,1,0,0,0,00)/(0,0,2,0,0,00)/(0,0,0,2,0,00)/$ $(0,0,0,0,2,00)/(0,0,0,0,0,10)/(0,0,0,0,0,00)^2$
442	$\bar{A}_1^2 A_1^2 G_2$	$(1,1,0,0,00)/(0,0,2,0,00)/(0,0,0,2,00)/$ $(0,0,0,0,10)/(0,0,0,0,00)^2$
355b	$\bar{A}_1^2 A_1^2 \bar{A}_3$	
443	$A_1 A_1^2 B_3$ via $(1_a, 1_a^{[r]}, 1_b, 1_c, 100)$ $(r \neq 0)$	
444	$A_1 \bar{A}_1 A_1 B_3$ via $(1_a^{[r]}, 1_b, 1_a^{[s]}, 1_c, 100)$ $(rs = 0)$	
445	$\bar{A}_1^2 A_1 B_3$ via $(1_a, 1_b, 1_c, 1_c^{[r]}, 100)$ $(r \neq 0)$	
446	$A_1 A_1 B_3$ via $(1_a^{[r]}, 1_a^{[s]}, 1_a^{[t]}, 1_b, 100)$ $(rt = 0; r < s)$	
447	$A_1 \bar{A}_1 B_3$ via $(1_a^{[r]}, 1_b, 1_a^{[s]}, 1_a^{[t]}, 100)$ $(rs = 0; s < t)$	
448	$A_1 A_1 B_3$ via $(1_a, 1_a^{[r]}, 1_b, 1_b^{[s]}, 100)$ $(rs \neq 0)$	
449	$A_1 A_1 B_3$ via $(1_a^{[r]}, 1_b^{[t]}, 1_a^{[s]}, 1_b^{[u]}, 100)$ $(rs = tu = 0;\ r \leq t;$ if $r = t$ then $s \leq u)$	
450	$A_1 B_3$ via $(1^{[r]}, 1^{[s]}, 1^{[t]}, 1^{[u]}, 100)$ $(rt = 0; r < s; t < u)$	
	In $M_4 = \bar{A}_1^2 A_1^3 B_2$ $(p = 2)$ $(E_8(\#441))$	
451	$\bar{A}_1^2 A_1^5$	$(1,1,0,0,0,0,0)/(0,0,2,0,0,0,0)/(0,0,0,2,0,0,0)/$ $(0,0,0,0,2,0,0)/(0,0,0,0,0,2,0)/(0,0,0,0,0,0,2)/$ $(0,0,0,0,0,0,0)^2$
190	$\bar{A}_1^4 A_1^3$	
	See Table 11.35	
	In $M_5 = \bar{A}_1^2 A_1^5$ $(p = 2)$ $(E_8(\#451))$	
	See Table 11.36	
	In $M_4 = \bar{A}_1^2 A_1^2 G_2$ $(p = 2)$ $(E_8(\#442))$	
263$^{\{0\}}$	$\bar{A}_1^3 A_1^2 A_1$	
512	$A_1 A_1^2 G_2$ via $(1_a, 1_a^{[r]}, 1_b, 1_c, 10)$ $(r \neq 0)$	
513	$A_1 \bar{A}_1 A_1 G_2$ via $(1_a^{[r]}, 1_b, 1_a^{[s]}, 1_c, 10)$ $(rs = 0)$	
514	$\bar{A}_1^2 A_1 G_2$ via $(1_a, 1_b, 1_c, 1_c^{[r]}, 10)$ $(r \neq 0)$	
515	$A_1 A_1 G_2$ via $(1_a^{[r]}, 1_a^{[s]}, 1_a^{[t]}, 1_b, 10)$ $(rt = 0; r < s)$	
516	$A_1 \bar{A}_1 G_2$ via $(1_a^{[r]}, 1_b, 1_a^{[s]}, 1_a^{[t]}, 10)$ $(rs = 0; s < t)$	
517	$A_1 A_1 G_2$ via $(1_a, 1_a^{[r]}, 1_b, 1_b^{[s]}, 10)$ $(rs \neq 0)$	
518	$A_1 A_1 G_2$ via $(1_a^{[r]}, 1_b^{[t]}, 1_a^{[s]}, 1_b^{[u]}, 10)$ $(rs = tu = 0;\ r \leq t;$ if $r = t$ then $s \leq u)$	
519	$A_1 G_2$ via $(1^{[r]}, 1^{[s]}, 1^{[t]}, 1^{[u]}, 10)$ $(rt = 0; r < s; t < u)$	
	In $M_3 = \bar{A}_1^2 A_1 B_2^2$ $(p = 2)$ $(E_8(\#404))$	
126b	$\bar{A}_1^4 A_1 B_2$	
441	$\bar{A}_1^2 A_1^3 B_2$	
	See Table 11.37	
	In $M_2 = \bar{A}_1^2 B_2 B_3$ $(E_8(\#121))$	
125	$\bar{A}_1^4 B_3$	
531	$\bar{A}_1^2 A_1 B_3$ $(p \geq 5)$	$(1,1,0,000)/(0,0,4,000)/(0,0,0,100)$

403	$\bar{A}_1^2 A_1^2 B_3$ $(p = 2)$	
350	$\bar{A}_1^2 B_2 \bar{A}_3$	
126a	$\bar{A}_1^4 B_2 A_1$ $(p \neq 2)$	
532	$\bar{A}_1^2 B_2 G_2$	$(1,1,00,00)/(0,0,W(10),00)/(0,0,00,W(10))$
404	$\bar{A}_1^2 A_1 B_2^2$ $(p = 2)$	
533	$A_1 B_2 B_3$ via $(1_a, 1_a^{[r]}, 10, 100)$ (if $p = 2$ then $r \neq 0$)	
	In $M_3 = \bar{A}_1^2 A_1 B_3$ $(p \geq 5)$ $(E_8(\#531))$	
360	$\bar{A}_1^2 A_1 \bar{A}_3$	
189	$\bar{A}_1^4 A_1 A_1$	
534	$\bar{A}_1^2 A_1 G_2$	$(1,1,0,00)/(0,0,4,00)/(0,0,0,10)$
535	$A_1 A_1 B_3$ via $(1_a, 1_a^{[r]}, 1_b, 100)$	
536	$A_1 A_1 B_3$ via $(1_a^{[r]}, 1_b, 1_a^{[s]}, 100)$ $(rs = 0)$	
537	$A_1 B_3$ via $(1^{[r]}, 1^{[s]}, 1^{[t]}, 100)$ $(rt = 0;\ r \leq s)$	
	In $M_4 = \bar{A}_1^2 A_1 G_2$ $(p \geq 5)$ $(E_8(\#534))$	
209$^{\{\underline{0}\}}$	$\bar{A}_1^3 A_1 A_1$	
538	$\bar{A}_1^2 A_1 A_1$ $(p \geq 7)$	$(1,1,0,0)/(0,0,4,0)/(0,0,0,6)$
539	$A_1 A_1 G_2$ via $(1_a, 1_a^{[r]}, 1_b, 10)$	
540	$\bar{A}_1 A_1 G_2$ via $(1_a, 1_b^{[r]}, 1_b^{[s]}, 10)$ $(rs = 0)$	
541	$A_1 G_2$ via $(1^{[r]}, 1^{[s]}, 1^{[t]}, 10)$ $(rt = 0; r \leq s)$	
	In $M_5 = \bar{A}_1^2 A_1 A_1$ $(p \geq 7)$ $(E_8(\#538))$	
	See Table 11.38	
	In $M_3 = \bar{A}_1^2 B_2 G_2$ $(E_8(\#532))$	
167	$\bar{A}_1^4 G_2$	
534	$\bar{A}_1^2 A_1 G_2$ $(p \geq 5)$	
442	$\bar{A}_1^2 A_1^2 G_2$ $(p = 2)$	
551	$\bar{A}_1^2 B_2 A_2$ $(p = 3)$	$(1,1,00,00)/(0,0,10,00)/(0,0,00,11)$
193$^{\{\underline{0}\}}$	$\bar{A}_1^3 B_2 A_1$	
552	$\bar{A}_1^2 B_2 A_1$ $(p \geq 7)$	$(1,1,00,0)/(0,0,10,0)/(0,0,00,6)$
553	$A_1 B_2 G_2$ via $(1, 1^{[r]}, 10, 10)$ (if $p = 2$ then $r \neq 0$)	
	In $M_4 = \bar{A}_1^2 B_2 A_2$ $(p = 3)$ $(E_8(\#551))$	
127b	$\bar{A}_1^4 A_2$	
195$^{\{\delta_2\}}$	$\bar{A}_1^2 B_2 A_1$	
554	$A_1 B_2 A_2$ via $(1, 1^{[r]}, 10, 10)$	
	In $M_4 = \bar{A}_1^2 B_2 A_1$ $(p \geq 7)$ $(E_8(\#552))$	
538	$\bar{A}_1^2 A_1 A_1$	
174	$\bar{A}_1^4 A_1$	
555	$A_1 B_2 A_1$ via $(1_a, 1_a^{[r]}, 10, 1_b)$	
556	$A_1 \bar{A}_1 B_2$ via $(1_a^{[r]}, 1_b, 10, 1_a^{[s]})$ $(rs = 0)$	
557	$A_1 B_2$ via $(1^{[r]}, 1^{[s]}, 10, 1^{[t]})$ $(rt = 0; r \leq s)$	
	In $M_2 = \bar{A}_1^2 A_1 C_3$ $(E_8(\#122))$	
193$^{\{\underline{0}\}}$	$\bar{A}_1^3 A_1 B_2$	

558	$\bar{A}_1^2A_1^2A_1$ $(p \neq 2)$	$(1,1,0,0,0)/(0,0,1,1,2)$
559	$\bar{A}_1^2A_1A_1$ $(p \geq 7)$	$(1,1,0,0)/(0,0,1,5)$
560	$\bar{A}_1^2A_1A_3$ $(p = 2)$	$(1,1,0,000)/(0,0,1,010)$
561	$\bar{A}_1^2A_1G_2$ $(p = 2)$	$(1,1,0,00)/(0,0,1,10)$
562	$A_1A_1C_3$ via $(1_a^{[r]},1_a^{[s]},1_b,100)$ $(rs=0)$	
563	$A_1\bar{A}_1C_3$ via $(1_a^{[r]},1_b,1_a^{[s]},100)$ $(rs=0$; if $p=2$ then $r \neq s)$	
564	$\bar{A}_1A_1C_3$ via $(1_a,1_b^{[r]},1_b^{[s]},100)$ $(rs=0)$	
565	A_1C_3 via $(1^{[r]},1^{[s]},1^{[t]},100)$ $(rst=0$; if $p=2$ then $r \neq t)$	
	In $M_3 = \bar{A}_1^2A_1^2A_1$ $(p \neq 2)$ $(E_8(\#558))$	
	See Table 11.39	
	In $M_3 = \bar{A}_1^2A_1A_1$ $(p \geq 7)$ $(E_8(\#559))$	
	See Table 11.40	
	In $M_3 = \bar{A}_1^2A_1A_3$ $(p = 2)$ $(E_8(\#560))$	
611	$A_1A_1A_3$ via $(1_a^{[r]},1_a^{[s]},1_b,100)$ $(rs=0)$	
612	$A_1\bar{A}_1A_3$ via $(1_a^{[r]},1_b,1_a^{[s]},100)$ $(rs=0;\ r \neq s)$	
613	$\bar{A}_1A_1A_3$ via $(1_a,1_b^{[r]},1_b^{[s]},100)$ $(rs=0)$	
614	A_1A_3 via $(1^{[r]},1^{[s]},1^{[t]},100)$ $(rst=0; r \neq t)$	
	In $M_3 = \bar{A}_1^2A_1G_2$ $(p = 2)$ $(E_8(\#561))$	
276$^{\{\delta_3\}}$	$\bar{A}_1^2A_1A_1A_1$	
615	$A_1A_1G_2$ via $(1_a^{[r]},1_a^{[s]},1_b,10)$ $(rs=0)$	
616	$A_1\bar{A}_1G_2$ via $(1_a^{[r]},1_b,1_a^{[s]},10)$ $(rs=0; r \neq s)$	
617	$\bar{A}_1A_1G_2$ via $(1_a,1_b^{[r]},1_b^{[s]},10)$ $(rs=0)$	
618	A_1G_2 via $(1^{[r]},1^{[s]},1^{[t]},10)$ $(rst=0; r \neq t)$	
	In $M_1 = \bar{D}_4^2$ $(E_8(\#108))$	
619	$B_3\bar{D}_4$	$(W(100),0)/(000,\lambda_1)/(000,0)$
620a	$A_2\bar{D}_4$ $(p \neq 3)$	$(11,0)/(00,\lambda_1)$
621a	$A_1B_2\bar{D}_4$ $(p \neq 2)$	$(2,00,0)/(0,10,0)/(0,00,\lambda_1)$
117	$\bar{A}_1^4\bar{D}_4$	
48	D_4 via $(\lambda_1,\lambda_1^{[r]})$ $(r \neq 0)$	
49	D_4 via $(\lambda_1,\lambda_1^{[\tau]})$	
50	D_4 via $(\lambda_1,\lambda_1^{[\tau r]})$ $(r \neq 0)$	
51	D_4 via $(\lambda_1,\lambda_1^{[\iota r]})$ $(r \neq 0)$	
	In $M_2 = B_3\bar{D}_4$ $(E_8(\#619))$	
128$^{\{0\}}$	$\bar{A}_1^2A_1\bar{D}_4$	
621b	$A_1B_2\bar{D}_4$ $(p = 2)$	$(2,00,0)/(0,10,0)/(0,00,\lambda_1)/(0,00,0)^2$
622	$G_2\bar{D}_4$	$(W(10),0)/(00,\lambda_1)/(00,0)$
623	B_3^2	$(W(100),000)/(000,001)/(000,000)$
624	B_3A_2 $(p \neq 3)$	$(W(100),00)/(000,11)/(000,00)$
533$^{\{0\}}$	$B_3A_1B_2$ $(p \neq 2)$	
625a	$B_3A_1B_2$ $(p \neq 2)$	$(100,0,00)/(000,1,01)/(000,0,00)$
125	$B_3\bar{A}_1^4$	

	In $M_3 = G_2\bar{D}_4$ ($E_8(\#622)$)	
620b	$A_2\bar{D}_4$ $(p = 3)$	$(11, 0)/(00, \lambda_1)/(00, 0)$
129$^{\{\underline{0}\}}$	$\bar{A}_1A_1\bar{D}_4$	
626	$A_1\bar{D}_4$ $(p \geq 7)$	$(6, 0)/(0, \lambda_1)/(0, 0)$
627	G_2A_2 $(p \neq 3)$	$(W(10), 00)/(00, 11)/(00, 00)$
553$^{\{0\}}$	$G_2A_1B_2$ $(p \neq 2)$	
167	$G_2\bar{A}_1^4$	
	In $M_4 = A_1\bar{D}_4$ $(p \geq 7)$ ($E_8(\#626)$)	
628	$A_1A_1B_2$	$(6, 0, 00)/(0, 2, 00)/(0, 0, 10)/(0, 0, 00)$
629	A_1A_2	$(6, 00)/(0, 11)/(0, 00)$
174	$A_1\bar{A}_1^4$	
	In $M_5 = A_1A_1B_2$ $(p \geq 7)$ ($E_8(\#628)$)	
630	$A_1A_1A_1$	$(6, 0, 0)/(0, 2, 0)/(0, 0, 4)/(0, 0, 0)$
631	A_1B_2 via $(1^{[r]}, 1^{[s]}, 10)$ $(rs = 0)$	
	In $M_6 = A_1A_1A_1$ $(p \geq 7)$ ($E_8(\#630)$)	
632	A_1A_1 via $(1_a^{[r]}, 1_a^{[s]}, 1_b)$ $(rs = 0)$	
633	A_1A_1 via $(1_a^{[r]}, 1_b, 1_a^{[s]})$ $(rs = 0)$	
634	A_1A_1 via $(1_a, 1_b^{[r]}, 1_b^{[s]})$ $(rs = 0)$	
17$^{\{s,s,t,r\}}$	A_1 via $(1^{[r]}, 1^{[s]}, 1^{[t]})$ $(rst = 0)$	
	In $M_5 = A_1A_2$ $(p \geq 7)$ ($E_8(\#629)$)	
542$^{\{0\}}$	A_1A_1	
	In $M_4 = G_2A_2$ $(p \neq 3)$ ($E_8(\#627)$)	
347$^{\{\underline{0}\}}$	$\bar{A}_1A_1A_2$	
629	A_1A_2 $(p \geq 7)$	
541$^{\{\underline{0}\}}$	G_2A_1 $(p \geq 5)$	
	In $M_3 = B_3^2$ ($E_8(\#623)$)	
169$^{\{\underline{0}\}}$	$\bar{A}_1^2A_1B_3$	
625b	$B_3A_1B_2$ $(p = 2)$	$(100, 0, 00)/(000, 1, 10)/(000, 0, 00)^2$
52	B_3 via $(100, 100^{[r]})$ $(r \neq 0)$	
	In $M_3 = B_3A_2$ $(p \neq 3)$ ($E_8(\#624)$)	
627	G_2A_2	
346$^{\{0\}}$	$\bar{A}_1^2A_1A_2$	
537$^{\{\underline{0}\}}$	B_3A_1 $(p \geq 5)$	
	In $M_3 = B_3A_1B_2$ ($E_8(\#625)$)	
553$^{\{0\}}$	$G_2A_1B_2$ $(p \neq 2)$	
191$^{\{0\}}$	$\bar{A}_1^2A_1^2B_2$ $(p \neq 2)$	
676b	$A_1^2B_2^2$ $(p = 2)$	$(2, 0, 00, 00)/(0, 1, 00, 01)/(0, 0, 10, 00)/(0, 0, 00, 00)^2$
635	$B_3A_1A_1$ $(p \geq 5)$	$(100, 0, 0)/(000, 1, 3)/(000, 0, 0)$
169$^{\{\underline{0}\}}$	$B_3A_1\bar{A}_1^2$	
636	$B_3A_1^3$ $(p = 2)$	$(100, 0, 0, 0)/(000, 1, 1, 1)/(000, 0, 0, 0)^2$
	In $M_4 = B_3A_1A_1$ $(p \geq 5)$ ($E_8(\#635)$)	

207$^{\{0\}}$	$\bar{A}_1^2 A_1^2 A_1$	
539$^{\{0\}}$	$G_2 A_1 A_1$	
637	$B_3 A_1$ via $(100, 1^{[r]}, 1^{[s]})$ $(rs = 0; r \neq s)$	
535$^{\{0\}}$	$B_3 A_1$ via $(100, 1, 1)$	
	In $M_4 = B_3 A_1^3$ $(p = 2)$ $(E_8(\#636))$	
638	$A_1^4 B_2$	$(2,0,0,0,00)/(0,1,1,1,00)/(0,0,0,0,10)/$ $(0,0,0,0,00)^2$
639	$B_3 A_1 A_1$ via $(100, 1_a, 1_a^{[r]}, 1_b)$ $(r \neq 0)$	
640	$B_3 A_1$ via $(100, 1, 1^{[r]}, 1^{[s]})$ $(0 < r < s)$	
	In $M_5 = A_1^4 B_2$ $(p = 2)$ $(E_8(\#638))$	
641	A_1^6	$(2,0,0,0,0,0)/(0,2,0,0,0,0)/(0,0,2,0,0,0)/$ $(0,0,0,1,1,1)/(0,0,0,0,0,0)^2$
262$^{\{0\}}$	$A_1^4 \bar{A}_1^2$	
642	$A_1 A_1^2 B_2$ via $(1_a^{[r]}, 1_a^{[s]}, 1_b, 1_c, 10)$ $(rs = 0)$	
643	$A_1 A_1 A_1 B_2$ via $(1_a, 1_b, 1_b^{[r]}, 1_c, 10)$ $(r \neq 0)$	
644	$A_1 A_1 B_2$ via $(1_a^{[r]}, 1_a^{[s]}, 1_a^{[t]}, 1_b, 10)$ $(rs = 0; s < t)$	
645	$A_1 A_1 B_2$ via $(1_a, 1_b, 1_b^{[r]}, 1_b^{[s]}, 10)$ $(0 < r < s)$	
646	$A_1 A_1 B_2$ via $(1_a^{[r]}, 1_a^{[s]}, 1_b, 1_b^{[t]}, 10)$ $(rs = 0 \neq t)$	
647	$A_1 B_2$ via $(1^{[r]}, 1^{[s]}, 1^{[t]}, 1^{[u]}, 10)$ $(rs = 0; s < t < u)$	
	In $M_6 = A_1^6$ $(p = 2)$ $(E_8(\#641))$	
	See Table 11.41	
	In $M_2 = A_2 \bar{D}_4$ $(E_8(\#620))$	
682$^{\{\underline{0}\}}$	$A_1 \bar{D}_4$ $(p \geq 5)$	
131$^{\{\delta_3\}}$	$A_1 \bar{D}_4$ $(p = 3)$	
624	$A_2 B_3$ $(p \neq 3)$	
667	$A_2 A_1 B_2$	$(W(11), 0, 00)/(00, W(2), 00)/(00, 0, W(10))$
127	$A_2 \bar{A}_1^4$	
668	A_2^2 $(p \neq 3)$	$(11, 00)/(00, 11)$
	In $M_3 = A_2 A_1 B_2$ $(E_8(\#667))$	
685$^{\{\underline{0}\}}$	$A_1 A_1 B_2$ $(p \geq 5)$	
201$^{\{\delta_3\}}$	$A_1 A_1 B_2$ $(p = 3)$	
346$^{\{0\}}$	$A_2 A_1 \bar{A}_1^2$ $(p \neq 3)$	
669	$A_2 A_1 A_1$ $(p \geq 5)$	$(11, 0, 0)/(00, 2, 0)/(00, 0, 4)$
670	$A_2 A_1^3$ $(p = 2)$	$(11,0,0,0)/(00,2,0,0)/(00,0,2,0)/(00,0,0,2)/$ $(00,0,0,0)^2$
	In $M_4 = A_2 A_1 A_1$ $(p \geq 5)$ $(E_8(\#669))$	
688$^{\{\underline{0}\}}$	$A_1 A_1 A_1$ $(p \geq 5)$	
672	$A_2 A_1$ via $(10, 1^{[r]}, 1^{[s]})$ $(rs = 0)$	
	In $M_4 = A_2 A_1^3$ $(p = 2)$ $(E_8(\#670))$	
672	$A_2 A_1 A_1$ via $(10, 1_a, 1_a^{[r]}, 1_b)$ $(r \neq 0)$	
673	$A_2 A_1$ via $(10, 1, 1^{[r]}, 1^{[s]})$ $(0 < r < s)$	
	In $M_3 = A_2^2$ $(p \neq 3)$ $(E_8(\#668))$	

672$^{\{\underline{0}\}}$	A_1A_2 $(p \geq 5)$	
53	A_2 via $(10, 10^{[r]})$ $(r \neq 0)$	
54	A_2 via $(10, 01^{[r]})$ $(r \neq 0)$	
	In $M_2 = A_1B_2\bar{D}_4$ $(E_8(\#621))$	
128$^{\{0\}}$	$A_1\bar{A}_1^2\bar{D}_4$	
674	$A_1^3\bar{D}_4$ $(p = 2)$	$(2,0,0,0)/(0,2,0,0)/(0,0,2,0)/(0,0,0,\lambda_1)/$ $(0,0,0,0)^2$
675	$A_1A_1\bar{D}_4$ $(p \geq 5)$	$(2,0,0)/(0,4,0)/(0,0,\lambda_1)$
676a	$A_1^2B_2^2$ $(p \neq 2)$	$(2,0,00,00)/(0,1,00,01)/(0,0,10,00)$
677	$A_1^2B_2^2$ $(p \neq 2)$	$(2,0,00,00)/(0,2,00,00)/(0,0,10,00)/(0,00,00,10)$
625	$A_1B_2B_3$	
533$^{\{0\}}$	$A_1B_2B_3$ $(p \neq 2)$	
667	$A_1B_2A_2$ $(p \neq 3)$	
126	$A_1B_2\bar{A}_1^4$	
	In $M_3 = A_1^3\bar{D}_4$ $(p = 2)$ $(E_8(\#674))$	
636	$A_1^3B_3$	
670	$A_1^3A_2$	
190	$A_1^3\bar{A}_1^4$	
678	$A_1A_1\bar{D}_4$ via $(1_a, 1_a^{[r]}, 1_b, \lambda_1)$ $(r \neq 0)$	
679	$A_1\bar{D}_4$ via $(1, 1^{[r]}, 1^{[s]}, \lambda_1)$ $(0 < r < s)$	
	In $M_3 = A_1A_1\bar{D}_4$ $(p \geq 5)$ $(E_8(\#675))$	
535$^{\{0\}}$	$A_1A_1B_3$	
635	$A_1A_1B_3$	
680	$A_1^2A_1B_2$	$(2,0,0,00)/(0,2,0,00)/(0,0,4,00)/(0,0,0,10)$
681	$A_1^2A_1B_2$	$(2,0,0,00)/(0,0,4,00)/(0,1,0,01)$
189	$A_1A_1\bar{A}_1^4$	
669	$A_1A_1A_2$	
682	$A_1\bar{D}_4$ via $(1^{[r]}, 1^{[s]}, \lambda_1)$ $(rs = 0)$	
	In $M_4 = A_1^2A_1B_2$ $(p \geq 5)$ $(E_8(\#680))$	
207$^{\{0\}}$	$A_1^2A_1\bar{A}_1^2$	
683	$A_1^2A_1^2$	$(2,0,0,0)/(0,2,0,0)/(0,0,4,0)/(0,0,0,4)$
684	$A_1A_1B_2$ via $(1_a, 1_a^{[r]}, 1_b, 10)$ $(r \neq 0)$	
685	$A_1A_1B_2$ via $(1_a^{[r]}, 1_b, 1_a^{[s]}, 10)$ $(rs = 0)$	
687	A_1B_2 via $(1^{[r]}, 1^{[s]}, 1^{[t]}, 10)$ $(rt = 0; r < s)$	
	In $M_5 = A_1^2A_1^2$ $(p \geq 5)$ $(E_8(\#683))$	
	See Table 11.42	
	In $M_4 = A_1^2A_1B_2$ $(p \geq 5)$ $(E_8(\#681))$	
208$^{\{\underline{0}\}}$	$A_1^2A_1\bar{A}_1^2$	
694	$A_1^2A_1^2$	$(2,0,0,0)/(0,1,0,3)/(0,0,4,0)$
695	$A_1A_1B_2$ via $(1_a^{[r]}, 1_a^{[s]}, 1_b, 10)$ $(rs = 0)$	
696	$A_1A_1B_2$ via $(1_a^{[r]}, 1_b, 1_a^{[s]}, 10)$ $(rs = 0; r \neq s)$	
685$^{\{\underline{0}\}}$	$A_1A_1B_2$ via $(1_a, 1_b, 1_a, 10)$	

697	$A_1A_1B_2$ via $(1_a, 1_b^{[r]}, 1_b^{[s]}, 10)$ $(rs = 0)$	
698	A_1B_2 via $(1^{[r]}, 1^{[s]}, 1^{[t]}, 10)$ $(rst = 0; r \neq t)$	
	In $M_5 = A_1^2A_1^2$ $(p \geq 5)$ $(E_8(\#694))$	
	See Table 11.43	
	In $M_3 = A_1^2B_2^2$ $(E_8(\#676))$	
192$^{\{0\}}$	$A_1^2\bar{A}_1^2B_2$	
638	$A_1^4B_2$ $(p = 2)$	
681	$A_1^2A_1B_2$ $(p \geq 5)$	
708	$A_1B_2^2$ via $(1, 1^{[r]}, 10, 10)$	
709	$A_1^2B_2$ via $(1, 1, 10, 10^{[r]})$	
710	$A_1^2B_2$ via $(1, 1, 10, 02^{[r]})$ $(p = 2)$	
711	A_1B_2 via $(1, 1^{[r]}, 10^{[s]}, 10^{[t]})$ $(st = 0;$ if $r = 0$ then $0 < s \leq t)$	
712	A_1B_2 via $(1, 1^{[r]}, 10^{[s]}, 02^{[t]})$ $(p = 2; st = 0)$	
	In $M_3 = A_1^2B_2^2$ $(p \neq 2)$ $(E_8(\#677))$	
191$^{\{0\}}$	$A_1^2\bar{A}_1^2B_2$	
680	$A_1^2A_1B_2$ $(p \geq 5)$	
713	$A_1B_2^2$ via $(1, 1^{[r]}, 10, 10)$ $(r \neq 0)$	
714	$A_1^2B_2$ via $(1, 1, 10, 10^{[r]})$ $(r \neq 0)$	
715	A_1B_2 via $(1, 1^{[r]}, 10, 10^{[s]})$ $(rs \neq 0)$	
	In $M_1 = \bar{A}_3D_5$ $(E_8(\#109))$	
716	B_2D_5	$(W(10), 0)/(00, \lambda_1)/(00, 0)$
717a	$A_1^2D_5$ $(p \neq 2)$	$(2, 0, 0)/(0, 2, 0)/(0, 0, \lambda_1)$
718	$\bar{A}_3B_4$	$(010, 0)/(000, W(\lambda_1))/(000, 0)$
719a	$\bar{A}_3A_1B_3$ $(p \neq 2)$	$(010, 0, 000)/(000, 2, 000)/(000, 0, 100)$
119	$\bar{A}_3^2\bar{A}_1^2$	
720a	$\bar{A}_3B_2^2$ $(p \neq 2)$	$(010, 00, 00)/(000, 10, 00)/(000, 00, 10)$
721	$\bar{A}_3B_2$ $(p \neq 2)$	$(010, 00)/(000, 02)$
	In $M_2 = B_2D_5$ $(E_8(\#716))$	
722	A_1D_5 $(p \geq 5)$	$(4, 0)/(0, \lambda_1)/(0, 0)$
717b	$A_1^2D_5$ $(p = 2)$	$(2, 0, 0)/(0, 2, 0)/(0, 0, \lambda_1)/(0, 0, 0)^2$
533$^{\{0\}}$	$B_2A_1B_3$ $(p \neq 2)$	
350	$B_2\bar{A}_1^2\bar{A}_3$	
723	B_2^3 $(p \neq 2)$	$(10, 00, 00)/(00, 10, 00)/(00, 00, 10)/(00, 00, 00)$
724	B_2B_2 $(p \neq 2)$	$(10, 00)/(00, 02)/(00, 00)$
	In $M_3 = A_1D_5$ $(p \geq 5)$ $(E_8(\#722))$	
535$^{\{0\}}$	$A_1A_1B_3$	
360	$A_1\bar{A}_1^2\bar{A}_3$	
725	$A_1B_2^2$	$(4, 00, 00)/(0, 10, 00)/(0, 00, 10)/(0, 00, 00)$
726	A_1B_2	$(4, 00)/(0, 02)/(0, 00)$
	In $M_4 = A_1B_2^2$ $(p \geq 5)$ $(E_8(\#725))$	
727	$A_1^2B_2$	$(4, 0, 00)/(0, 4, 00)/(0, 0, 10)/(0, 0, 00)$
728	A_1B_2 via $(1, 10, 10^{[r]})$ $(r \neq 0)$	

	In $M_5 = A_1^2B_2$ $(p \geq 5)$ $(E_8(\#727))$	
729	A_1^3	$(4,0,0)/(0,4,0)/(0,0,4)/(0,0,0)$
730	A_1B_2 via $(1,1^{[r]},10)$ $(r \neq 0)$	
	In $M_6 = A_1^3$ $(p \geq 5)$ $(E_8(\#729))$	
731	A_1A_1 via $(1_a,1_a^{[r]},1_b)$ $(r \neq 0)$	
20	A_1 via $(1,1^{[r]},1^{[s]})$ $(0 < r < s)$	
	In $M_4 = A_1B_2$ $(p \geq 5)$ $(E_8(\#726))$	
233$^{\{\underline{0}\}}$	$A_1A_1^2$	
547$^{\{\underline{0}\}}$	A_1A_1 $(p \geq 7)$	
	In $M_3 = B_2^3$ $(p \neq 2)$ $(E_8(\#723))$	
725	$A_1B_2^2$ $(p \geq 5)$	
732	B_2B_2 via $(10_a,10_a^{[r]},10_b)$ $(r \neq 0)$	
61	B_2 via $(10,10^{[r]},10^{[s]})$ $(0 < r < s)$	
	In $M_3 = B_2B_2$ $(p \neq 2)$ $(E_8(\#724))$	
726	A_1B_2 $(p \geq 5)$	
203$^{\{\underline{0}\}}$	$B_2A_1^2$	
557$^{\{\underline{0}\}}$	B_2A_1 $(p \geq 7)$	
59	B_2 via $(10,10)$	
60	B_2 via $(10^{[r]},10^{[s]})$ $(rs = 0; r \neq s)$	
	In $M_2 = A_1^2D_5$ $(E_8(\#717))$	
405$^{\{0\}}$	$A_1^2B_4$ $(p \neq 2)$	
733	$A_1^3B_3$ $(p \neq 2)$	$(2,0,0,000)/(0,2,0,000)/(0,0,2,000)/(0,0,0,100)$
355	$A_1^2\bar{A}_1^2\bar{A}_3$	
677	$A_1^2B_2^2$ $(p \neq 2)$	
734	$A_1^2B_2$ $(p \neq 2)$	$(2,0,00)/(0,2,00)/(0,0,02)$
735	A_1D_5 via $(1,1^{[r]},\lambda_1)$ $(r \neq 0)$	
	In $M_3 = A_1^3B_3$ $(p \neq 2)$ $(E_8(\#733))$	
365	$A_1^4\bar{A}_1^2$	
736	$A_1^3G_2$	$(2,0,0,00)/(0,2,0,00)/(0,0,2,00)/(0,0,0,10)$
737	$A_1A_1B_3$ via $(1_a,1_a^{[r]},1_b,100)$ $(r \neq 0)$	
738	A_1B_3 via $(1,1^{[r]},1^{[s]},100)$ $(0 < r < s)$	
	In $M_4 = A_1^3G_2$ $(p \neq 2)$ $(E_8(\#736))$	
739	$A_1^3A_2$ $(p = 3)$	$(2,0,0,00)/(0,2,0,00)/(0,0,2,00)/(0,0,0,11)$
375$^{\{\underline{0}\}}$	$A_1^3\bar{A}_1A_1$	
740	$A_1^3A_1$ $(p \geq 7)$	$(2,0,0,0)/(0,2,0,0)/(0,0,2,0)/(0,0,0,6)$
741	$A_1A_1G_2$ via $(1_a,1_a^{[r]},1_b,10)$ $(r \neq 0)$	
742	A_1G_2 via $(1,1^{[r]},1^{[s]},10)$ $(0 < r < s)$	
	In $M_5 = A_1^3A_2$ $(p = 3)$ $(E_8(\#739))$	
377$^{\{\delta_2\}}$	$A_1^3A_1$	
743	$A_1A_1A_2$ via $(1_a,1_a^{[r]},1_b,10)$ $(r \neq 0)$	
744	A_1A_2 via $(1,1^{[r]},1^{[s]},10)$ $(0 < r < s)$	

	In $M_5 = A_1^3A_1$ $(p \geq 7)$ $(E_8(\#740))$	
745	$A_1A_1A_1$ via $(1_a, 1_a^{[r]}, 1_b, 1_c)$ $(r \neq 0)$	
746	$A_1A_1^2$ via $(1_a^{[r]}, 1_b, 1_c, 1_a^{[s]})$ $(rs = 0)$	
747	A_1A_1 via $(1_a, 1_a^{[r]}, 1_a^{[s]}, 1_b)$ $(0 < r < s)$	
748	A_1A_1 via $(1_a^{[r]}, 1_a^{[s]}, 1_b, 1_a^{[t]})$ $(rt = 0; r < s)$	
749	A_1A_1 via $(1_a, 1_a^{[r]}, 1_b^{[s]}, 1_b^{[t]})$ $(r \neq 0; st = 0)$	
19	A_1 via $(1^{[r]}, 1^{[s]}, 1^{[t]}, 1^{[u]})$ $(ru = 0; r < s < t)$	
	In $M_3 = A_1^2B_2$ $(p \neq 2)$ $(E_8(\#734))$	
$\mathit{382}^{\{\underline{0}\}}$	$A_1^2A_1^2$	
$\mathit{746}^{\{\underline{0}\}}$	$A_1^2A_1$ $(p \geq 7)$	
750	A_1B_2 via $(1, 1^{[r]}, 10)$ $(r \neq 0)$	
	In $M_2 = \bar{A}_3B_4$ $(E_8(\#718))$	
$\mathit{405}^{\{0\}}$	$A_1^2B_4$ $(p \neq 2)$	
$\mathit{356}^{\{0\}}$	$\bar{A}_3^2A_1$ $(p \neq 2)$	
$\mathit{350}$	$\bar{A}_3\bar{A}_1^2B_2$ $(p \neq 2)$	
751	$\bar{A}_3A_1^2$ $(p \neq 2)$	$(010,0,0)/(000,2,2)/(000,0,0)$
752	$\bar{A}_3A_1$ $(p \geq 11)$	$(010,0)/(000,8)/(000,0)$
719b	$\bar{A}_3A_1B_3$ $(p = 2)$	$(010,0,000)/(000,2,000)/(000,0,100)/(000,0,000)^2$
720b	$\bar{A}_3B_2^2$ $(p = 2)$	$(010,00,00)/(000,10,00)/(000,00,10)/(000,00,00)^2$
	In $M_3 = \bar{A}_3A_1^2$ $(p \neq 2)$ $(E_8(\#751))$	
$\mathit{408}^{\{0\}}$	$A_1^2A_1^2$	
753	$\bar{A}_3A_1$ via $(100, 1, 1^{[r]})$ $(r \neq 0)$	
	In $M_3 = \bar{A}_3A_1$ $(p \geq 11)$ $(E_8(\#752))$	
$\mathit{432}^{\{0\}}$	$A_1^2A_1$	
	In $M_2 = \bar{A}_3A_1B_3$ $(E_8(\#719))$	
$\mathit{733}$	$A_1^3B_3$ $(p \neq 2)$	
$\mathit{533}^{\{0\}}$	$B_2A_1B_3$ $(p \neq 2)$	
$\mathit{356}^{\{0\}}$	$\bar{A}_3^2A_1$	
$\mathit{355a}$	$\bar{A}_3A_1^2\bar{A}_1^2$ $(p \neq 2)$	
754	$\bar{A}_3A_1G_2$	$(010,0,00)/(000,W(2),00)/(000,0,W(10))$
	In $M_3 = \bar{A}_3A_1G_2$ $(E_8(\#754))$	
$\mathit{736}$	$A_1^3G_2$ $(p \neq 2)$	
$\mathit{553}^{\{0\}}$	$B_2A_1G_2$ $(p \neq 2)$	
$\mathit{367}^{\{\underline{0}\}}$	$\bar{A}_3A_1\bar{A}_1A_1$	
755	$\bar{A}_3A_1A_2$ $(p = 3)$	$(010,0,00)/(000,2,00)/(000,0,11)$
756	$\bar{A}_3A_1A_1$ $(p \geq 7)$	$(010,0,0)/(000,2,0)/(000,0,6)$
	In $M_4 = \bar{A}_3A_1A_2$ $(p = 3)$ $(E_8(\#755))$	
$\mathit{739}$	$A_1^3A_2$	
$\mathit{554}^{\{0\}}$	$B_2A_1A_2$	
$\mathit{369}^{\{\delta_2\}}$	$\bar{A}_3A_1A_1$	
	In $M_4 = \bar{A}_3A_1A_1$ $(p \geq 7)$ $(E_8(\#756))$	

740	$A_1^3A_1$	
555$^{\{0\}}$	$B_2A_1A_1$	
757	$\bar{A}_3A_1$ via $(100, 1^{[r]}, 1^{[s]})$ $(rs = 0)$	
	In $M_2 = \bar{A}_3B_2^2$ $(E_8(\#720))$	
677	$A_1^2B_2^2$ $(p \neq 2)$	
723	B_2^3 $(p \neq 2)$	
350	$\bar{A}_3\bar{A}_1^2B_2$	
758	$\bar{A}_3A_1B_2$ $(p \geq 5)$	$(010, 0, 00)/(000, 4, 00)/(000, 0, 10)$
759	$\bar{A}_3A_1^2B_2$ $(p = 2)$	$(010, 0, 0, 00)/(000, 2, 0, 00)/(000, 0, 2, 00)/$ $(000, 0, 0, 10)/(000, 0, 0, 00)^2$
760	$\bar{A}_3B_2$ via $(100, 10, 10^{[r]})$ $(r \neq 0)$	
761	$\bar{A}_3B_2$ via $(100, 10, 02^{[r]})$ $(p = 2)$	
	In $M_3 = \bar{A}_3A_1B_2$ $(p \geq 5)$ $(E_8(\#758))$	
680	$A_1^2A_1B_2$	
725	$B_2A_1B_2$	
360	$\bar{A}_3A_1\bar{A}_1^2$	
762	$\bar{A}_3A_1^2$	$(010, 0, 0)/(000, 4, 0)/(000, 0, 4)$
	In $M_4 = \bar{A}_3A_1^2$ $(p \geq 5)$ $(E_8(\#762))$	
683	$A_1^2A_1^2$	
727	$B_2A_1^2$	
763	$\bar{A}_3A_1$ via $(100, 1, 1^{[r]})$ $(r \neq 0)$	
	In $M_3 = \bar{A}_3A_1^2B_2$ $(p = 2)$ $(E_8(\#759))$	
355b	$\bar{A}_3A_1^2\bar{A}_1^2$	
764	$\bar{A}_3A_1^4$	$(010, 0, 0, 0, 0)/(000, 2, 0, 0, 0)/(000, 0, 2, 0, 0)/$ $(000, 0, 0, 2, 0)/(000, 0, 0, 0, 2)/(000, 0, 0, 0, 0)^2$
765	$\bar{A}_3A_1B_2$ via $(100, 1, 1^{[r]}, 10)$ $(r \neq 0)$	
	In $M_4 = \bar{A}_3A_1^4$ $(p = 2)$ $(E_8(\#764))$	
766	$\bar{A}_3A_1A_1^2$ via $(100, 1_a, 1_a^{[r]}, 1_b, 1_c)$ $(r \neq 0)$	
767	$\bar{A}_3A_1A_1$ via $(100, 1_a, 1_a^{[r]}, 1_a^{[s]}, 1_b)$ $(0 < r < s)$	
768	$\bar{A}_3A_1A_1$ via $(100, 1_a, 1_a^{[r]}, 1_b, 1_b^{[s]})$ $(0 < r \leq s)$	
769	$\bar{A}_3A_1$ via $(100, 1, 1^{[r]}, 1^{[s]}, 1^{[t]})$ $(0 < r < s < t)$	
	In $M_2 = \bar{A}_3B_2$ $(p \neq 2)$ $(E_8(\#721))$	
724	B_2B_2	
734	$A_1^2B_2$	
372$^{\{\underline{0}\}}$	$\bar{A}_3A_1^2$	
757$^{\{\underline{0}\}}$	$\bar{A}_3A_1$ $(p \geq 7)$	
	In $M_1 = B_7$ $(E_8(\#44))$	
123$^{\{0\}}$	A_1D_6 $(p \neq 2)$	
118	$\bar{A}_1^2B_5$ $(p \neq 2)$	
716	B_2D_5 $(p \neq 2)$	
718	$\bar{A}_3B_4$ $(p \neq 2)$	
619	$B_3\bar{D}_4$ $(p \neq 2)$	

47	A_3 $(p \neq 2)$	$101/000$
770	A_1B_2 $(p \neq 2)$	$(2,10)/(0,00)$
10	A_1 $(p \geq 17)$	$14/0$
110b	A_1B_6 $(p = 2)$	$(2,0)/(0,\lambda_1)/(0,0)^2$
111b	B_2B_5 $(p = 2)$	$(10,0)/(00,\lambda_1)/(00,0)^2$
112b	B_3B_4 $(p = 2)$	$(100,0)/(000,\lambda_1)/(000,0)^2$
	In $M_2 = A_3$ $(p \neq 2)$ $(E_8(\#47))$	
428$^{\{\underline{0}\}}$	A_1^2	
59	B_2	
	In $M_2 = A_1B_2$ $(p \neq 2)$ $(E_8(\#770))$	
573$^{\{\underline{0}\}}$	$A_1A_1^2$	
771	A_1A_1 $(p \geq 5)$	$(2,4)/(0,0)$
	In $M_3 = A_1A_1$ $(p \geq 5)$ $(E_8(\#771))$	
7	A_1 via $(1^{[r]}, 1^{[s]})$ $(rs = 0; r \neq s)$	
17$^{\{\underline{0}\}}$	A_1 via $(1,1)$ $(p \geq 7)$	
	In $M_1 = A_1B_6$ $(E_8(\#110))$	
123$^{\{0\}}$	A_1D_6	
717a	$A_1^2D_5$ $(p \neq 2)$	
120a	$A_1\bar{A}_1^2B_4$ $(p \neq 2)$	
621a	$A_1B_2\bar{D}_4$ $(p \neq 2)$	
719a	$A_1\bar{A}_3B_3$ $(p \neq 2)$	
772	A_1A_1 $(p \geq 13)$	$(2,0)/(0,12)$
773	A_1B_2 $(p = 5)$	$(2,00)/(0,20)$
774	A_1C_3 $(p = 3)$	$(2,000)/(0,010)$
775	$A_1^2B_5$ $(p = 2)$	$(2,0,0)/(0,2,0)/(0,0,\lambda_1)/(0,0,0)^2$
776	$A_1B_2B_4$ $(p = 2)$	$(2,00,0)/(0,10,0)/(0,00,\lambda_1)/(0,00,0)^2$
777	$A_1B_3^2$ $(p = 2)$	$(2,000,000)/(0,100,000)/(0,000,100)/(0,000,000)^2$
	In $M_2 = A_1A_1$ $(p \geq 13)$ $(E_8(\#772))$	
11	A_1 via $(1^{[r]}, 1^{[s]})$ $(rs = 0)$	
	In $M_2 = A_1B_2$ $(p = 5)$ $(E_8(\#773))$	
423$^{\{\underline{0}\}}$	$A_1A_1^2$	
704$^{\{\delta_1\}}$	A_1A_1	
	In $M_2 = A_1C_3$ $(p = 3)$ $(E_8(\#774))$	
709$^{\{0\}}$	$A_1^2B_2$	
423$^{\{\delta_3\}}$	$A_1A_1A_1$	
	In $M_2 = A_1^2B_5$ $(p = 2)$ $(E_8(\#775))$	
717b	$A_1^2D_5$	
778	$A_1^3B_4$	$(2,0,0,0)/(0,2,0,0)/(0,0,2,0)/(0,0,0,\lambda_1)/$ $(0,0,0,0)^2$
779	$A_1^2B_2B_3$	$(2,0,00,000)/(0,2,00,000)/(0,0,10,000)/$ $(0,0,00,100)/(0,0,00,000)^2$
780	A_1B_5 via $(1, 1^{[r]}, \lambda_1)$ $(r \neq 0)$	

	In $M_3 = A_1^3B_4$ $(p=2)$ $(E_8(\#778))$	
674	$A_1^3\bar{D}_4$	
781	$A_1^4B_3$	$(2,0,0,0,000)/(0,2,0,0,000)/(0,0,2,0,000)/$ $(0,0,0,2,000)/(0,0,0,0,100)/(0,0,0,0,000)^2$
782	$A_1^3B_2^2$	$(2,0,0,00,00)/(0,2,0,00,00)/(0,0,2,00,00)/$ $(0,0,0,10,00)/(0,0,0,00,10)/(0,0,0,00,00)^2$
783	$A_1A_1B_4$ via $(1_a, 1_a^{[r]}, 1_b, \lambda_1)$ $(r \neq 0)$	
784	A_1B_4 via $(1, 1^{[r]}, 1^{[s]}, \lambda_1)$ $(0 < r < s)$	
	In $M_4 = A_1^4B_3$ $(p=2)$ $(E_8(\#781))$	
764	$A_1^4\bar{A}_3$	
785	$A_1^5B_2$	$(2,0,0,0,0,00)/(0,2,0,0,0,00)/(0,0,2,0,0,00)/$ $(0,0,0,2,0,00)/(0,0,0,0,2,00)/(0,0,0,0,0,10)/$ $(0,0,0,0,0,00)^2$
786	$A_1^4G_2$	$(2,0,0,0,00)/(0,2,0,0,00)/(0,0,2,0,00)/$ $(0,0,0,2,00)/(0,0,0,0,10)/(0,0,0,0,00)^2$
787	$A_1A_1^2B_3$ via $(1_a, 1_a^{[r]}, 1_b, 1_c, 100)$ $(r \neq 0)$	
788	$A_1A_1B_3$ via $(1_a, 1_a^{[r]}, 1_a^{[s]}, 1_b, 100)$ $(0 < r < s)$	
789	$A_1A_1B_3$ via $(1_a, 1_a^{[r]}, 1_b, 1_b^{[s]}, 100)$ $(rs \neq 0)$	
790	A_1B_3 via $(1, 1^{[r]}, 1^{[s]}, 1^{[t]}, 100)$ $(0 < r < s < t)$	
	In $M_5 = A_1^5B_2$ $(p=2)$ $(E_8(\#785))$	
451	$A_1^5\bar{A}_1^2$	
791	A_1^7	$(2,0,0,0,0,0,0)/(0,2,0,0,0,0,0)/(0,0,2,0,0,0,0)/$ $(0,0,0,2,0,0,0)/(0,0,0,0,2,0,0)/(0,0,0,0,0,2,0)/$ $(0,0,0,0,0,0,2)/(0,0,0,0,0,0,0)^2$
792	$A_1A_1^3B_2$ via $(1_a, 1_a^{[r]}, 1_b, 1_c, 1_d, 10)$ $(r \neq 0)$	
793	$A_1A_1^2B_2$ via $(1_a, 1_a^{[r]}, 1_a^{[s]}, 1_b, 1_c, 10)$ $(0 < r < s)$	
794	$A_1A_1A_1B_2$ via $(1_a, 1_a^{[r]}, 1_b, 1_b^{[s]}, 1_c, 10)$ $(rs \neq 0)$	
795	$A_1A_1B_2$ via $(1_a, 1_a^{[r]}, 1_a^{[s]}, 1_a^{[t]}, 1_b, 10)$ $(rs \neq 0)$	
796	$A_1A_1B_2$ via $(1_a, 1_a^{[r]}, 1_a^{[s]}, 1_b, 1_b^{[t]}, 10)$ $(rs \neq 0)$	
797	A_1B_2 via $(1, 1^{[r]}, 1^{[s]}, 1^{[t]}, 1^{[u]}, 10)$ $(0 < r < s < t < u)$	
	In $M_6 = A_1^7$ $(p=2)$ $(E_8(\#791))$	
	See Table 11.44	
	In $M_5 = A_1^4G_2$ $(p=2)$ $(E_8(\#786))$	
468$^{\{\underline{0}\}}$	$A_1^4\bar{A}_1A_1$	
811	$A_1A_1^2G_2$ via $(1_a, 1_a^{[r]}, 1_b, 1_c, 10)$ $(r \neq 0)$	
812	$A_1A_1G_2$ via $(1_a, 1_a^{[r]}, 1_a^{[s]}, 1_b, 10)$ $(0 < r < s)$	
813	$A_1A_1G_2$ via $(1_a, 1_a^{[r]}, 1_b, 1_b^{[s]}, 10)$ $(rs \neq 0)$	
814	A_1G_2 via $(1, 1^{[r]}, 1^{[s]}, 1^{[t]}, 10)$ $(0 < r < s < t)$	
	In $M_4 = A_1^3B_2^2$ $(p=2)$ $(E_8(\#782))$	
441	$A_1^3\bar{A}_1^2B_2$	
785	$A_1^5B_2$	

815	$A_1A_1B_2^2$ via $(1_a, 1_a^{[r]}, 1_b, 10, 10)$ $(r \neq 0)$	
816	$A_1^3B_2$ via $(1_a, 1_b, 1_c, 10, 10^{[r]})$ $(r \neq 0)$	
817	$A_1^3B_2$ via $(1_a, 1_b, 1_c, 10^{[r]}, 02^{[s]})$ $(rs = 0)$	
818	$A_1B_2^2$ via $(1_a, 1_a^{[r]}, 1_a^{[s]}, 10, 10)$ $(0 < r < s)$	
819	$A_1A_1B_2$ via $(1_a, 1_a^{[r]}, 1_b, 10, 10^{[s]})$ $(rs \neq 0)$	
820	$A_1A_1B_2$ via $(1_a, 1_a^{[r]}, 1_b, 10, 02^{[s]})$ $(r \neq 0)$	
821	A_1B_2 via $(1, 1^{[r]}, 1^{[s]}, 10, 10^{[t]})$ $(0 < r < s; t \neq 0)$	
822	A_1B_2 via $(1, 1^{[r]}, 1^{[s]}, 10, 02^{[t]})$ $(0 < r < s)$	
	In $M_3 = A_1^2B_2B_3$ $(p = 2)$ $(E_8(\#779))$	
403	$A_1^2\bar{A}_1^2B_3$	
781	$A_1^4B_3$	
782	$A_1^3B_2^2$	
759	$A_1^2B_2\bar{A}_3$	
823	$A_1^2B_2G_2$	$(2,0,00,00)/(0,2,00,00)/(0,0,10,00)/(0,0,00,10)/$ $(0,0,00,00)^2$
824	$A_1B_2B_3$ via $(1_a, 1_a^{[r]}, 10, 100)$ $(r \neq 0)$	
	In $M_4 = A_1^2B_2G_2$ $(p = 2)$ $(E_8(\#823))$	
442	$A_1^2\bar{A}_1^2G_2$	
786	$A_1^4G_2$	
453$^{\{\underline{0}\}}$	$A_1^2B_2\bar{A}_1A_1$	
825	$A_1B_2G_2$ via $(1_a, 1_a^{[r]}, 10, 10)$ $(r \neq 0)$	
	In $M_2 = A_1B_2B_4$ $(p = 2)$ $(E_8(\#776))$	
120b	$A_1\bar{A}_1^2B_4$	
778	$A_1^3B_4$	
621b	$A_1B_2\bar{D}_4$	
779	$A_1^2B_2B_3$	
826	$A_1B_2^3$	$(2,00,00,00)/(0,10,00,00)/(0,00,10,00)/$ $(0,00,00,10)/(0,00,00,00)^2$
	In $M_3 = A_1B_2^3$ $(p = 2)$ $(E_8(\#826))$	
404	$A_1\bar{A}_1^2B_2^2$	
782	$A_1^3B_2^2$	
827	$A_1B_2B_2$ via $(1, 10_a, 10_a^{[r]}, 10_b)$ $(r \neq 0)$	
828	$A_1B_2B_2$ via $(1, 10_a^{[r]}, 02_a^{[s]}, 10_b)$ $(rs = 0)$	
829	A_1B_2 via $(1, 10, 10^{[r]}, 10^{[s]})$ $(0 < r < s)$	
830	A_1B_2 via $(1, 10^{[r]}, 10^{[s]}, 02^{[t]})$ $(rt = 0; r < s)$	
	In $M_2 = A_1B_3^2$ $(p = 2)$ $(E_8(\#777))$	
719b	$A_1\bar{A}_3B_3$	
779	$A_1^2B_2B_3$	
831	$A_1G_2B_3$	$(2,00,000)/(0,10,000)/(0,00,100)/(0,00,000)^2$
832	A_1B_3 via $(1, 100, 100^{[r]})$ $(r \neq 0)$	
	In $M_3 = A_1G_2B_3$ $(p = 2)$ $(E_8(\#831))$	
444$^{\{\underline{0}\}}$	$A_1\bar{A}_1A_1B_3$	

754	$A_1G_2\bar{A}_3$	
823	$A_1^2G_2B_2$	
833	$A_1G_2^2$	$(2,00,00)/(0,10,00)/(0,00,10)/(0,00,00)^2$
	In $M_4 = A_1G_2^2$ $(p=2)$ $(E_8(\#833))$	
513$^{\{\underline{0}\}}$	$A_1\bar{A}_1A_1G_2$	
834	A_1G_2 via $(1,10,10^{[r]})$ $(r \neq 0)$	
	In $M_1 = B_2B_5$ $(E_8(\#111))$	
118	$\bar{A}_1^2B_5$	
835	A_1B_5 $(p \geq 5)$	$(4,0)/(0,\lambda_1)$
775	$A_1^2B_5$ $(p=2)$	
716	B_2D_5	
621a	$B_2A_1\bar{D}_4$ $(p \neq 2)$	
121a	$B_2\bar{A}_1^2B_3$ $(p \neq 2)$	
720a	$B_2^2\bar{A}_3$ $(p \neq 2)$	
836	B_2A_1 $(p \geq 11)$	$(10,0)/(00,10)$
776	$B_2A_1B_4$ $(p=2)$	
837	$B_2^2B_3$ $(p=2)$	$(10,00,000)/(00,10,000)/(00,00,100)/(00,00,000)^2$
	In $M_2 = A_1B_5$ $(p \geq 5)$ $(E_8(\#835))$	
722	A_1D_5	
675	$A_1A_1\bar{D}_4$	
531	$A_1\bar{A}_1^2B_3$	
758	$A_1B_2\bar{A}_3$	
838	A_1A_1 $(p \geq 11)$	$(4,0)/(0,10)$
	In $M_3 = A_1A_1$ $(p \geq 11)$ $(E_8(\#838))$	
12	A_1 via $(1^{[r]},1^{[s]})$ $(rs=0)$	
	In $M_2 = B_2A_1$ $(p \geq 11)$ $(E_8(\#836))$	
351	$\bar{A}_1^2A_1$	
838	A_1A_1 $(p \geq 11)$	
	In $M_2 = B_2^2B_3$ $(p=2)$ $(E_8(\#837))$	
121a	$\bar{A}_1^2B_2B_3$	
779	$A_1^2B_2B_3$	
826	$A_1B_2^3$	
720b	$B_2^2\bar{A}_3$	
839	$B_2^2G_2$	$(10,00,00)/(00,10,00)/(00,00,10)/(00,00,00)^2$
840	B_2B_3 via $(10,10^{[r]},100)$ $(r \neq 0)$	
841	B_2B_3 via $(10^{[r]},02^{[s]},100)$ $(rs=0)$	
	In $M_3 = B_2^2G_2$ $(p=2)$ $(E_8(\#839))$	
532b	$\bar{A}_1^2B_2G_2$	
823	$A_1^2B_2G_2$	
521$^{\{\underline{0}\}}$	$B_2^2\bar{A}_1A_1$	
842	B_2G_2 via $(10,10^{[r]},10)$ $(r \neq 0)$	
843	B_2G_2 via $(10^{[r]},02^{[s]},10)$ $(rs=0)$	

	In $M_1 = B_3B_4$ (E_8(#112))	
718	$\bar{A}_3B_4$	
120a	$\bar{A}_1^2A_1B_4$ ($p \neq 2$)	
844	G_2B_4	$(W(10), 0)/(00, \lambda_1)$
776	$A_1B_2B_4$ ($p = 2$)	
619	$B_3\bar{D}_4$	
719a	$B_3A_1\bar{A}_3$ ($p \neq 2$)	
121a	$B_3\bar{A}_1^2B_2$ ($p \neq 2$)	
845	$B_3A_1^2$ ($p \neq 2$)	$(100, 0, 0)/(000, 2, 2)$
846	B_3A_1 ($p \geq 11$)	$(100, 0)/(000, 8)$
777	$B_3^2A_1$ ($p = 2$)	
837	$B_3B_2^2$ ($p = 2$)	
	In $M_2 = G_2B_4$ (E_8(#844))	
406$^{\{\underline{0}\}}$	$\bar{A}_1A_1B_4$	
847	A_2B_4 ($p = 3$)	$(11, 0)/(00, \lambda_1)$
848	A_1B_4 ($p \geq 7$)	$(6, 0)/(0, \lambda_1)$
622	$G_2\bar{D}_4$	
754	$G_2A_1\bar{A}_3$ ($p \neq 2$)	
532a	$G_2\bar{A}_1^2B_2$ ($p \neq 2$)	
849	$G_2A_1^2$ ($p \neq 2$)	$(10, 0, 0)/(00, 2, 2)$
850	G_2A_1 ($p \geq 11$)	$(10, 0)/(00, 8)$
831	$G_2B_3A_1$ ($p = 2$)	
839	$G_2B_2^2$ ($p = 2$)	
	In $M_3 = A_2B_4$ ($p = 3$) (E_8(#847))	
407$^{\{\delta_2\}}$	A_1B_4	
620b	$A_2\bar{D}_4$	
755	$A_2A_1\bar{A}_3$	
551	$A_2\bar{A}_1^2B_2$	
851	$A_2A_1^2$	$(11, 0, 0)/(00, 2, 2)$
	In $M_4 = A_2A_1^2$ ($p = 3$) (E_8(#851))	
413$^{\{\delta_2\}}$	$A_1A_1^2$	
852	A_2A_1 via $(10, 1, 1^{[r]})$ ($r \neq 0$)	
	In $M_3 = A_1B_4$ ($p \geq 7$) (E_8(#848))	
626	$A_1\bar{D}_4$	
756	$A_1A_1\bar{A}_3$	
552	$A_1\bar{A}_1^2B_2$	
853	$A_1A_1^2$	$(6, 0, 0)/(0, 2, 2)$
854	A_1A_1 ($p \geq 11$)	$(6, 0)/(0, 8)$
	In $M_4 = A_1A_1^2$ ($p \geq 7$) (E_8(#853))	
855	A_1A_1 via $(1_a^{[r]}, 1_a^{[s]}, 1_b)$ ($rs = 0$)	
856	A_1A_1 via $(1_a, 1_b, 1_b^{[r]})$ ($r \neq 0$)	

546$^{\{\underline{0}\}}$	A_1A_1 via $(1_a, 1_b, 1_b)$	
16	A_1 via $(1^{[r]}, 1^{[s]}, 1^{[t]})$ $(rs = 0; s < t)$	
	In $M_4 = A_1A_1$ $(p \geq 11)$ $(E_8(\#854))$	
14	A_1 via $(1^{[r]}, 1^{[s]})$	
	In $M_3 = G_2A_1^2$ $(p \neq 2)$ $(E_8(\#849))$	
409$^{\{\underline{0}\}}$	$\bar{A}_1A_1A_1^2$	
851	$A_2A_1^2$ $(p = 3)$	
853	$A_1A_1^2$ $(p \geq 7)$	
857	G_2A_1 via $(10, 1, 1^{[r]})$ $(r \neq 0)$	
541$^{\{\underline{0}\}}$	G_2A_1 via $(10, 1, 1)$ $(p \geq 5)$	
	In $M_3 = G_2A_1$ $(p \geq 11)$ $(E_8(\#850))$	
433$^{\{\underline{0}\}}$	$\bar{A}_1A_1A_1$	
854	A_1A_1	
	In $M_2 = B_3A_1^2$ $(p \neq 2)$ $(E_8(\#845))$	
751	$\bar{A}_3A_1^2$	
401	$\bar{A}_1^2A_1A_1^2$	
849	$G_2A_1^2$	
858	B_3A_1 via $(100, 1, 1^{[r]})$ $(r \neq 0)$	
537$^{\{\underline{0}\}}$	B_3A_1 via $(100, 1, 1)$ $(p \geq 5)$	
	In $M_2 = B_3A_1$ $(p \geq 11)$ $(E_8(\#846))$	
752	$\bar{A}_3A_1$	
402	$\bar{A}_1^2A_1A_1$	
850	G_2A_1	
	In $M_1 = B_2^2$ $(E_8(\#113))$	
859	$A_1^2B_2$ $(p = 2)$	$(1, 1, 01)$
860	A_1B_2 $(p \geq 5)$	$(3, 01)$
55	B_2 via $(10, 10^{[r]})$ $(r \neq 0)$	
56	B_2 via $(10, 02)$ $(p = 2)$	
57	B_2 via $(10, 02^{[r]})$ $(p = 2; r \neq 0)$	
59	B_2 via $(10, 10)$ $(p \neq 2)$	
	In $M_2 = A_1^2B_2$ $(p = 2)$ $(E_8(\#859))$	
865b	A_1^4	$(1, 1, 1, 1)$
861	A_1B_2 via $(1, 1^{[r]}, 10)$ $(r \neq 0)$	
	In $M_2 = A_1B_2$ $(p \geq 5)$ $(E_8(\#860))$	
862	A_1^2	$(3, 3)$
	In $M_3 = A_1^2$ $(p \geq 5)$ $(E_8(\#862))$	
4	A_1 via $(1, 1^{[r]})$ $(r \neq 0)$	
17$^{\{\underline{0}\}}$	A_1 via $(1, 1)$ $(p \geq 7)$	
	In $M_1 = B_2^2$ $(p \neq 2)$ $(E_8(\#114))$	
863	A_1B_2 $(p \geq 5)$	$(3, 01)$
58	B_2 via $(10, 10^{[r]})$ $(r \neq 0)$	

59	B_2 via $(10,10)$	
	In $M_2 = A_1B_2$ $(p \geq 5)$ $(E_8(\#863))$	
864	A_1^2	$(3,3)$
	In $M_3 = A_1^2$ $(p \geq 5)$ $(E_8(\#864))$	
3	A_1 via $(1, 1^{[r]})$ $(r \neq 0)$	
17$^{\{\underline{0}\}}$	A_1 via $(1,1)$ $(p \geq 7)$	
	In $M_1 = A_1C_4$ $(E_8(\#115))$	
564$^{\{\underline{0}\}}$	$A_1\bar{A}_1C_3$	
708$^{\{0\}}$	$A_1B_2^2$	
865a	A_1^4 $(p \neq 2)$	$(1,1,1,1)$
866	A_1A_1 $(p \geq 11)$	$(1,7)$
867	A_1D_4 $(p = 2)$	$(1,\lambda_1)$
	In $M_2 = A_1^4$ $(E_8(\#865))$	
868	$A_1A_1^2$ via $(1_a, 1_a^{[r]}, 1_b, 1_c)$ $(r \neq 0)$	
583$^{\{\underline{0}\}}$	$A_1A_1^2$ via $(1_a, 1_a, 1_b, 1_c)$ $(p \neq 2)$	
869	$A_1A_1^2$ via $(1_a, 1_a^{[r]}, 1_a^{[s]}, 1_b)$ $(0 < r < s)$	
870	A_1A_1 via $(1_a, 1_a^{[r]}, 1_b, 1_b^{[s]})$ $(rs \neq 0; r \leq s)$	
8	A_1 via $(1, 1^{[r]}, 1^{[s]}, 1^{[t]})$ $(0 < r < s < t)$	
	In $M_2 = A_1A_1$ $(p \geq 11)$ $(E_8(\#866))$	
2	A_1 via $(1^{[r]}, 1^{[s]})$ $(rs = 0; r \neq s)$	
14$^{\{\underline{0}\}}$	A_1 via $(1,1)$	
	In $M_2 = A_1D_4$ $(p = 2)$ $(E_8(\#867))$	
871	A_1B_3	$(1,001)$
649$^{\{\underline{0}\}}$	$A_1A_1^4$	
872	A_1A_2	$(1,11)$
	In $M_3 = A_1B_3$ $(p = 2)$ $(E_8(\#871))$	
859	$A_1^2B_2$	
	In $M_1 = A_1C_4$ $(p \neq 2)$ $(E_8(\#116))$	
563$^{\{\underline{0}\}}$	$A_1\bar{A}_1C_3$	
873	A_1^4	$(1,1,1,1)$
874	A_1A_1 $(p \geq 11)$	$(1,7)$
	In $M_2 = A_1^4$ $(p \neq 2)$ $(E_8(\#873))$	
875	$A_1A_1^2$ via $(1_a, 1_a^{[r]}, 1_b, 1_c)$ $(r \neq 0)$	
580$^{\{\underline{0}\}}$	$A_1A_1^2$ via $(1_a, 1_a, 1_b, 1_c)$	
876	$A_1A_1^2$ via $(1_a, 1_a^{[r]}, 1_a^{[s]}, 1_b)$ $(0 < r < s)$	
877	A_1A_1 via $(1_a, 1_a^{[r]}, 1_b, 1_b^{[s]})$ $(rs \neq 0; r \leq s)$	
9	A_1 via $(1, 1^{[r]}, 1^{[s]}, 1^{[t]})$ $(0 < r < s < t)$	
	In $M_2 = A_1A_1$ $(p \geq 11)$ $(E_8(\#874))$	
1	A_1 via $(1^{[r]}, 1^{[s]})$ $(rs = 0; r \neq s)$	
14$^{\{\underline{0}\}}$	A_1 via $(1,1)$	
	In $M_1 = B_4$ $(E_8(\#45))$	

49	D_4	
113b	B_2^2 $(p = 2)$	$(01, 01)$
$709^{\{0\}}$	$A_1^2 B_2$ $(p \neq 2)$	
$707^{\{\underline{0}\}}$	A_1^2 $(p \geq 5)$	
$12^{\{\underline{0}\}}$	A_1 $(p \geq 11)$	
	In $M_1 = B_4$ $(p \neq 2)$ $(E_8(\#46))$	
$693^{\{\underline{0}\}}$	A_1^2 $(p \geq 5)$	
$12^{\{\underline{0}\}}$	A_1 $(p \geq 11)$	
	In $M = \bar{A}_1 E_7$ $(E_8(\#102))$	
107	$\bar{A}_1^2 D_6$	$(1, E_7(\#30))$
878	$\bar{A}_1 \bar{A}_2 A_5$	$(1, E_7(\#31))$
879	$\bar{A}_1 A_7$	$(1, E_7(\#22))$
880	$\bar{A}_1 G_2 C_3$	$(1, E_7(\#32))$
881	$\bar{A}_1 A_1 F_4$	$(1, E_7(\#33))$
882	$\bar{A}_1 A_1 G_2$ $(p \neq 2)$	$(1, E_7(\#34))$
883	$\bar{A}_1 A_1 A_1$ $(p \geq 5)$	$(1, E_7(\#35))$
884	$\bar{A}_1 A_2$ $(p \geq 5)$	$(1, E_7(\#29))$
885	$\bar{A}_1 A_1$ $(p \geq 17)$	$(1, E_7(\#20))$
886	$\bar{A}_1 A_1$ $(p \geq 19)$	$(1, E_7(\#21))$
	In $M_1 = \bar{A}_1 \bar{A}_2 A_5$ $(E_8(\#878))$	
887	$\bar{A}_1 \bar{A}_2 A_2 A_1$	$(1, E_7(\#300))$
888	$\bar{A}_1 \bar{A}_2 C_3$	$(1, E_7(\#301))$
889a	$\bar{A}_1 \bar{A}_2 A_3$ $(p \neq 2)$	$(1, E_7(\#302a))$
890	$\bar{A}_1 \bar{A}_2 A_2$ $(p \neq 2)$	$(1, E_7(\#303))$
891	$\bar{A}_1 A_1 A_5$ $(p \neq 2)$	$(1, E_7(\#304))$
	In $M_2 = \bar{A}_1 \bar{A}_2 A_2 A_1$ $(E_8(\#887))$	
892	$\bar{A}_1 A_1 A_2 A_1$ $(p \neq 2)$	$(1, E_7(\#305))$
893	$\bar{A}_1 \bar{A}_2 A_1 A_1$ $(p \neq 2)$	$(1, E_7(\#306))$
894	$A_1 \bar{A}_2 A_2$ via $(1^{[r]}, 10, 10, 1^{[s]})$ $(rs = 0)$	
895	$\bar{A}_1 A_2 A_1$ via $(1, 10^{[r]}, 10^{[s]}, 1)$ $(rs = 0)$	
896	$\bar{A}_1 A_2 A_1$ via $(1, 10^{[r]}, 01^{[s]}, 1)$ $(rs = 0; r \neq s)$	
$347^{\{\underline{0}\}}$	$\bar{A}_1 A_2 A_1$ via $(1, 10, 01, 1)$ $(p \neq 3)$	
897	$A_1 A_2$ via $(1^{[r]}, 10^{[t]}, 10^{[u]}, 1^{[s]})$ $(rs = tu = 0)$	
898	$A_1 A_2$ via $(1^{[r]}, 10^{[t]}, 01^{[u]}, 1^{[s]})$ $(rs = tu = 0; t \neq u)$	
	In $M_3 = \bar{A}_1 A_1 A_2 A_1$ $(p \neq 2)$ $(E_8(\#892))$	
$567^{\{\underline{0}\}}$	$\bar{A}_1 A_1 A_1 A_1$	$(1, E_7(\#275^{\{\underline{0}\}}))$
899	$A_1 A_2 A_1$ via $(1_a^{[r]}, 1_a^{[s]}, 10, 1_b)$ $(rs = 0)$	
900	$A_1 A_1 A_2$ via $(1_a^{[r]}, 1_b, 10, 1_a^{[s]})$ $(rs = 0)$	
901	$\bar{A}_1 A_1 A_2$ via $(1_a, 1_b^{[r]}, 10, 1_b^{[s]})$ $(rs = 0)$	
902	$A_1 A_2$ via $(1^{[r]}, 1^{[s]}, 10, 1^{[t]})$ $(rst = 0)$	

	In $M_3 = \bar{A}_1\bar{A}_2A_1A_1$ $(p \neq 2)$ $(E_8(\#893))$	
567$^{\{\underline{0}\}}$	$\bar{A}_1A_1A_1A_1$	
903	$A_1\bar{A}_2A_1$ via $(1_a^{[r]}, 10, 1_a^{[s]}, 1_b)$ $(rs = 0)$	
904	$A_1\bar{A}_2A_1$ via $(1_a^{[r]}, 10, 1_b, 1_a^{[s]})$ $(rs = 0)$	
905	$\bar{A}_1A_1\bar{A}_2$ via $(1_a, 10, 1_b^{[r]}, 1_b^{[s]})$ $(rs = 0)$	
906	A_1A_2 via $(1^{[r]}, 10, 1^{[s]}, 1^{[t]})$ $(rst = 0)$	
	In $M_2 = \bar{A}_1\bar{A}_2C_3$ $(E_8(\#888))$	
563$^{\{\underline{0}\}}$	$\bar{A}_1A_1C_3$ $(p \neq 2)$	$(1, E_7(\#294^{\{\underline{0}\}}))$
893	$\bar{A}_1\bar{A}_2A_1A_1$ $(p \neq 2)$	
907	$\bar{A}_1\bar{A}_2A_1$ $(p \geq 7)$	$(1, E_7(\#311))$
889b	$\bar{A}_1\bar{A}_2A_3$ $(p = 2)$	$(1, E_7(\#302\text{b}))$
908	$\bar{A}_1\bar{A}_2G_2$ $(p = 2)$	$(1, E_7(\#312))$
	In $M_3 = \bar{A}_1\bar{A}_2A_1$ $(p \geq 7)$ $(E_8(\#907))$	
599$^{\{\underline{0}\}}$	$\bar{A}_1A_1A_1$ $(p \geq 7)$	$(1, E_7(\#295^{\{\underline{0}\}}))$
909	$A_1\bar{A}_2$ via $(1^{[r]}, 10, 1^{[s]})$ $(rs = 0)$	
	In $M_2 = \bar{A}_1\bar{A}_2A_3$ $(E_8(\#889))$	
910	$\bar{A}_1A_1A_3$ $(p \neq 2)$	$(1, E_7(\#313))$
	In $M_3 = \bar{A}_1A_1A_3$ $(p \neq 2)$ $(E_8(\#910))$	
911	A_1A_3 via $(1^{[r]}, 1^{[s]}, 100)$	
	In $M_2 = \bar{A}_1\bar{A}_2A_2$ $(p \neq 2)$ $(E_8(\#890))$	
912	$\bar{A}_1A_1A_2$	$(1, E_7(\#314))$
913	$\bar{A}_1A_2$ via $(1, 10, 10)$ $(p \geq 5)$	
914	$\bar{A}_1A_2$ via $(1, 10^{[r]}, 10^{[s]})$ $(rs = 0; r \neq s)$	
915	$\bar{A}_1A_2$ via $(1, 10^{[r]}, 01^{[s]})$ $(rs = 0)$	
	In $M_3 = \bar{A}_1A_1A_2$ $(p \neq 2)$ $(E_8(\#912))$	
916	A_1A_2 via $(1^{[r]}, 1^{[s]}, 10)$ $(rs = 0)$	
	In $M_2 = \bar{A}_1A_1A_5$ $(p \neq 2)$ $(E_8(\#891))$	
892	$\bar{A}_1A_1A_1A_2$	
563$^{\{\underline{0}\}}$	$\bar{A}_1A_1C_3$	
910	$\bar{A}_1A_1A_3$	
912	$\bar{A}_1A_1A_2$	
	In $M_1 = \bar{A}_1A_7$ $(E_8(\#879))$	
917	$\bar{A}_1D_4$ $(p \neq 2)$	$(1, E_7(\#23))$
910	$\bar{A}_1A_1A_3$ $(p \neq 2)$	
	In $M_2 = \bar{A}_1D_4$ $(p \neq 2)$ $(E_8(\#917))$	
913	$\bar{A}_1A_2$ $(p \geq 5)$	
	In $M_1 = \bar{A}_1G_2C_3$ $(E_8(\#880))$	
888	$\bar{A}_1\bar{A}_2C_3$	
122	$\bar{A}_1^2A_1C_3$	$(1, E_7(\#42))$
918	$\bar{A}_1A_2C_3$ $(p = 3)$	$(1, E_7(\#315))$

919	$\bar{A}_1A_1C_3$ $(p \geq 7)$	$(1, E_7(\#316))$
532	$\bar{A}_1^2G_2B_2$	$(1, E_7(\#253))$
921	$\bar{A}_1G_2A_1A_1$ $(p \neq 2)$	$(1, E_7(\#317))$
921	$\bar{A}_1G_2A_1$ $(p \geq 7)$	$(1, E_7(\#318))$
922	$\bar{A}_1G_2A_3$ $(p = 2)$	$(1, E_7(\#319))$
923	$\bar{A}_1G_2G_2$ $(p = 2)$	$(1, E_7(\#320))$
	In $M_2 = \bar{A}_1A_2C_3$ $(p = 3)$ $(E_8(\#918))$	
$563^{\{\delta_1\}}$	$\bar{A}_1A_1C_3$	$(1, E_7(\#294^{\{\delta_1\}}))$
551	$\bar{A}_1^2A_2B_2$	$(1, E_7(\#263))$
924	$\bar{A}_1A_2A_1A_1$	$(1, E_7(\#321))$
	In $M_3 = \bar{A}_1A_2A_1A_1$ $(p = 3)$ $(E_8(\#924))$	
$567^{\{\delta_1\}}$	$\bar{A}_1A_1A_1A_1$	$(1, E_7(\#275^{\{\delta_1\}}))$
925	$A_1A_2A_1$ via $(1_a^{[r]}, 10, 1_a^{[s]}, 1_b)$ $(rs = 0)$	
926	$A_1A_2A_1$ via $(1_a^{[r]}, 10, 1_b, 1_a^{[s]})$ $(rs = 0)$	
927	$\bar{A}_1A_2A_1$ via $(1_a, 10, 1_b^{[r]}, 1_b^{[s]})$ $(rs = 0; r \neq s)$	
928	A_1A_2 via $(1^{[r]}, 10, 1^{[s]}, 1^{[t]})$ $(rst = 0; s \neq t)$	
	In $M_2 = \bar{A}_1A_1C_3$ $(p \geq 7)$ $(E_8(\#919))$	
552	$\bar{A}_1^2A_1B_2$	$(1, E_7(\#264))$
929	$\bar{A}_1A_1A_1A_1$	$(1, E_7(\#323))$
930	$\bar{A}_1A_1A_1$	$(1, E_7(\#324))$
931	A_1C_3 via $(1^{[r]}, 1^{[s]}, 100)$ $(rs = 0)$	
	In $M_3 = \bar{A}_1A_1A_1A_1$ $(p \geq 7)$ $(E_8(\#929))$	
	See Table 11.45	
	In $M_3 = \bar{A}_1A_1A_1$ $(p \geq 7)$ $(E_8(\#930))$	
945	A_1A_1 via $(1_a^{[r]}, 1_a^{[s]}, 1_b)$ $(rs = 0)$	
946	A_1A_1 via $(1_a^{[r]}, 1_b, 1_a^{[s]})$ $(rs = 0)$	
947	$\bar{A}_1A_1$ via $(1_a, 1_b^{[r]}, 1_b^{[s]})$ $(rs = 0; r \neq s)$	
$438^{\{\underline{0}\}}$	$\bar{A}_1A_1$ via $(1_a, 1_b, 1_b)$ $(p \geq 11)$	
31	A_1 via $(1^{[r]}, 1^{[s]}, 1^{[t]})$ $(s \neq t)$	
	In $M_2 = \bar{A}_1G_2A_1A_1$ $(p \neq 2)$ $(E_8(\#921))$	
893	$\bar{A}_1\bar{A}_2A_1A_1$	
558	$\bar{A}_1^2A_1^2A_1$	$(1, E_7(\#268))$
924	$\bar{A}_1A_2A_1A_1$ $(p = 3)$	
929	$\bar{A}_1A_1A_1A_1$ $(p \geq 7)$	
948	$A_1A_1G_2$ via $(1_a^{[r]}, 10, 1_a^{[s]}, 1_b)$ $(rs = 0)$	
949	$A_1A_1G_2$ via $(1_a^{[r]}, 10, 1_b, 1_a^{[s]})$ $(rs = 0)$	
950	$\bar{A}_1A_1G_2$ via $(1_a, 10, 1_b^{[r]}, 1_b^{[s]})$ $(rs = 0; r \neq s)$	
$540^{\{\underline{0}\}}$	$\bar{A}_1A_1G_2$ via $(1_a, 10, 1_b, 1_b)$ $(p \geq 5)$	

951	A_1G_2 via $(1^{[r]}, 10, 1^{[s]}, 1^{[t]})$ $(rst = 0; s \neq t)$	
	In $M_2 = \bar{A}_1G_2A_1$ $(p \geq 7)$ $(E_8(\#921))$	
907	$\bar{A}_1\bar{A}_2A_1$	
559	$\bar{A}_1^2A_1A_1$	$(1, E_7(\#291))$
930	$\bar{A}_1A_1A_1$	
952	A_1G_2 via $(1^{[r]}, 10, 1^{[s]})$ $(rs = 0)$	
	In $M_2 = \bar{A}_1G_2A_3$ $(p = 2)$ $(E_8(\#922))$	
889b	$\bar{A}_1\bar{A}_2A_3$	
560	$\bar{A}_1^2A_1A_3$	$(1, E_7(\#292))$
	In $M_2 = \bar{A}_1G_2G_2$ $(p = 2)$ $(E_8(\#923))$	
908	$\bar{A}_1\bar{A}_2G_2$	
561	$\bar{A}_1^2A_1G_2$	$(1, E_7(\#293))$
513$^{\{\underline{0}\}}$	$\bar{A}_1G_2A_1A_1$	$(1, E_7(\#92^{\{\underline{0}\}}))$
953	A_1G_2 via $(1, 10^{[r]}, 10^{[s]})$ $(rs = 0; r \neq s)$	
	In $M_1 = \bar{A}_1A_1F_4$ $(E_8(\#881))$	
406$^{\{\underline{0}\}}$	$\bar{A}_1A_1B_4$	$(1, E_7(\#190^{\{\underline{0}\}}))$
954	$\bar{A}_1A_1C_4$ $(p = 2)$	$(1, E_7(\#329))$
122	$\bar{A}_1^2A_1C_3$ $(p \neq 2)$	
887	$\bar{A}_1A_1\bar{A}_2A_2$	
921	$\bar{A}_1G_2A_1A_1$ $(p \neq 2)$	
955	$\bar{A}_1A_1G_2$ $(p = 7)$	$(1, E_7(\#330))$
956	$\bar{A}_1A_1A_1$ $(p \geq 13)$	$(1, E_7(\#331))$
957	A_1F_4 via $(1^{[r]}, 1^{[s]}, \lambda_1)$ $(rs = 0;$ if $p = 2$ then $r \neq s)$	
	In $M_2 = \bar{A}_1A_1C_4$ $(p = 2)$ $(E_8(\#954))$	
521$^{\{\underline{0}\}}$	$\bar{A}_1A_1B_2^2$	$(1, E_7(\#218^{\{\underline{0}\}}))$
122	$\bar{A}_1^2A_1C_3$	
958	$\bar{A}_1A_1D_4$	$(1, E_7(\#332))$
959	A_1C_4 via $(1^{[r]}, 1^{[s]}, \lambda_1)$ $(rs = 0; r \neq s)$	
	In $M_3 = \bar{A}_1A_1D_4$ $(p = 2)$ $(E_8(\#958))$	
468$^{\{\underline{0}\}}$	$\bar{A}_1A_1A_1^4$	$(1, E_7(\#230^{\{\underline{0}\}}))$
895$^{\{\delta_1\}}$	$\bar{A}_1A_1A_2$	
960	A_1D_4 via $(1^{[r]}, 1^{[s]}, \lambda_1)$ $(rs = 0; r \neq s)$	
	In $M_2 = \bar{A}_1A_1G_2$ $(p = 7)$ $(E_8(\#955))$	
895$^{\{\underline{0}\}}$	$\bar{A}_1A_1A_2$	
569$^{\{\underline{0}\}}$	$\bar{A}_1A_1A_1A_1$	$(1, E_7(\#274^{\{\underline{0}\}}))$
600$^{\{\delta_1\}}$	$\bar{A}_1A_1A_1$	$(1, E_7(\#287^{\{\delta_1\}}))$
961	A_1G_2 via $(1^{[r]}, 1^{[s]}, 10)$ $(rs = 0)$	
	In $M_2 = \bar{A}_1A_1A_1$ $(p \geq 13)$ $(E_8(\#956))$	
962	A_1A_1 via $(1_a^{[r]}, 1_a^{[s]}, 1_b)$ $(rs = 0)$	
963	A_1A_1 via $(1_a^{[r]}, 1_b, 1_a^{[s]})$ $(rs = 0)$	
964	$\bar{A}_1A_1$ via $(1_a, 1_b^{[r]}, 1_b^{[s]})$ $(rs = 0)$	

34	A_1 via $(1^{[r]}, 1^{[s]}, 1^{[t]})$ $(rst = 0)$	
	In $M_1 = \bar{A}_1 A_1 G_2$ $(p \neq 2)$ $(E_8(\#882))$	
901$^{\{\underline{0}\}}$	$\bar{A}_1 A_1 A_2$	
965	$\bar{A}_1 A_1 A_2$ $(p = 3)$	$(1, E_7(\#333))$
410$^{\{\underline{0}\}}$	$\bar{A}_1 A_1 A_1 A_1$	$(1, E_7(\#206^{\{\underline{0}\}}))$
966	$\bar{A}_1 A_1 A_1$ $(p \geq 7)$	$(1, E_7(\#334))$
967	$A_1 G_2$ via $(1^{[r]}, 1^{[s]}, 10)$ $(rs = 0)$	
	In $M_2 = \bar{A}_1 A_1 A_2$ $(p = 3)$ $(E_8(\#965))$	
422$^{\{\delta_3\}}$	$\bar{A}_1 A_1 A_1$	$(1, E_7(\#213^{\{\delta_3\}}))$
968	$A_1 A_2$ via $(1^{[r]}, 1^{[s]}, 10)$ $(rs = 0)$	
	In $M_2 = \bar{A}_1 A_1 A_1$ $(p \geq 7)$ $(E_8(\#966))$	
969	$A_1 A_1$ via $(1_a^{[r]}, 1_a^{[s]}, 1_b)$ $(rs = 0)$	
970	$A_1 A_1$ via $(1_a^{[r]}, 1_b, 1_a^{[s]})$ $(rs = 0)$	
971	$\bar{A}_1 A_1$ via $(1_a, 1_b^{[r]}, 1_b^{[s]})$ $(rs = 0; r \neq s)$	
435$^{\{\underline{0}\}}$	$\bar{A}_1 A_1$ via $(1_a, 1_b, 1_b)$ $(p \geq 11)$	
33	A_1 via $(1^{[r]}, 1^{[s]}, 1^{[t]})$ $(rst = 0; s \neq t)$	
	In $M_1 = \bar{A}_1 A_1 A_1$ $(p \geq 5)$ $(E_8(\#883))$	
972	$A_1 A_1$ via $(1_a^{[r]}, 1_a^{[s]}, 1_b)$ $(rs = 0)$	
973	$A_1 A_1$ via $(1_a^{[r]}, 1_b, 1_a^{[s]})$ $(rs = 0)$	
974	$\bar{A}_1 A_1$ via $(1_a, 1_b^{[r]}, 1_b^{[s]})$ $(rs = 0; r \neq s)$	
35	A_1 via $(1^{[r]}, 1^{[s]}, 1^{[t]})$ $(rst = 0; s \neq t)$	
	In $M_1 = \bar{A}_1 A_1$ $(p \geq 17)$ $(E_8(\#885))$	
36	A_1 via $(1^{[r]}, 1^{[s]})$ $(rs = 0)$	
	In $M_1 = \bar{A}_1 A_1$ $(p \geq 19)$ $(E_8(\#886))$	
37	A_1 via $(1^{[r]}, 1^{[s]})$ $(rs = 0)$	
	In $M = \bar{A}_2 E_6$ $(E_8(\#103))$	
878	$\bar{A}_2 \bar{A}_1 A_5$	$(10, E_6(\#24))$
975	$\bar{A}_2^4$	$(10, E_6(\#25))$
976	$\bar{A}_2 F_4$	$(10, E_6(\#7))$
977a	$\bar{A}_2 C_4$ $(p \neq 2)$	$(10, E_6(\#8\text{a}))$
978	$\bar{A}_2 A_2 G_2$	$(10, E_6(\#26))$
979a	$\bar{A}_2 G_2$ $(p \neq 7)$	$(10, E_6(\#10\text{a}))$
980a	$\bar{A}_2 A_2$ $(p \geq 5)$	$(10, E_6(\#11\text{a}))$
981	$A_1 E_6$ $(p \neq 2)$	$(2, E_6(\#0))$
	In $M_1 = \bar{A}_2^4$ $(E_8(\#975))$	
982	$A_1 \bar{A}_2^3$ $(p \neq 2)$	$(2, E_6(\#25))$
	See Table 11.46	
	In $M_2 = A_1 \bar{A}_2^3$ $(p \neq 2)$ $(E_8(\#982))$	
989	$A_1^2 \bar{A}_2^2$	$(2, E_6(\#38))$
990	$A_1 A_2 \bar{A}_2$ via $(1, 10_a, 10_a^{[r]}, 10_b)$	
991	$A_1 A_2 \bar{A}_2$ via $(1, 10_a, 01_a^{[r]}, 10_b)$	

992	A_1A_2 via $(1,10,10^{[r]},10^{[s]})$ $(r \leq s$; if $r=0$ then $s \neq 0)$	
993	A_1A_2 via $(1,10,10^{[r]},01^{[s]})$	
994	A_1A_2 via $(1,10,01^{[r]},01^{[s]})$ $(0<r \leq s)$	
	In $M_3 = A_1^2\bar{A}_2^2$ $(p \neq 2)$ $(E_8(\#989))$	
995	$A_1^3\bar{A}_2$	$(2, E_6(\#41))$
996	$A_1\bar{A}_2^2$ via $(1,1^{[r]},10,10)$	
997	$A_1^2A_2$ via $(1,1,10,10^{[r]})$	
998	A_1A_2 via $(1,1^{[r]},10,10^{[s]})$	
999	A_1A_2 via $(1,1^{[r]},10,01^{[s]})$ $(r \neq 0)$	
	In $M_4 = A_1^3\bar{A}_2$ $(p \neq 2)$ $(E_8(\#995))$	
873	A_1^4	$(2, E_6(\#44))$
1000	$A_1A_1\bar{A}_2$ via $(1_a,1_a^{[r]},1_b,10)$ $(r \neq 0)$	
903$^{\{\underline{0}\}}$	$A_1A_1\bar{A}_2$ via $(1_a,1_a,1_b,10)$	
1001	$A_1A_1\bar{A}_2$ via $(1,1^{[r]},1^{[s]},10)$ $(0<r<s)$	
	In $M_1 = \bar{A}_2F_4$ $(E_8(\#976))$	
977b	$\bar{A}_2C_4$ $(p=2)$	$(10, E_6(\#8\text{b}))$
888	$\bar{A}_2\bar{A}_1C_3$ $(p \neq 2)$	$(10, E_6(\#27))$
983$^{\{0\}}$	$\bar{A}_2^2A_2$	
979b	$\bar{A}_2G_2$ $(p=7)$	$(10, E_6(\#10\text{b}))$
1002	$\bar{A}_2A_1G_2$ $(p \neq 2)$	$(10, E_6(\#47))$
1003	$\bar{A}_2A_1$ $(p \geq 13)$	$(10, E_6(\#5))$
957$^{\{\underline{0}\}}$	A_1F_4 $(p \neq 2)$	$(2, E_6(\#7))$
	In $M_2 = \bar{A}_2A_1G_2$ $(p \neq 2)$ $(E_8(\#1002))$	
949$^{\{\underline{0}\}}$	$A_1A_1G_2$	$(2, E_6(\#47))$
996$^{\{0\}}$	$\bar{A}_2^2A_1$	
893	$\bar{A}_2A_1\bar{A}_1A_1$	$(10, E_6(\#31))$
1004	$\bar{A}_2A_1A_2$ $(p=3)$	$(10, E_6(\#50))$
1005	$\bar{A}_2A_1A_1$ $(p \geq 7)$	$(10, E_6(\#51))$
	In $M_3 = \bar{A}_2A_1A_2$ $(p=3)$ $(E_8(\#1004))$	
926$^{\{\underline{0}\}}$	$A_1A_1A_2$	$(2, E_6(\#50))$
903$^{\{\delta_1\}}$	$\bar{A}_2A_1A_1$	$(10, E_6(\#35^{\{\delta_1\}}))$
1006	A_2A_1 via $(10^{[r]},1,10^{[s]})$ $(rs=0)$	
1007	A_2A_1 via $(10^{[r]},1,01^{[s]})$ $(rs=0)$	
	In $M_3 = \bar{A}_2A_1A_1$ $(p \geq 7)$ $(E_8(\#1005))$	
933$^{\{\underline{0}\}}$	$A_1A_1A_1$	$(2, E_6(\#51))$
1008	$\bar{A}_2A_1$ via $(10,1^{[r]},1^{[s]})$ $(rs=0; r \neq s)$	
909$^{\{\underline{0}\}}$	$\bar{A}_2A_1$ via $(10,1,1)$	
	In $M_2 = \bar{A}_2A_1$ $(p \geq 13)$ $(E_8(\#1003))$	
962$^{\{\underline{0}\}}$	A_1A_1	$(2, E_6(\#5))$
	In $M_1 = \bar{A}_2C_4$ $(E_8(\#977))$	
116	A_1C_4 $(p \neq 2)$	$(2, E_6(\#8\text{a}))$
888	$\bar{A}_2\bar{A}_1C_3$	

995	$\bar{A}_2A_1^3$ $(p \neq 2)$	
1009	$\bar{A}_2A_1$ $(p \geq 11)$	$(10, E_6(\#6))$
1010	$\bar{A}_2D_4$ $(p = 2)$	$(10, E_6(\#9))$
	In $M_2 = \bar{A}_2A_1$ $(p \geq 11)$ $(E_8(\#1009))$	
874	A_1A_1	$(10, E_6(\#6))$
	In $M_2 = \bar{A}_2D_4$ $(p = 2)$ $(E_8(\#1010))$	
985$^{\{\delta_2\}}$	$\bar{A}_2A_2$	
	In $M_1 = \bar{A}_2A_2G_2$ $(E_8(\#978))$	
1011	$A_1A_2G_2$ $(p \neq 2)$	$(10, E_6(\#26))$
1002	$\bar{A}_2A_1G_2$ $(p \neq 2)$	
983$^{\{0\}}$	$\bar{A}_2^2A_2$	
1012	$\bar{A}_2A_2A_2$ $(p = 3)$	$(10, E_6(\#48))$
887	$\bar{A}_2A_2\bar{A}_1A_1$	$(10, E_6(\#28))$
1013	$\bar{A}_2A_2A_1$ $(p \geq 7)$	$(10, E_6(\#49))$
1014	A_2G_2 via $(10^{[r]}, 10^{[s]}, 10)$ $(rs = 0$; if $p = 3$ then $r \neq s)$	
1015	A_2G_2 via $(10^{[r]}, 01^{[s]}, 10)$ $(rs = 0)$	
	In $M_2 = A_1A_2G_2$ $(p \neq 2)$ $(E_8(\#1011))$	
949$^{\{\underline{0}\}}$	$A_1A_1G_2$ $(p \neq 2)$	$(2, E_6(\#47))$
991$^{\{0\}}$	$A_1A_2\bar{A}_2$	
1016	$A_1A_2A_2$ $(p = 3)$	$(10, E_6(\#48))$
892	$A_1A_2\bar{A}_1A_1$	$(2, E_6(\#28))$
1017	$A_1A_2A_1$ $(p \geq 7)$	$(10, E_6(\#49))$
	In $M_3 = A_1A_2A_2$ $(p = 3)$ $(E_8(\#1016))$	
926$^{\{\underline{0}\}}$	$A_1A_1A_2$	
900$^{\{\delta_1\}}$	$A_1A_2A_1$	$(2, E_6(\#37^{\{\delta_1\}}))$
1018	A_1A_2 via $(1, 10^{[r]}, 10^{[s]})$ $(rs = 0; r \neq s)$	
	In $M_3 = A_1A_2A_1$ $(p \geq 7)$ $(E_8(\#1017))$	
933$^{\{\underline{0}\}}$	$A_1A_1A_1$	
1019	A_1A_2 via $(1^{[r]}, 10, 1^{[s]})$ $(rs = 0)$	
	In $M_2 = \bar{A}_2A_2A_2$ $(p = 3)$ $(E_8(\#1012))$	
1016	$A_1A_2A_2$	
1004	$\bar{A}_2A_1A_2$	
894$^{\{\delta_1\}}$	$\bar{A}_2A_2A_1$	$(10, E_6(\#37^{\{\delta_1\}}))$
	See Table 11.47	
	In $M_2 = \bar{A}_2A_2A_1$ $(p \geq 7)$ $(E_8(\#1013))$	
1017	$A_1A_2A_1$	
1005	$\bar{A}_2A_1A_1$	
1024	A_2A_1 via $(10^{[r]}, 10^{[s]}, 1)$ $(rs = 0$; if $p = 3$ then $r \neq s)$	
1025	A_2A_1 via $(10^{[r]}, 01^{[s]}, 1)$ $(rs = 0)$	
	In $M_1 = \bar{A}_2G_2$ $(E_8(\#979))$	
1026	A_1G_2 $(p \neq 2)$	$(2, E_6(\#10))$
985$^{\{\underline{0}\}}$	$\bar{A}_2A_2$	

904$^{\{\underline{0}\}}$	$\bar{A}_2A_1A_1$ $(p \neq 2)$	$(10, E_6(\#34^{\{\underline{0}\}}))$
980b	$\bar{A}_2A_2$ $(p = 3)$	$(10, E_6(\#11\text{b}))$
1009	$\bar{A}_2A_1$ $(p \geq 11)$	
909$^{\{\delta_1\}}$	$\bar{A}_2A_1$ $(p = 7)$	$(10, E_6(\#1^{\{\delta_1\}}))$
	In $M_2 = A_1G_2$ $(p \neq 2)$ $(E_8(\#1026))$	
993$^{\{\underline{0}\}}$	A_1A_2	
581$^{\{\underline{0}\}}$	$A_1A_1A_1$	$(2, E_6(\#34^{\{\underline{0}\}}))$
1027	A_1A_2 $(p = 3)$	
874	A_1A_1 $(p \geq 11)$	
	In $M_1 = \bar{A}_2A_2$ $(p \neq 2)$ $(E_8(\#980))$	
1027	A_1A_2	$(2, E_6(\#11))$
909$^{\{\underline{0}\}}$	$\bar{A}_2A_1$ $(p \geq 7)$	
906$^{\{\delta_3\}}$	$\bar{A}_2A_1$ $(p = 3)$	$(10, E_6(\#2^{\{\delta_3\}}))$
90	A_2 via $(10, 10)$	
91	A_2 via $(10^{[r]}, 10^{[s]})$ $(rs = 0; r \neq s)$	
	In $M_2 = A_1A_2$ $(p \neq 2)$ $(E_8(\#1027))$	
609$^{\{\underline{0}\}}$	A_1A_1 $(p \geq 7)$	$(2, E_6(\#1^{\{\underline{0}\}}))$
595$^{\{\delta_2\}}$	A_1A_1 $(p = 3)$	$(2, E_6(\#2^{\{\delta_3\}}))$
	In $M_1 = A_1E_6$ $(p \neq 2)$ $(E_8(\#981))$	
891	$A_1\bar{A}_1A_5$	$(2, E_6(\#24))$
982	$A_1\bar{A}_2^3$	
957$^{\{\underline{0}\}}$	A_1F_4	
1011	$A_1A_2G_2$	
1026	A_1G_2 $(p \neq 7)$	
1027	A_1A_2 $(p \geq 5)$	

	In $M = A_8$ $(E_8(\#62))$	
46	B_4 $(p \neq 2)$	λ_1
668	A_2^2 $(p \neq 3)$	$(10, 10)$

	In $M = \bar{A}_4^2$ $(E_8(\#104))$	
1028	$B_2\bar{A}_4$ $(p \neq 2)$	$(10, \lambda_1)$
63	A_4 via (λ_1, λ_1)	
64	A_4 via $(\lambda_1, \lambda_1^{[r]})$ $(r \neq 0)$	
65	A_4 via $(\lambda_1, \lambda_4^{[r]})$ $(r \neq 0)$	
	In $M_1 = B_2\bar{A}_4$ $(p \neq 2)$ $(E_8(\#1028))$	
1029	$A_1\bar{A}_4$ $(p \geq 5)$	$(4, \lambda_1)$
114	B_2^2	$(10, 10)$
	In $M_2 = A_1\bar{A}_4$ $(p \geq 5)$ $(E_8(\#1029))$	
863	A_1B_2	$(4, 10)$

	In $M = G_2F_4$ $(E_8(\#105))$	
976	$\bar{A}_2F_4$	$(G_2(\#4), F_4(\#0))$
881	$\bar{A}_1A_1F_4$	$(G_2(\#6), F_4(\#0))$

1030	A_2F_4 $(p=3)$	$(G_2(\#5), F_4(\#0))$
1031	A_1F_4 $(p\geq 7)$	$(G_2(\#3), F_4(\#0))$
844	G_2B_4	$(G_2(\#0), F_4(\#12))$
1032	G_2C_4 $(p=2)$	$(G_2(\#0), F_4(\#14))$
880a	$G_2\bar{A}_1C_3$ $(p\neq 2)$	$(G_2(\#0), F_4(\#24\text{a}))$
1033	$G_2^2A_1$ $(p\neq 2)$	$(G_2(\#0), F_4(\#25))$
978	$G_2\bar{A}_2A_2$	$(G_2(\#0), F_4(\#26))$
1034	G_2A_1 $(p\geq 13)$	$(G_2(\#0), F_4(\#10))$
1035	G_2G_2 $(p=7)$	$(G_2(\#0), F_4(\#16))$
	In $M_1 = A_2F_4$ $(p=3)$ $(E_8(\#1030))$	
957$^{\{\delta_1\}}$	A_1F_4	$(G_2(\#1^{\{\delta_1\}}), F_4(\#0))$
847	A_2B_4	$(G_2(\#5), F_4(\#12))$
918	$A_2\bar{A}_1C_3$	$(G_2(\#5), F_4(\#24))$
1036	$A_2A_1G_2$	$(G_2(\#5), F_4(\#25))$
1012	$A_2\bar{A}_2A_2$	$(G_2(\#5), F_4(\#26))$
	In $M_2 = A_2A_1G_2$ $(p=3)$ $(E_8(\#1036))$	
949$^{\{\delta_1\}}$	$A_1A_1G_2$	$(G_2(\#1^{\{\delta_1\}}), F_4(\#25))$
1004	$A_2A_1\bar{A}_2$	$(G_2(\#5), F_4(\#64))$
924	$A_2A_1\bar{A}_1A_1$	$(G_2(\#5), F_4(\#57))$
1037	$A_2^2A_1$	$(G_2(\#5), F_4(\#65))$
	In $M_3 = A_2^2A_1$ $(p=3)$ $(E_8(\#1037))$	
926$^{\{\delta_1\}}$	$A_1A_2A_1$	$(G_2(\#1^{\{\delta_1\}}), F_4(\#65))$
1038	A_2A_1 via $(10, 10^{[r]}, 1)$	
	In $M_1 = A_1F_4$ $(p\geq 7)$ $(E_8(\#1031))$	
848	A_1B_4	$(G_2(\#3), F_4(\#12))$
919	$A_1\bar{A}_1C_3$	$(G_2(\#3), F_4(\#24))$
1039	$A_1A_1G_2$	$(G_2(\#3), F_4(\#25))$
1013	$A_1\bar{A}_2A_2$	$(G_2(\#3), F_4(\#26))$
1040	A_1A_1 $(p\geq 13)$	$(G_2(\#3), F_4(\#10))$
1041	A_1G_2 $(p=7)$	$(G_2(\#3), F_4(\#16))$
	In $M_2 = A_1A_1G_2$ $(p\geq 7)$ $(E_8(\#1039))$	
1005	$A_1A_1\bar{A}_2$	$(G_2(\#3), F_4(\#64))$
929	$A_1A_1\bar{A}_1A_1$	$(G_2(\#3), F_4(\#57))$
1042	$A_1^2A_1$	$(G_2(\#3), F_4(\#66))$
1043	A_1G_2 via $(1^{[r]}, 10, 1^{[s]})$ $(rs=0)$	
952$^{\{\underline{0}\}}$	A_1G_2 via $(1, 10, 1)$	
	In $M_3 = A_1^2A_1$ $(p\geq 7)$ $(E_8(\#1042))$	
1044	A_1A_1 via $(1_a, 1_a^{[r]}, 1_b)$ $(r\neq 0)$	
969$^{\{\underline{0}\}}$	A_1A_1 via $(1_a, 1_a, 1_b)$	
1045	A_1A_1 via $(1_a, 1_b^{[r]}, 1_b^{[s]})$ $(rs=0)$	
946$^{\{\underline{0}\}}$	A_1A_1 via $(1_a, 1_b, 1_b)$	

39	A_1 via $(1^{[r]}, 1^{[s]}, 1^{[t]})$ $(rt = 0; r < s; t \notin \{r, s\})$	
	In $M_2 = A_1A_1$ $(p \geq 13)$ $(E_8(\#1040))$	
38	A_1 via $(1^{[r]}, 1^{[s]})$ $(rs = 0; r \neq s)$	
11$^{\{\underline{0}\}}$	A_1 via $(1, 1)$	
	In $M_2 = A_1G_2$ $(p = 7)$ $(E_8(\#1041))$	
994$^{\{\underline{0}\}}$	A_1A_2	$(G_2(\#3), F_4(\#17))$
933$^{\{\underline{0}\}}$	$A_1A_1A_1$	$(G_2(\#3), F_4(\#61^{\{\underline{0}\}}))$
946$^{\{\delta_1\}}$	A_1A_1	$(G_2(\#3), F_4(\#8^{\{\delta_1\}}))$
	In $M_1 = G_2C_4$ $(p = 2)$ $(E_8(\#1032))$	
977b	$\bar{A}_2C_4$	$(G_2(\#4), F_4(\#14))$
954	$\bar{A}_1A_1C_4$	$(G_2(\#6), F_4(\#14))$
1046	G_2D_4	$(G_2(\#0), F_4(\#15))$
880b	$G_2\bar{A}_1C_3$	$(G_2(\#0), F_4(\#24\text{b}))$
839	$G_2B_2^2$	$(G_2(\#0), F_4(\#30))$
	In $M_2 = G_2D_4$ $(p = 2)$ $(E_8(\#1046))$	
1010	$\bar{A}_2D_4$	$(G_2(\#4), F_4(\#15))$
958	$\bar{A}_1A_1D_4$	$(G_2(\#6), F_4(\#15))$
786	$G_2A_1^4$	$(G_2(\#0), F_4(\#51))$
1014$^{\{\delta_1\}}$	G_2A_2	$(G_2(\#0), F_4(\#18^{\{\delta_1\}}))$
	In $M_1 = G_2^2A_1$ $(p \neq 2)$ $(E_8(\#1033))$	
1002	$\bar{A}_2G_2A_1$	$(G_2(\#4), F_4(\#25))$
921	$\bar{A}_1A_1G_2A_1$	$(G_2(\#6), F_4(\#25))$
1036	$A_2G_2A_1$ $(p = 3)$	
1039	$A_1G_2A_1$ $(p \geq 7)$	
1047	G_2A_1 via $(10, 10^{[r]}, 1)$ $(r \neq 0)$	
967$^{\{\underline{0}\}}$	G_2A_1 via $(10, 10, 1)$	
1048	G_2A_1 via $(10, 03^{[r]}, 1)$ $(p = 3)$	
	In $M_1 = G_2A_1$ $(p \geq 13)$ $(E_8(\#1034))$	
1003	$\bar{A}_2A_1$	$(G_2(\#4), F_4(\#10))$
956	$\bar{A}_1A_1A_1$	$(G_2(\#6), F_4(\#10))$
1040	A_1A_1	
	In $M_1 = G_2G_2$ $(p = 7)$ $(E_8(\#1035))$	
979b	$\bar{A}_2G_2$	$(G_2(\#4), F_4(\#16))$
955	$\bar{A}_1A_1G_2$	$(G_2(\#6), F_4(\#16))$
1041	A_1G_2	
1014$^{\{\underline{0}\}}$	G_2A_2	$(G_2(\#0), F_4(\#17))$
948$^{\{\underline{0}\}}$	$G_2A_1A_1$	$(G_2(\#0), F_4(\#61^{\{\underline{0}\}}))$
952$^{\{\delta_1\}}$	G_2A_1	$(G_2(\#0), F_4(\#8^{\{\delta_1\}}))$
100	G_2 via $(10^{[r]}, 10^{[s]})$ $(rs = 0; r \neq s)$	
	In $M = F_4$ $(p = 3)$ $(E_8(\#1049))$	
45	B_4	$F_4(\#(12))$
774	A_1C_3	$F_4(\#(24))$

967$^{\{\delta_1\}}$	A_1G_2	$F_4(\#(25))$
	In $M = B_2$ $(p \geq 5)$ $(E_8(\#101))$	
707$^{\{\underline{0}\}}$	A_1^2	$(1,1)/(0,0)$
12$^{\{\underline{0}\}}$	A_1 $(p \geq 11)$	4
	In $M = A_1A_2$ $(p \geq 5)$ $(E_8(\#106))$	
973$^{\{\underline{0}\}}$	A_1A_1	$(1,2)$

Table 11.5A: The connected overgroups for irreducible subgroups of E_8.

ID	X	p	Immediate connected overgroups
10	A_1	≥ 17	$B_7[\#44]$
$11^{\{\underline{0}\}}$	A_1	≥ 13	$A_1A_1[\#772]$, $A_1A_1[\#1040]$
$12^{\{\underline{0}\}}$	A_1	≥ 11	$B_4[\#45]$, $B_4[\#46]$, $B_2[\#101]$, $A_1A_1[\#838]$
$14^{\{\underline{0}\}}$	A_1	≥ 11	$A_1A_1[\#854]$, $A_1A_1[\#866]$, $A_1A_1[\#874]$
$17^{\{\underline{0}\}}$	A_1	≥ 7	immediate overgroups in $\bar{A}_1^2A_1A_1[\#538]$, $A_1A_1[\#771]$, $A_1^2[\#862]$, $A_1^2[\#864]$
$17^{\{s,s,t,r\}}$	A_1	≥ 7	immediate overgroups in $\bar{A}_1^2A_1A_1[\#538]$, $A_1A_1A_1[\#630]$
40	A_1	≥ 23	$E_8[\#0]$
41	A_1	≥ 29	$E_8[\#0]$
42	A_1	≥ 31	$E_8[\#0]$
43	D_8		$E_8[\#0]$
44	B_7		$D_8[\#43]$
45	B_4		$D_8[\#43]$, $F_4[\#1049]$ $(p = 3)$
46	B_4	$\neq 2$	$D_8[\#43]$, $A_8[\#62]$
47	A_3	$\neq 2$	$B_7[\#44]$
49	D_4		$B_4[\#45]$, $\bar{D}_4^2[\#108]$
59	B_2	$\neq 2$	$A_3[\#47]$, $B_2^2[\#113]$, $B_2^2[\#114]$, $B_2B_2[\#724]$
62	A_8		$E_8[\#0]$
101	B_2	≥ 5	$E_8[\#0]$
102	$\bar{A}_1E_7$		$E_8[\#0]$
103	$\bar{A}_2E_6$		$E_8[\#0]$
104	$\bar{A}_4^2$		$E_8[\#0]$
105	G_2F_4		$E_8[\#0]$
106	A_1A_2	≥ 5	$E_8[\#0]$
107	$\bar{A}_1^2D_6$		$D_8[\#43]$, $\bar{A}_1E_7[\#102]$
108	$\bar{D}_4^2$		$D_8[\#43]$
109	$\bar{A}_3D_5$		$D_8[\#43]$
110	A_1B_6		$D_8[\#43]$ $(p \neq 2)$, $B_7[\#44]$ $(p = 2)$
111	B_2B_5		$D_8[\#43]$ $(p \neq 2)$, $B_7[\#44]$ $(p = 2)$
112	B_3B_4		$D_8[\#43]$ $(p \neq 2)$, $B_7[\#44]$ $(p = 2)$

113	B_2^2		$D_8[\#43]$ $(p \neq 2)$, $B_4[\#45]$ $(p = 2)$
114	B_2^2	$\neq 2$	$D_8[\#43]$, $B_2\bar{A}_4[\#1028]$
115	A_1C_4		$D_8[\#43]$
116	A_1C_4	$\neq 2$	$D_8[\#43]$, $\bar{A}_2C_4[\#977]$
117	$\bar{A}_1^4\bar{D}_4$		$\bar{A}_1^2D_6[\#107]$, $\bar{D}_4^2[\#108]$
118	$\bar{A}_1^2B_5$		$B_7[\#44]$ $(p \neq 2)$, $\bar{A}_1^2D_6[\#107]$, $B_2B_5[\#111]$
119	$\bar{A}_1^2\bar{A}_3^2$		$\bar{A}_1^2D_6[\#107]$, $\bar{A}_3D_5[\#109]$
120	$\bar{A}_1^2A_1B_4$		$\bar{A}_1^2D_6[\#107]$ $(p \neq 2)$, $A_1B_6[\#110]$ $(p \neq 2)$, $B_3B_4[\#112]$ $(p \neq 2)$, $\bar{A}_1^2B_5[\#118]$ $(p = 2)$, $A_1B_2B_4[\#776]$ $(p = 2)$
121	$\bar{A}_1^2B_2B_3$		$\bar{A}_1^2D_6[\#107]$ $(p \neq 2)$, $B_2B_5[\#111]$ $(p \neq 2)$, $B_3B_4[\#112]$ $(p \neq 2)$, $\bar{A}_1^2B_5[\#118]$ $(p = 2)$, $B_2^2B_3[\#837]$ $(p = 2)$
122	$\bar{A}_1^2A_1C_3$		$\bar{A}_1^2D_6[\#107]$, $\bar{A}_1G_2C_3[\#880]$, $\bar{A}_1A_1F_4[\#881]$ $(p \neq 2)$, $\bar{A}_1A_1C_4[\#954]$ $(p = 2)$
$123^{\{0\}}$	A_1D_6		$B_7[\#44]$ $(p \neq 2)$, $\bar{A}_1^2D_6[\#107]$, $A_1B_6[\#110]$
124	$\bar{A}_1^8$		$\bar{A}_1^2D_6[\#107]$
125	$\bar{A}_1^4B_3$		$\bar{A}_1^4\bar{D}_4[\#117]$, $\bar{A}_1^2B_5[\#118]$ $(p \neq 2)$, $\bar{A}_1^2B_2B_3[\#121]$, $B_3\bar{D}_4[\#619]$
126	$\bar{A}_1^4A_1B_2$		$\bar{A}_1^4\bar{D}_4[\#117]$ $(p \neq 2)$, $\bar{A}_1^2A_1B_4[\#120]$ $(p \neq 2)$, $\bar{A}_1^2B_2B_3[\#121]$ $(p \neq 2)$, $\bar{A}_1^4B_3[\#125]$ $(p = 2)$, $\bar{A}_1^2A_1B_2^2[\#404]$ $(p = 2)$, $A_1B_2\bar{D}_4[\#621]$
127	$\bar{A}_1^4A_2$		$\bar{A}_1^4\bar{D}_4[\#117]$ $(p \neq 3)$, $\bar{A}_1^4G_2[\#167]$ $(p = 3)$, $\bar{A}_1^2B_2A_2[\#551]$ $(p = 3)$, $A_2\bar{D}_4[\#620]$
$128^{\{0\}}$	$\bar{A}_1^2A_1\bar{D}_4$		$\bar{A}_1^4\bar{D}_4[\#117]$, $\bar{A}_1^2B_5[\#118]$ $(p \neq 2)$, $\bar{A}_1^2A_1B_4[\#120]$, $B_3\bar{D}_4[\#619]$, $A_1B_2\bar{D}_4[\#621]$
$129^{\{\underline{0}\}}$	$\bar{A}_1A_1\bar{D}_4$		$\bar{A}_1^2A_1\bar{D}_4[\#128^0]$, $G_2\bar{D}_4[\#622]$
$131^{\{\delta_3\}}$	A_1D_4		immediate overgroups in $\bar{A}_1^4\bar{D}_4[\#117]$, $A_2\bar{D}_4[\#620]$ $(p = 3)$
$132^{\{0\}}$	$\bar{A}_1^6A_1$	$\neq 2$	$\bar{A}_1^8[\#124]$, $\bar{A}_1^4B_3[\#125]$ $(p \neq 2)$, $\bar{A}_1^4A_1B_2[\#126]$
$133^{\{\underline{0}\}}$	$\bar{A}_1^5A_1$		$\bar{A}_1^6A_1[\#132^{\{0\}}]$, $\bar{A}_1^4G_2[\#167]$
$136^{\{\delta_3\}}$	$\bar{A}_1^4A_1$		immediate overgroups in $\bar{A}_1^8[\#124]$, $\bar{A}_1^4A_2[\#127]$ $(p = 3)$
167	$\bar{A}_1^4G_2$		$\bar{A}_1^4B_3[\#125]$, $\bar{A}_1^2B_2G_2[\#532]$, $G_2\bar{D}_4[\#622]$
$169^{\{\underline{0}\}}$	$\bar{A}_1^2A_1B_3$		$\bar{A}_1^4B_3[\#125]$, $B_3^2[\#623]$, $B_3A_1B_2[\#625]$
174	$\bar{A}_1^4A_1$	≥ 7	$\bar{A}_1^4G_2[\#167]$, $\bar{A}_1^2B_2A_1[\#552]$, $A_1\bar{D}_4[\#626]$
189	$\bar{A}_1^4A_1A_1$	≥ 5	$\bar{A}_1^4A_1B_2[\#126]$, $\bar{A}_1^2A_1B_3[\#531]$, $A_1A_1\bar{D}_4[\#675]$
190	$\bar{A}_1^4A_1^3$	2	$\bar{A}_1^4A_1B_2[\#126]$, $\bar{A}_1^2A_1^3B_2[\#441]$, $A_1^3\bar{D}_4[\#674]$
$191^{\{0\}}$	$\bar{A}_1^2A_1^2B_2$	$\neq 2$	$\bar{A}_1^4A_1B_2[\#126]$, $\bar{A}_1^2B_2\bar{A}_3[\#350]$, $\bar{A}_1^2A_1^2\bar{A}_3[\#355]$, $B_3A_1B_2[\#625]$, $A_1^2B_2^2[\#677]$
$192^{\{0\}}$	$\bar{A}_1^2A_1A_1B_2$		$\bar{A}_1^4A_1B_2[\#126]$, $A_1^2B_2^2[\#676]$
$193^{\{\underline{0}\}}$	$\bar{A}_1^3A_1B_2$		$\bar{A}_1^2A_1C_3[\#122]$, $\bar{A}_1^4A_1B_2[\#126]$, $\bar{A}_1^2B_2G_2[\#532]$
$195^{\{\delta_2\}}$	$\bar{A}_1^2A_1B_2$		immediate overgroups in $\bar{A}_1^4A_1B_2[\#126]$, $\bar{A}_1^2B_2A_2[\#551]$ $(p = 3)$

$201^{\{\delta_3\}}$	$A_1A_1B_2$		immediate overgroups in $\bar{A}_1^4A_1B_2[\#126]$, $A_2A_1B_2[\#667]$ $(p=3)$
$203^{\{\underline{0}\}}$	$A_1^2B_2$		immediate overgroups in $\bar{A}_1^4A_1B_2[\#126]$, $B_2B_2[\#724]$
$207^{\{0\}}$	$\bar{A}_1^2A_1^2A_1$	≥ 5	$\bar{A}_1^4A_1A_1[\#189]$, $\bar{A}_1^2A_1\bar{A}_3[\#360]$, $B_3A_1A_1[\#635]$, $A_1^2A_1B_2[\#680]$
$208^{\{0\}}$	$\bar{A}_1^2A_1A_1^2$	≥ 5	$\bar{A}_1^4A_1A_1[\#189]$, $A_1^2A_1B_2[\#681]$
$209^{\{\underline{0}\}}$	$\bar{A}_1^3A_1A_1$	≥ 5	$\bar{A}_1^4A_1A_1[\#189]$, $\bar{A}_1^2A_1G_2[\#534]$
$211^{\{\underline{0}\}}$	$\bar{A}_1^4A_1$	≥ 5	$\bar{A}_1^4A_2[\#127]$, $\bar{A}_1^4A_1A_1[\#189]$
$214^{\{\underline{0}\}}$	$\bar{A}_1^2A_1A_1$	≥ 5	$\bar{A}_1^2A_1^2A_1[\#207^{\{\underline{0}\}}]$, $\bar{A}_1^3A_1A_1[\#210^{\{\underline{0}\}}]$, $\bar{A}_1^2A_1A_1^2[\#401]$
$227^{\{\underline{0}\}}$	$\bar{A}_1^2A_1A_1$	≥ 5	$\bar{A}_1^3A_1A_1[\#209^{\{\underline{0}\}}]$, $\bar{A}_1^3A_1A_1[\#210^{\{\underline{0}\}}]$, $\bar{A}_1^2A_1^2A_1[\#558]$
$233^{\{\underline{0}\}}$	$A_1A_1^2$	≥ 5	immediate overgroups in $\bar{A}_1^4A_1A_1[\#189]$, $A_1B_2[\#726]$
$262^{\{0\}}$	$\bar{A}_1^2A_1^4$	2	$\bar{A}_1^4A_1^3[\#190]$, $A_1^4B_2[\#638]$
$263^{\{\underline{0}\}}$	$\bar{A}_1^3A_1^2A_1$	2	$\bar{A}_1^4A_1^3[\#190]$, $\bar{A}_1^2A_1^2G_2[\#442]$
$276^{\{\delta_3\}}$	$\bar{A}_1^2A_1A_1A_1$	2	$\bar{A}_1^3A_1^2A_1[\#263^{\{\underline{0}\}}]$, $\bar{A}_1^3A_1A_1A_1[\#263^{\{\delta_1\}}]$, $\bar{A}_1^2A_1G_2[\#561]$
$346^{\{0\}}$	$\bar{A}_1^2A_1A_2$	$\neq 3$	$\bar{A}_1^4A_2[\#127]$, $B_3A_2[\#624]$, $A_2A_1B_2[\#667]$
$347^{\{\underline{0}\}}$	$\bar{A}_1A_1A_2$	$\neq 3$	$\bar{A}_1^2A_1A_2[\#346^{\{0\}}]$, $G_2A_2[\#627]$, $\bar{A}_1\bar{A}_2A_2A_1[\#887]$
350	$\bar{A}_1^2B_2\bar{A}_3$		$\bar{A}_1^2B_5[\#118]$ $(p\neq 2)$, $\bar{A}_1^2\bar{A}_3^2[\#119]$, $\bar{A}_1^2B_2B_3[\#121]$, $B_2D_5[\#716]$, $\bar{A}_3B_4[\#718]$ $(p\neq 2)$, $\bar{A}_3B_2^2[\#720]$
351	$\bar{A}_1^2A_1$	≥ 11	$\bar{A}_1^2B_5[\#118]$, $B_2A_1[\#836]$
355	$\bar{A}_1^2A_1^2\bar{A}_3$		$\bar{A}_1^2\bar{A}_3^2[\#119]$ $(p\neq 2)$, $\bar{A}_1^2A_1B_4[\#120]$ $(p\neq 2)$, $\bar{A}_1^2B_2\bar{A}_3[\#350]$ $(p=2)$, $\bar{A}_1^2A_1^2B_3[\#403]$ $(p=2)$, $A_1^2D_5[\#717]$, $\bar{A}_3A_1B_3[\#719]$ $(p\neq 2)$, $\bar{A}_3A_1^2B_2[\#759]$ $(p=2)$
$356^{\{0\}}$	$A_1\bar{A}_3^2$		$\bar{A}_1^2\bar{A}_3^2[\#119]$, $\bar{A}_3B_4[\#718]$ $(p\neq 2)$, $\bar{A}_3A_1B_3[\#719]$
360	$\bar{A}_1^2A_1\bar{A}_3$	≥ 5	$\bar{A}_1^2B_2\bar{A}_3[\#350]$, $\bar{A}_1^2A_1B_3[\#531]$, $A_1D_5[\#722]$, $\bar{A}_3A_1B_2[\#758]$
365	$\bar{A}_1^2A_1^4$	$\neq 2$	$\bar{A}_1^2A_1^2\bar{A}_3[\#355]$, $A_1^3B_3[\#733]$
$367^{\{\underline{0}\}}$	$\bar{A}_1A_1A_1\bar{A}_3$		$\bar{A}_1^2A_1^2\bar{A}_3[\#355]$, $\bar{A}_3A_1G_2[\#754]$
$369^{\{\delta_2\}}$	$A_1A_1\bar{A}_3$		immediate overgroups in $\bar{A}_1^2A_1^2\bar{A}_3[\#355]$, $\bar{A}_3A_1A_2[\#755]$ $(p=3)$
$372^{\{\underline{0}\}}$	$A_1^2\bar{A}_3$		$\bar{A}_1A_1A_1\bar{A}_3[\#367^{\{\underline{0}\}}]$, $\bar{A}_3B_2[\#721]$ $(p\neq 2)$
$375^{\{\underline{0}\}}$	$\bar{A}_1A_1A_1^3$	$\neq 2$	$\bar{A}_1^2A_1^4[\#365]$, $A_1^3G_2[\#736]$
$377^{\{\delta_2\}}$	$A_1A_1^3$	$\neq 2$	immediate overgroups in $\bar{A}_1^2A_1^4[\#365]$, $A_1^3A_2[\#739]$ $(p=3)$
$382^{\{\underline{0}\}}$	$A_1^2A_1^2$	$\neq 2$	$\bar{A}_1A_1A_1^3[\#375^{\{\underline{0}\}}]$, $A_1^2B_2[\#734]$
401	$\bar{A}_1^2A_1A_1^2$	$\neq 2$	$\bar{A}_1^2A_1B_4[\#120]$, $B_3A_1^2[\#845]$
402	$\bar{A}_1^2A_1A_1$	≥ 11	$\bar{A}_1^2A_1B_4[\#120]$, $B_3A_1[\#846]$
403	$\bar{A}_1^2A_1^2B_3$	2	$\bar{A}_1^2A_1B_4[\#120]$, $\bar{A}_1^2B_2B_3[\#121]$, $A_1^2B_2B_3[\#779]$
404	$\bar{A}_1^2A_1B_2^2$	2	$\bar{A}_1^2A_1B_4[\#120]$, $\bar{A}_1^2B_2B_3[\#121]$, $A_1B_2^3[\#826]$

$405^{\{0\}}$	$A_1A_1B_4$	$\neq 2$	$\bar{A}_1^2A_1B_4[\#120]$, $A_1^2D_5[\#717]$, $\bar{A}_3B_4[\#718]$
$406^{\{\underline{0}\}}$	$\bar{A}_1A_1B_4$		$\bar{A}_1^2A_1B_4[\#120]$, $G_2B_4[\#844]$, $\bar{A}_1A_1F_4[\#881]$
$407^{\{\delta_2\}}$	A_1B_4		$A_1A_1B_4[\#405^{\{1\}}]$, $\bar{A}_1A_1B_4[\#406^{\{\delta_1\}}]$, $A_2B_4[\#847]$ $(p=3)$
$408^{\{0\}}$	$A_1^2A_1^2$	$\neq 2$	$\bar{A}_1^2A_1A_1^2[\#401]$, $\bar{A}_3A_1^2[\#751]$
$409^{\{\underline{0}\}}$	$\bar{A}_1A_1A_1^2$	$\neq 2$	$\bar{A}_1^2A_1A_1^2[\#401]$, $G_2A_1^2[\#849]$
$410^{\{\underline{0}\}}$	$\bar{A}_1A_1A_1A_1$	$\neq 2$	$\bar{A}_1^2A_1A_1^2[\#401]$, $\bar{A}_1A_1G_2[\#882]$
$411^{\{\underline{0}\}}$	$\bar{A}_1^2A_1A_1$	$\neq 2$	$\bar{A}_1^2A_1A_1^2[\#401]$, $\bar{A}_1^2A_1^2A_1[\#558]$
$413^{\{\delta_2\}}$	$A_1A_1^2$	$\neq 2$	immediate overgroups in $\bar{A}_1^2A_1A_1^2[\#401]$, $A_2A_1^2[\#851]$ $(p=3)$
$422^{\{\delta_3\}}$	$\bar{A}_1A_1A_1$	$\neq 2$	immediate overgroups in $\bar{A}_1^2A_1A_1^2[\#401]$, $\bar{A}_1A_1A_2[\#965]$ $(p=3)$
$423^{\{\underline{0}\}}$	$A_1A_1^2$	$\neq 2$	$\bar{A}_1A_1A_1A_1[\#410^{\{\underline{0}\}}]$, $A_1B_2[\#773]$ $(p=5)$
$423^{\{\delta_3\}}$	$A_1A_1A_1$	$\neq 2$	$\bar{A}_1A_1A_1A_1[\#410^{\{\underline{0}\}}]$, $\bar{A}_1A_1A_1A_1[\#410^{\{\delta_1\}}]$, $A_1C_3[\#774]$ $(p=3)$
$428^{\{\underline{0}\}}$	A_1^2	$\neq 2$	$A_3[\#47]$, immediate overgroups in $\bar{A}_1^2A_1A_1^2[\#401]$
$432^{\{0\}}$	$A_1^2A_1$	≥ 11	$\bar{A}_1^2A_1A_1[\#402]$, $\bar{A}_3A_1[\#752]$
$433^{\{\underline{0}\}}$	$\bar{A}_1A_1A_1$	≥ 11	$\bar{A}_1^2A_1A_1[\#402]$, $G_2A_1[\#850]$
$435^{\{\underline{0}\}}$	$\bar{A}_1A_1$	≥ 11	$\bar{A}_1^2A_1A_1[\#402]$, $\bar{A}_1A_1A_1[\#966]$
$438^{\{\underline{0}\}}$	$\bar{A}_1A_1$	≥ 11	immediate overgroups in $\bar{A}_1^2A_1A_1[\#402]$, $\bar{A}_1A_1A_1[\#930]$
441	$\bar{A}_1^2A_1^3B_2$	2	$\bar{A}_1^2A_1^2B_3[\#403]$, $\bar{A}_1^2A_1B_2^2[\#404]$, $A_1^3B_2^2[\#782]$,
442	$\bar{A}_1^2A_1^2G_2$	2	$\bar{A}_1^2A_1^2B_3[\#403]$, $\bar{A}_1^2B_2G_2[\#532]$, $A_1^2B_2G_2[\#823]$
$444^{\{\underline{0}\}}$	$\bar{A}_1A_1A_1B_3$	2	$\bar{A}_1^2A_1^2B_3[\#403]$, $A_1G_2B_3[\#831]$
451	$\bar{A}_1^2A_1^5$	2	$\bar{A}_1^2A_1^3B_2[\#441]$, $A_1^5B_2[\#785]$
$453^{\{\underline{0}\}}$	$\bar{A}_1A_1^2A_1B_2$	2	$\bar{A}_1^2A_1^3B_2[\#441]$, $A_1^2B_2G_2[\#823]$
$468^{\{\underline{0}\}}$	$\bar{A}_1A_1A_1^4$	2	$\bar{A}_1^2A_1^5[\#451]$, $A_1^4G_2[\#786]$, $\bar{A}_1A_1D_4[\#958]$
$513^{\{\underline{0}\}}$	$\bar{A}_1G_2A_1A_1$	2	$\bar{A}_1^2A_1^2G_2[\#442]$, $A_1G_2^2[\#833]$, $\bar{A}_1G_2G_2[\#923]$
$521^{\{\underline{0}\}}$	$\bar{A}_1A_1B_2^2$	2	$\bar{A}_1^2A_1B_2^2[\#404]$, $B_2^2G_2[\#839]$, $\bar{A}_1A_1C_4[\#954]$
531	$\bar{A}_1^2A_1B_3$	≥ 5	$\bar{A}_1^2B_2B_3[\#121]$, $A_1B_5[\#835]$
532	$\bar{A}_1^2B_2G_2$		$\bar{A}_1^2B_2B_3[\#121]$, $B_2^2G_2[\#839]$ $(p=2)$, $G_2B_4[\#844]$ $(p\neq 2)$, $\bar{A}_1G_2C_3[\#880]$
$533^{\{0\}}$	$A_1B_2B_3$	$\neq 2$	$\bar{A}_1^2B_2B_3[\#121]$, $B_3\bar{D}_4[\#619]$, $A_1B_2\bar{D}_4[\#621]$, $B_2D_5[\#716]$, $\bar{A}_3A_1B_3[\#719]$
534	$\bar{A}_1^2A_1G_2$	≥ 5	$\bar{A}_1^2A_1B_3[\#531]$, $\bar{A}_1^2B_2G_2[\#532]$
$535^{\{0\}}$	$A_1A_1B_3$	≥ 5	$\bar{A}_1^2A_1B_3[\#531]$, $B_3A_1A_1[\#635]$, $A_1A_1\bar{D}_4[\#675]$, $A_1D_5[\#722]$
$537^{\{\underline{0}\}}$	A_1B_3	≥ 5	$A_1A_1B_3[\#535^{\{0\}}]$, $A_1A_1B_3[\#536^{\{\underline{0}\}}]$, $B_3A_2[\#624]$, $B_3A_1^2[\#845]$
538	$\bar{A}_1^2A_1A_1$	≥ 7	$\bar{A}_1^2A_1G_2[\#534]$, $\bar{A}_1^2B_2A_1[\#552]$
$539^{\{0\}}$	$A_1A_1G_2$	≥ 5	$\bar{A}_1^2A_1G_2[\#534]$, $B_3A_1A_1[\#635]$
$540^{\{\underline{0}\}}$	$\bar{A}_1A_1G_2$	≥ 5	$\bar{A}_1^2A_1G_2[\#534]$, $\bar{A}_1G_2A_1A_1[\#921]$

$541^{\{\underline{0}\}}$	A_1G_2	≥ 5	$A_1A_1G_2[\#539^{\{0\}}]$, $\bar{A}_1A_1G_2[\#540^{\{\underline{0}\}}]$, $G_2A_2[\#627]$, $G_2A_1^2[\#849]$
$542^{\{0\}}$	$A_1A_1A_1$	≥ 7	$\bar{A}_1^2A_1A_1[\#538]$, $A_1A_2[\#629]$
$543^{\{\underline{0}\}}$	$\bar{A}_1A_1A_1$	≥ 7	$\bar{A}_1^2A_1A_1[\#538]$, $\bar{A}_1A_1A_1A_1[\#929]$
$545^{\{\underline{0}\}}$	$\bar{A}_1^2A_1$	≥ 7	$\bar{A}_1^2A_1A_1[\#538]$, $\bar{A}_1^2A_1A_1[\#559]$
$546^{\{\underline{0}\}}$	A_1A_1	≥ 7	$A_1A_1A_1[\#542^{\{0\}}]$, $\bar{A}_1A_1A_1[\#543^{\{\underline{0}\}}]$, $A_1A_1^2[\#853]$
$547^{\{\underline{0}\}}$	A_1A_1	≥ 7	$A_1A_1A_1[\#542^{\{0\}}]$, $\bar{A}_1A_1A_1[\#544^{\{\underline{0}\}}]$, $A_1B_2[\#726]$
551	$\bar{A}_1^2B_2A_2$	3	$\bar{A}_1^2B_2G_2[\#532]$, $A_2B_4[\#847]$, $\bar{A}_1A_2C_3[\#918]$
552	$\bar{A}_1^2B_2A_1$	≥ 7	$\bar{A}_1^2B_2G_2[\#532]$, $A_1B_4[\#848]$, $\bar{A}_1A_1C_3[\#919]$
$553^{\{0\}}$	$A_1B_2G_2$	$\neq 2$	$\bar{A}_1^2B_2G_2[\#532]$, $G_2\bar{D}_4[\#622]$, $B_3A_1B_2[\#625]$, $\bar{A}_3A_1G_2[\#754]$
$554^{\{0\}}$	$A_1B_2A_2$	3	$\bar{A}_1^2B_2A_2[\#551]$, $\bar{A}_3A_1A_2[\#755]$
$555^{\{0\}}$	$A_1A_1B_2$	≥ 7	$\bar{A}_1^2B_2A_1[\#552]$, $\bar{A}_3A_1A_1[\#756]$
$557^{\{\underline{0}\}}$	A_1B_2	≥ 7	$A_1A_1B_2[\#555^{\{0\}}]$, $\bar{A}_1A_1B_2[\#556^{\{\underline{0}\}}]$, $B_2B_2[\#724]$
558	$\bar{A}_1^2A_1^2A_1$	$\neq 2$	$\bar{A}_1^2A_1C_3[\#122]$, $\bar{A}_1G_2A_1A_1[\#921]$
559	$\bar{A}_1^2A_1A_1$	≥ 7	$\bar{A}_1^2A_1C_3[\#122]$, $\bar{A}_1G_2A_1[\#921]$
560	$\bar{A}_1^2A_1A_3$	2	$\bar{A}_1^2A_1C_3[\#122]$, $\bar{A}_1G_2A_3[\#922]$
561	$\bar{A}_1^2A_1G_2$	2	$\bar{A}_1^2A_1C_3[\#122]$, $\bar{A}_1G_2G_2[\#923]$
$563^{\{\underline{0}\}}$	$\bar{A}_1A_1C_3$	$\neq 2$	$A_1C_4[\#116]$, $\bar{A}_1^2A_1C_3[\#122]$, $\bar{A}_1\bar{A}_2C_3[\#888]$, $\bar{A}_1A_1A_5[\#891]$
$563^{\{\delta_1\}}$	$\bar{A}_1A_1C_3$		$\bar{A}_1^2A_1C_3[\#122]$, $\bar{A}_1A_2C_3[\#918]$ $(p=3)$
$564^{\{\underline{0}\}}$	$\bar{A}_1A_1C_3$		$A_1C_4[\#115]$, $\bar{A}_1^2A_1C_3[\#122]$
$567^{\{\underline{0}\}}$	$\bar{A}_1A_1A_1A_1$	$\neq 2$	$\bar{A}_1^2A_1^2A_1[\#558]$, $\bar{A}_1A_1A_2A_1[\#892]$, $\bar{A}_1\bar{A}_2A_1A_1[\#893]$
$567^{\{\delta_1\}}$	$\bar{A}_1A_1A_1A_1$	$\neq 2$	$\bar{A}_1^2A_1^2A_1[\#558]$, $\bar{A}_1A_2A_1A_1[\#924]$ $(p=3)$
$569^{\{\underline{0}\}}$	$\bar{A}_1A_1A_1A_1$	$\neq 2$	$\bar{A}_1^2A_1^2A_1[\#558]$, $\bar{A}_1A_1G_2[\#955]$ $(p=7)$
$573^{\{\underline{0}\}}$	$A_1A_1^2$	$\neq 2$	$A_1A_1^2A_1[\#566^{\{0\}}]$, $\bar{A}_1A_1A_1A_1[\#568^{\{\underline{0}\}}]$, $A_1B_2[\#770]$
$580^{\{\underline{0}\}}$	$A_1A_1^2$	$\neq 2$	$\bar{A}_1A_1A_1A_1[\#567^{\{\underline{0}\}}]$, $A_1^4[\#873]$
$581^{\{\underline{0}\}}$	$A_1A_1A_1$	$\neq 2$	$\bar{A}_1A_1A_1A_1[\#567^{\{\underline{0}\}}]$, $\bar{A}_1A_1^2A_1[\#569^{\{\underline{0}\}}]$, $A_1G_2[\#1026]$
$583^{\{\underline{0}\}}$	$A_1A_1^2$	$\neq 2$	$\bar{A}_1A_1A_1A_1[\#567^{\{\underline{0}\}}]$, $A_1^4[\#865]$
$595^{\{\delta_2\}}$	A_1A_1	$\neq 2$	immediate overgroups in $\bar{A}_1^2A_1^2A_1[\#558]$, $A_1A_2[\#1027]$ $(p=3)$
$599^{\{\underline{0}\}}$	$\bar{A}_1A_1A_1$	≥ 7	$\bar{A}_1^2A_1A_1[\#559]$, $\bar{A}_1\bar{A}_2A_1[\#907]$
$600^{\{\underline{0}\}}$	$\bar{A}_1A_1A_1$	≥ 7	$\bar{A}_1^2A_1A_1[\#559]$, $\bar{A}_1A_1A_1A_1[\#929]$
$600^{\{\delta_1\}}$	$\bar{A}_1A_1A_1$	≥ 7	$\bar{A}_1^2A_1A_1[\#559]$, $\bar{A}_1A_1G_2[\#955]$ $(p=7)$
$609^{\{\underline{0}\}}$	A_1A_1	≥ 7	$\bar{A}_1A_1A_1[\#600^{\{\underline{0}\}}]$, $A_1A_2[\#1027]$
619	$B_3\bar{D}_4$		$B_7[\#44]$ $(p\neq 2)$, $\bar{D}_4^2[\#108]$, $B_3B_4[\#112]$
620	$A_2\bar{D}_4$		$\bar{D}_4^2[\#108]$ $(p\neq 3)$, $G_2\bar{D}_4[\#622]$ $(p=3)$, $A_2B_4[\#847]$ $(p=3)$

621	$A_1B_2\bar{D}_4$		$\bar{D}_4^2$[#108] ($p \neq 2$), A_1B_6[#110] ($p \neq 2$), B_2B_5[#111] ($p \neq 2$), $B_3\bar{D}_4$[#619] ($p = 2$), $A_1B_2B_4$[#776] ($p = 2$)
622	G_2D_4		$B_3\bar{D}_4$[#619], G_2B_4[#844]
623	B_3^2		$B_3\bar{D}_4$[#619]
624	B_3A_2	$\neq 3$	$B_3\bar{D}_4$[#619], $A_2\bar{D}_4$[#620]
625	$B_3A_1B_2$		$B_3\bar{D}_4$[#619] ($p \neq 2$), $A_1B_2\bar{D}_4$[#621], B_3^2[#623] ($p = 2$)
626	$A_1\bar{D}_4$	≥ 7	$G_2\bar{D}_4$[#622], A_1B_4[#848]
627	G_2A_2	$\neq 3$	$G_2\bar{D}_4$[#622], B_3A_2[#624]
628	$A_1A_1B_2$	≥ 7	$A_1\bar{D}_4$[#626]
629	A_1A_2	≥ 7	$A_1\bar{D}_4$[#626], G_2A_2[#627]
630	$A_1A_1A_1$	≥ 7	$A_1A_1B_2$[#628]
635	$B_3A_1A_1$	≥ 5	$B_3A_1B_2$[#625], $A_1A_1\bar{D}_4$[#675]
636	$B_3A_1^3$	2	$B_3A_1B_2$[#625], $A_1^3\bar{D}_4$[#674]
638	$A_1^4B_2$	2	$B_3A_1^3$[#636], $A_1^2B_2^2$[#676]
641	A_1^6	2	$A_1^4B_2$[#638]
$649^{\{\underline{0}\}}$	$A_1A_1^4$	2	A_1^6[#641], A_1D_4[#867]
667	$A_2A_1B_2$		$A_2\bar{D}_4$[#620], $A_1B_2\bar{D}_4$[#621]
668	A_2^2	$\neq 3$	$A_2\bar{D}_4$[#620], A_8[#62]
669	$A_2A_1A_1$	≥ 5	$A_2A_1B_2$[#667], $A_1A_1\bar{D}_4$[#675],
670	$A_2A_1^3$	2	$A_2A_1B_2$[#667], $A_1^3\bar{D}_4$[#674]
$672^{\{\underline{0}\}}$	A_1A_2	≥ 5	A_2^2[#668], $A_2A_1A_1$[#669]
674	$A_1^3\bar{D}_4$	2	$A_1B_2\bar{D}_4$[#621], $A_1^3B_4$[#778]
675	$A_1A_1\bar{D}_4$	≥ 5	$A_1B_2\bar{D}_4$[#621], A_1B_5[#835]
676	$A_1^2B_2^2$		$A_1B_2\bar{D}_4$[#621] ($p \neq 2$), $B_3A_1B_2$[#625] ($p = 2$)
677	$A_1^2B_2^2$	$\neq 2$	$A_1B_2\bar{D}_4$[#621], $A_1^2D_5$[#717], $\bar{A}_3B_2^2$[#720]
680	$A_1^2A_1B_2$	≥ 5	$A_1A_1\bar{D}_4$[#675], $A_1^2B_2^2$[#677], $\bar{A}_3A_1B_2$[#758]
681	$A_1^2A_1B_2$	≥ 5	$A_1A_1\bar{D}_4$[#675], $A_1^2B_2^2$[#676]
$682^{\{\underline{0}\}}$	$A_1\bar{D}_4$	≥ 5	$A_2\bar{D}_4$[#620], $A_1A_1\bar{D}_4$[#675]
683	$A_1^2A_1^2$	≥ 5	$A_1^2A_1B_2$[#680], $\bar{A}_3A_1^2$[#762]
$685^{\{\underline{0}\}}$	$A_1A_1B_2$	≥ 5	$A_2A_1B_2$[#667], $A_1^2A_1B_2$[#680], $A_1^2A_1B_2$[#681]
$688^{\{\underline{0}\}}$	$A_1A_1A_1$	≥ 5	$A_2A_1A_1$[#669], $A_1^2A_1^2$[#683], $A_1^2A_1^2$[#694]
$693^{\{\underline{0}\}}$	A_1^2	≥ 5	B_4[#46], $A_1A_1A_1$[#$688^{\{\underline{0}\}}$]
694	$A_1^2A_1^2$	≥ 5	$A_1^2A_1B_2$[#681]
704$^{\{\delta_1\}}$	A_1A_1	≥ 5	immediate overgroups in $A_1^2A_1^2$[#694] , A_1B_2[#773] ($p = 5$)
$707^{\{\underline{0}\}}$	A_1^2	≥ 5	B_4[#45], B_2[#101], $A_1A_1A_1$[#$701^{\{\underline{0}\}}$]
$708^{\{0\}}$	$A_1B_2^2$		A_1C_4[#115], $A_1^2B_2^2$[#676]
$709^{\{0\}}$	$A_1^2B_2$		B_4[#45], $A_1^2B_2^2$[#676], A_1C_3[#774] ($p = 3$)
716	B_2D_5		B_7[#44] ($p \neq 2$), $\bar{A}_3D_5$[#109], B_2B_5[#111]
717	$A_1^2D_5$		$\bar{A}_3D_5$[#109] ($p \neq 2$), A_1B_6[#110] ($p \neq 2$), B_2D_5[#716] ($p = 2$), $A_1^2B_5$[#775] ($p = 2$)

718	$\bar{A}_3B_4$		$B_7[\#44]$ $(p \neq 2)$, $\bar{A}_3D_5[\#109]$, $B_3B_4[\#112]$
719	$\bar{A}_3A_1B_3$		$\bar{A}_3D_5[\#109]$ $(p \neq 2)$, $A_1B_6[\#110]$ $(p \neq 2)$, $B_3B_4[\#112]$ $(p \neq 2)$, $\bar{A}_3B_4[\#718]$ $(p = 2)$, $A_1B_3^2[\#777]$ $(p = 2)$
720	$\bar{A}_3B_2^2$		$\bar{A}_3D_5[\#109]$ $(p \neq 2)$, $B_2B_5[\#111]$ $(p \neq 2)$, $\bar{A}_3B_4[\#718]$ $(p = 2)$, $B_2^2B_3[\#837]$ $(p = 2)$
721	$\bar{A}_3B_2$	$\neq 2$	$\bar{A}_3D_5[\#109]$
722	A_1D_5	≥ 5	$B_2D_5[\#716]$, $A_1B_5[\#835]$
723	B_2^3	$\neq 2$	$B_2D_5[\#716]$, $\bar{A}_3B_2^2[\#720]$
724	B_2B_2	$\neq 2$	$B_2D_5[\#716]$, $\bar{A}_3B_2[\#721]$
725	$A_1B_2^2$	≥ 5	$A_1D_5[\#722]$, $B_2^3[\#723]$, $\bar{A}_3A_1B_2[\#758]$
726	A_1B_2	≥ 5	$A_1D_5[\#722]$, $B_2B_2[\#724]$
727	$A_1^2B_2$	≥ 5	$A_1B_2^2[\#725]$, $\bar{A}_3A_1^2[\#762]$
729	A_1^3	≥ 5	$A_1^2B_2[\#727]$
733	$A_1^3B_3$	$\neq 2$	$A_1^2D_5[\#717]$, $\bar{A}_3A_1B_3[\#719]$
734	$A_1^2B_2$	$\neq 2$	$A_1^2D_5[\#717]$, $\bar{A}_3B_2[\#721]$
736	$A_1^3G_2$	$\neq 2$	$A_1^3B_3[\#733]$, $\bar{A}_3A_1G_2[\#754]$
739	$A_1^3A_2$	3	$A_1^3G_2[\#736]$, $\bar{A}_3A_1A_2[\#755]$
740	$A_1^3A_1$	≥ 7	$A_1^3G_2[\#736]$, $\bar{A}_3A_1A_1[\#756]$
$746^{\{\underline{0}\}}$	$A_1A_1^2$	≥ 7	$A_1^2B_2[\#734]$, $A_1^3A_1[\#740]$
751	$\bar{A}_3A_1^2$	$\neq 2$	$\bar{A}_3B_4[\#718]$, $B_3A_1^2[\#845]$
752	$\bar{A}_3A_1$	≥ 11	$\bar{A}_3B_4[\#718]$, $B_3A_1[\#846]$
754	$\bar{A}_3A_1G_2$		$\bar{A}_3A_1B_3[\#719]$, $A_1G_2B_3[\#831]$ $(p = 2)$, $G_2B_4[\#844]$ $(p \neq 2)$
755	$\bar{A}_3A_1A_2$	3	$\bar{A}_3A_1G_2[\#754]$, $A_2B_4[\#847]$
756	$\bar{A}_3A_1A_1$	≥ 7	$\bar{A}_3A_1G_2[\#754]$, $A_1B_4[\#848]$
$757^{\{\underline{0}\}}$	$\bar{A}_3A_1$	≥ 7	$\bar{A}_3B_2[\#721]$, $\bar{A}_3A_1A_1[\#756]$
758	$\bar{A}_3A_1B_2$	≥ 5	$\bar{A}_3B_2^2[\#720]$, $A_1B_5[\#835]$
759	$\bar{A}_3A_1^2B_2$	2	$\bar{A}_3B_2^2[\#720]$, $A_1^2B_2B_3[\#779]$
762	$\bar{A}_3A_1^2$	≥ 5	$\bar{A}_3A_1B_2[\#758]$
764	$\bar{A}_3A_1^4$	2	$\bar{A}_3A_1^2B_2[\#759]$, $A_1^4B_3[\#781]$
770	A_1B_2	$\neq 2$	$B_7[\#44]$
771	A_1A_1	≥ 5	$A_1B_2[\#770]$
772	A_1A_1	≥ 13	$A_1B_6[\#110]$
773	A_1B_2	5	$A_1B_6[\#110]$
774	A_1C_3	3	$A_1B_6[\#110]$, $F_4[\#1049]$ $(p = 3)$
775	$A_1^2B_5$	2	$A_1B_6[\#110]$, $B_2B_5[\#111]$
776	$A_1B_2B_4$	2	$A_1B_6[\#110]$, $B_2B_5[\#111]$, $B_3B_4[\#112]$
777	$A_1B_3^2$	2	$A_1B_6[\#110]$, $B_3B_4[\#112]$
778	$A_1^3B_4$	2	$A_1^2B_5[\#775]$, $A_1B_2B_4[\#776]$
779	$A_1^2B_2B_3$	2	$A_1^2B_5[\#775]$, $A_1B_2B_4[\#776]$, $A_1B_3^2[\#777]$, $B_2^2B_3[\#837]$
781	$A_1^4B_3$	2	$A_1^3B_4[\#778]$, $A_1^2B_2B_3[\#779]$

782	$A_1^3B_2^2$	2	$A_1^3B_4$[#778], $A_1^2B_2B_3$[#779], $A_1B_2^3$[#826]
785	$A_1^5B_2$	2	$A_1^4B_3$[#781], $A_1^3B_2^2$[#782]
786	$A_1^4G_2$	2	$A_1^4B_3$[#781], $A_1^2B_2G_2$[#823], G_2D_4[#1046]
791	A_1^7	2	$A_1^5B_2$[#785]
823	$A_1^2B_2G_2$	2	$A_1^2B_2B_3$[#779], $A_1G_2B_3$[#831], $B_2^2G_2$[#839]
826	$A_1B_2^3$	2	$A_1B_2B_4$[#776], $B_2^2B_3$[#837]
831	$A_1G_2B_3$	2	$A_1B_3^2$[#777], G_2B_4[#844]
833	$A_1G_2^2$	2	$A_1G_2B_3$[#831]
835	A_1B_5	≥ 5	B_2B_5[#111]
836	B_2A_1	≥ 11	B_2B_5[#111]
837	$B_2^2B_3$	2	B_2B_5[#111], B_3B_4[#112]
838	A_1A_1	≥ 11	A_1B_5[#835], B_2A_1[#836]
839	$B_2^2G_2$	2	$B_2^2B_3$[#837], G_2B_4[#844], G_2C_4[#1032]
844	G_2B_4		G_2F_4[#105], B_3B_4[#112]
845	$B_3A_1^2$	$\neq 2$	B_3B_4[#112]
846	B_3A_1	≥ 11	B_3B_4[#112]
847	A_2B_4	3	G_2B_4[#844], A_2F_4[#1030]
848	A_1B_4	≥ 7	G_2B_4[#844], A_1F_4[#1031]
849	$G_2A_1^2$	$\neq 2$	G_2B_4[#844], $B_3A_1^2$[#845]
850	G_2A_1	≥ 11	G_2B_4[#844], B_3A_1[#846]
851	$A_2A_1^2$	3	A_2B_4[#847], $G_2A_1^2$[#849]
853	$A_1A_1^2$	≥ 7	A_1B_4[#848], $G_2A_1^2$[#849]
854	A_1A_1	≥ 11	A_1B_4[#848], G_2A_1[#850]
859	$A_1^2B_2$	2	B_2^2[#113], A_1B_3[#871]
860	A_1B_2	≥ 5	B_2^2[#113]
862	A_1^2	≥ 5	A_1B_2[#860]
863	A_1B_2	≥ 5	B_2^2[#114], $A_1\bar{A}_4$[#1029]
864	A_1^2	≥ 5	A_1B_2[#863]
865	A_1^4		A_1C_4[#115] ($p \neq 2$), $A_1^2B_2$[#859] ($p = 2$)
866	A_1A_1	≥ 11	A_1C_4[#115]
867	A_1D_4	2	A_1C_4[#115]
871	A_1B_3	2	A_1D_4[#867]
872	A_1A_2	2	A_1D_4[#867]
873	A_1^4	$\neq 2$	A_1C_4[#116], $A_1^3\bar{A}_2$[#995]
874	A_1A_1	≥ 11	A_1C_4[#116], $\bar{A}_2A_1$[#1009], A_1G_2[#1026]
878	$\bar{A}_1\bar{A}_2A_5$		$\bar{A}_1E_7$[#102], $\bar{A}_2E_6$[#103]
879	$\bar{A}_1A_7$		$\bar{A}_1E_7$[#102]
880	$\bar{A}_1G_2C_3$		$\bar{A}_1E_7$[#102], G_2F_4[#105] ($p \neq 2$), G_2C_4[#1032] ($p = 2$)
881	$\bar{A}_1A_1F_4$		$\bar{A}_1E_7$[#102], G_2F_4[#105]
882	$\bar{A}_1A_1G_2$	$\neq 2$	$\bar{A}_1E_7$[#102]
883	$\bar{A}_1A_1A_1$	≥ 5	$\bar{A}_1E_7$[#102]

884	$\bar{A}_1A_2$	≥ 5	$\bar{A}_1E_7[\#102]$
885	$\bar{A}_1A_1$	≥ 17	$\bar{A}_1E_7[\#102]$
886	$\bar{A}_1A_1$	≥ 19	$\bar{A}_1E_7[\#102]$
887	$\bar{A}_1\bar{A}_2A_1A_2$		$\bar{A}_1\bar{A}_2A_5[\#878]$, $\bar{A}_1A_1F_4[\#881]$, $\bar{A}_2A_2G_2[\#978]$
888	$\bar{A}_1\bar{A}_2C_3$		$\bar{A}_1\bar{A}_2A_5[\#878]$, $\bar{A}_1G_2C_3[\#880]$, $\bar{A}_2F_4[\#976]$ $(p \neq 2)$, $\bar{A}_2C_4[\#977]$
889	$\bar{A}_1\bar{A}_2A_3$		$\bar{A}_1\bar{A}_2A_5[\#878]$ $(p \neq 2)$, $\bar{A}_1\bar{A}_2C_3[\#888]$ $(p = 2)$, $\bar{A}_1G_2A_3[\#922]$ $(p = 2)$
890	$\bar{A}_1\bar{A}_2A_2$	$\neq 2$	$\bar{A}_1\bar{A}_2A_5[\#878]$
891	$\bar{A}_1A_1A_5$	$\neq 2$	$\bar{A}_1\bar{A}_2A_5[\#878]$, $A_1E_6[\#981]$
892	$\bar{A}_1A_1A_1A_2$	$\neq 2$	$\bar{A}_1\bar{A}_2A_2A_1[\#887]$, $\bar{A}_1A_1A_5[\#891]$, $A_1A_2G_2[\#1011]$
893	$\bar{A}_1\bar{A}_2A_1A_1$	$\neq 2$	$\bar{A}_1\bar{A}_2A_2A_1[\#887]$, $\bar{A}_1\bar{A}_2C_3[\#888]$, $\bar{A}_1G_2A_1A_1[\#921]$, $\bar{A}_2A_1G_2[\#1002]$
$894^{\{\delta_1\}}$	$A_1\bar{A}_2A_2$		$\bar{A}_1\bar{A}_2A_2A_1[\#887]$, $\bar{A}_2A_2A_2[\#1012]$ $(p = 3)$
$895^{\{\underline{0}\}}$	$\bar{A}_1A_2A_1$		$\bar{A}_1\bar{A}_2A_2A_1[\#887]$, $\bar{A}_1A_1G_2[\#955]$ $(p = 7)$
$895^{\{\delta_1\}}$	$\bar{A}_1A_2A_1$		$\bar{A}_1\bar{A}_2A_2A_1[\#887]$, $\bar{A}_1A_1D_4[\#958]$ $(p = 2)$
$900^{\{\delta_1\}}$	$A_1A_1A_2$	$\neq 2$	$\bar{A}_1A_1A_2A_1[\#892]$, $A_1A_2A_2[\#1016]$ $(p = 3)$
$901^{\{\underline{0}\}}$	$\bar{A}_1A_1A_2$	$\neq 2$	$\bar{A}_1A_1G_2[\#882]$, $\bar{A}_1A_1A_2A_1[\#892]$
$903^{\{\underline{0}\}}$	$A_1A_1\bar{A}_2$	$\neq 2$	$\bar{A}_1\bar{A}_2A_1A_1[\#893]$, $A_1^3\bar{A}_2[\#995]$
$903^{\{\delta_1\}}$	$A_1A_1\bar{A}_2$	$\neq 2$	$\bar{A}_1\bar{A}_2A_1A_1[\#893]$, $\bar{A}_2A_1A_2[\#1004]$ $(p = 3)$
$904^{\{\underline{0}\}}$	$A_1A_1\bar{A}_2$	$\neq 2$	$\bar{A}_1\bar{A}_2A_1A_1[\#893]$, $\bar{A}_2G_2[\#979]$
$906^{\{\delta_2\}}$	A_1A_2	$\neq 2$	$\bar{A}_1\bar{A}_2A_1A_1[\#893]$, $\bar{A}_2A_2[\#980]$ $(p = 3)$
907	$\bar{A}_1\bar{A}_2A_1$	≥ 7	$\bar{A}_1\bar{A}_2C_3[\#888]$, $\bar{A}_1G_2A_1[\#921]$
908	$\bar{A}_1\bar{A}_2G_2$	2	$\bar{A}_1\bar{A}_2C_3[\#888]$, $\bar{A}_1G_2G_2[\#923]$
$909^{\{\underline{0}\}}$	$A_1\bar{A}_2$	≥ 7	$\bar{A}_1\bar{A}_2A_1[\#907]$, $\bar{A}_2A_2[\#980]$, $\bar{A}_2A_1A_1[\#1005]$
$909^{\{\delta_1\}}$	$A_1\bar{A}_2$	≥ 7	$\bar{A}_1\bar{A}_2A_1[\#907]$, $\bar{A}_2G_2[\#979]$ $(p = 7)$
910	$\bar{A}_1A_1A_3$	$\neq 2$	$\bar{A}_1A_7[\#879]$, $\bar{A}_1\bar{A}_2A_3[\#889]$, $\bar{A}_1A_1A_5[\#891]$
912	$\bar{A}_1A_1A_2$	$\neq 2$	$\bar{A}_1\bar{A}_2A_2[\#890]$, $\bar{A}_1A_1A_5[\#891]$
913	$\bar{A}_1A_2$	≥ 5	$\bar{A}_1\bar{A}_2A_2[\#890]$, $\bar{A}_1D_4[\#917]$
917	$\bar{A}_1D_4$	$\neq 2$	$\bar{A}_1A_7[\#879]$
918	$\bar{A}_1A_2C_3$	3	$\bar{A}_1G_2C_3[\#880]$, $A_2F_4[\#1030]$
919	$\bar{A}_1A_1C_3$	≥ 7	$\bar{A}_1G_2C_3[\#880]$, $A_1F_4[\#1031]$
921	$\bar{A}_1G_2A_1A_1$	$\neq 2$	$\bar{A}_1G_2C_3[\#880]$, $\bar{A}_1A_1F_4[\#881]$, $G_2^2A_1[\#1033]$
921	$\bar{A}_1G_2A_1$	≥ 7	$\bar{A}_1G_2C_3[\#880]$
922	$\bar{A}_1G_2A_3$	2	$\bar{A}_1G_2C_3[\#880]$
923	$\bar{A}_1G_2G_2$	2	$\bar{A}_1G_2C_3[\#880]$
924	$\bar{A}_1A_2A_1A_1$	3	$\bar{A}_1A_2C_3[\#918]$, $\bar{A}_1G_2A_1A_1[\#921]$, $A_2A_1G_2[\#1036]$
$926^{\{\underline{0}\}}$	$A_1A_2A_1$	3	$\bar{A}_1A_2A_1A_1[\#924]$, $\bar{A}_2A_1A_2[\#1004]$, $A_1A_2A_2[\#1016]$
$926^{\{\delta_1\}}$	$A_1A_2A_1$	3	$\bar{A}_1A_2A_1A_1[\#924]$, $A_2^2A_1[\#1037]$
929	$\bar{A}_1A_1A_1A_1$	≥ 7	$\bar{A}_1A_1C_3[\#919]$, $\bar{A}_1G_2A_1A_1[\#921]$, $A_1A_1G_2[\#1039]$
930	$\bar{A}_1A_1A_1$	≥ 7	$\bar{A}_1A_1C_3[\#919]$, $\bar{A}_1G_2A_1[\#921]$
$933^{\{\underline{0}\}}$	$A_1A_1A_1$	≥ 7	$\bar{A}_1A_1A_1A_1[\#929]$, $\bar{A}_2A_1A_1[\#1005]$, $A_1A_2A_1[\#1017]$, $A_1G_2[\#1041]$ $(p = 7)$

$946^{\{\underline{0}\}}$	A_1A_1	≥ 7	$\bar{A}_1A_1A_1[\#930]$, $A_1^2A_1[\#1042]$
$946^{\{\delta_1\}}$	A_1A_1	≥ 7	$\bar{A}_1A_1A_1[\#930]$, $A_1G_2[\#1041]$ $(p=7)$
$948^{\{\underline{0}\}}$	$A_1A_1G_2$	$\neq 2$	$\bar{A}_1G_2A_1A_1[\#921]$, $G_2G_2[\#1035]$ $(p=7)$
$949^{\{\underline{0}\}}$	$A_1A_1G_2$	$\neq 2$	$\bar{A}_1G_2A_1A_1[\#921]$, $\bar{A}_2A_1G_2[\#1002]$, $A_1A_2G_2[\#1011]$
$949^{\{\delta_1\}}$	$A_1A_1G_2$	$\neq 2$	$\bar{A}_1G_2A_1A_1[\#921]$, $A_2A_1G_2[\#1036]$ $(p=3)$
$952^{\{\underline{0}\}}$	A_1G_2	≥ 7	$\bar{A}_1G_2A_1A_1[\#921]$, $A_1A_1G_2[\#1039]$
$952^{\{\delta_1\}}$	A_1G_2	≥ 7	$\bar{A}_1G_2A_1[\#921]$, $G_2G_2[\#1035]$ $(p=7)$
954	$\bar{A}_1A_1C_4$	2	$\bar{A}_1A_1F_4[\#881]$, $G_2C_4[\#1032]$
955	$\bar{A}_1A_1G_2$	7	$\bar{A}_1A_1F_4[\#881]$, $G_2G_2[\#1035]$
956	$\bar{A}_1A_1A_1$	≥ 13	$\bar{A}_1A_1F_4[\#881]$, $G_2A_1[\#1034]$
$957^{\{\underline{0}\}}$	A_1F_4	$\neq 2$	$\bar{A}_1A_1F_4[\#881]$, $\bar{A}_2F_4[\#976]$, $A_1E_6[\#981]$,
$957^{\{\delta_1\}}$	A_1F_4	$\neq 2$	$\bar{A}_1A_1F_4[\#881]$, $A_2F_4[\#1030]$ $(p=3)$
958	$\bar{A}_1A_1D_4$	2	$\bar{A}_1A_1C_4[\#954]$, $G_2D_4[\#1046]$
$962^{\{\underline{0}\}}$	A_1A_1	≥ 13	$\bar{A}_1A_1A_1[\#956]$, $\bar{A}_2A_1[\#1003]$
965	$\bar{A}_1A_1A_2$	3	$\bar{A}_1A_1G_2[\#882]$
966	$\bar{A}_1A_1A_1$	≥ 7	$\bar{A}_1A_1G_2[\#882]$
$967^{\{\underline{0}\}}$	A_1G_2	$\neq 2$	$\bar{A}_1A_1G_2[\#882]$, $G_2^2A_1[\#1033]$
$967^{\{\delta_1\}}$	A_1G_2	$\neq 2$	$\bar{A}_1A_1G_2[\#882]$, $F_4[\#1049]$ $(p=3)$
$969^{\{\underline{0}\}}$	A_1A_1	≥ 7	$\bar{A}_1A_1A_1[\#966]$, $A_1^2A_1[\#1042]$
$973^{\{\underline{0}\}}$	A_1A_1	≥ 5	$A_1A_2[\#106]$, $\bar{A}_1A_1A_1[\#883]$
975	$\bar{A}_2^4$		$\bar{A}_2E_6[\#103]$
976	$\bar{A}_2F_4$		$\bar{A}_2E_6[\#103]$, $G_2F_4[\#105]$
977	$\bar{A}_2C_4$		$\bar{A}_2E_6[\#103]$ $(p\neq 2)$, $\bar{A}_2F_4[\#976]$ $(p=2)$, $G_2C_4[\#1032]$ $(p=2)$
978	$\bar{A}_2A_2G_2$		$\bar{A}_2E_6[\#103]$, $G_2F_4[\#105]$
979	$\bar{A}_2G_2$		$\bar{A}_2E_6[\#103]$ $(p\neq 7)$, $\bar{A}_2F_4[\#976]$ $(p=7)$, $G_2G_2[\#1035]$ $(p=7)$
980	$\bar{A}_2A_2$	$\neq 2$	$\bar{A}_2E_6[\#103]$ $(p\geq 5)$, $\bar{A}_2G_2[\#979]$ $(p=3)$
981	A_1E_6	$\neq 2$	$\bar{A}_2E_6[\#103]$
982	$A_1\bar{A}_2^3$	$\neq 2$	$\bar{A}_2^4[\#975]$, $A_1E_6[\#981]$
$983^{\{0\}}$	$\bar{A}_2^2A_2$		$\bar{A}_2^4[\#975]$, $\bar{A}_2F_4[\#976]$, $\bar{A}_2A_2G_2[\#978]$
$985^{\{\underline{0}\}}$	$\bar{A}_2A_2$		$\bar{A}_2^4[\#975]$, $\bar{A}_2G_2[\#979]$
$985^{\{\delta_2\}}$	$\bar{A}_2A_2$		$\bar{A}_2^4[\#975]$, $\bar{A}_2D_4[\#1010]$ $(p=2)$
989	$A_1^2\bar{A}_2^2$	$\neq 2$	$A_1\bar{A}_2^3[\#982]$
$991^{\{0\}}$	$A_1A_2\bar{A}_2$	$\neq 2$	$A_1\bar{A}_2^3[\#982]$, $A_1A_2G_2[\#1011]$
$993^{\{\underline{0}\}}$	A_1A_2	$\neq 2$	$A_1A_2\bar{A}_2[\#990^{\{0\}}]$, $A_1A_2\bar{A}_2[\#991^{\{0\}}]$, $A_1G_2[\#1026]$
$994^{\{\underline{0}\}}$	A_1A_2	$\neq 2$	$A_1A_2\bar{A}_2[\#990^{\{0\}}]$, $A_1A_2\bar{A}_2[\#991^{\{0\}}]$, $A_1G_2[\#1041]$ $(p=7)$
995	$A_1^3\bar{A}_2$	$\neq 2$	$\bar{A}_2C_4[\#977]$, $A_1^2\bar{A}_2^2[\#989]$
$996^{\{0\}}$	$A_1\bar{A}_2^2$	$\neq 2$	$A_1^2\bar{A}_2^2[\#989]$, $\bar{A}_2A_1G_2[\#1002]$
1002	$\bar{A}_2A_1G_2$	$\neq 2$	$\bar{A}_2F_4[\#976]$, $\bar{A}_2A_2G_2[\#978]$, $G_2^2A_1[\#1033]$
1003	$\bar{A}_2A_1$	≥ 13	$\bar{A}_2F_4[\#976]$, $G_2A_1[\#1034]$

1004	$\bar{A}_2A_1A_2$	3	$\bar{A}_2A_1G_2$[#1002], $\bar{A}_2A_2A_2$[#1012], $A_2A_1G_2$[#1036]
1005	$\bar{A}_2A_1A_1$	≥ 7	$\bar{A}_2A_1G_2$[#1002], $\bar{A}_2A_2A_1$[#1013], $A_1A_1G_2$[#1039]
1009	$\bar{A}_2A_1$	≥ 11	$\bar{A}_2C_4$[#977], $\bar{A}_2G_2$[#979]
1010	$\bar{A}_2D_4$	2	$\bar{A}_2C_4$[#977], G_2D_4[#1046]
1011	$A_1A_2G_2$	$\neq 2$	$\bar{A}_2A_2G_2$[#978], A_1E_6[#981]
1012	$\bar{A}_2A_2A_2$	3	$\bar{A}_2A_2G_2$[#978], A_2F_4[#1030]
1013	$\bar{A}_2A_2A_1$	≥ 7	$\bar{A}_2A_2G_2$[#978], A_1F_4[#1031]
$1014^{\{0\}}$	A_2G_2	$\neq 3$	$\bar{A}_2A_2G_2$[#978], G_2G_2[#1035] $(p=7)$
$1014^{\{\delta_1\}}$	A_2G_2		$\bar{A}_2A_2G_2$[#978], G_2D_4[#1046] $(p=2)$
1016	$A_1A_2A_2$	3	$A_1A_2G_2$[#1011], $\bar{A}_2A_2A_2$[#1012]
1017	$A_1A_2A_1$	≥ 7	$A_1A_2G_2$[#1011], $\bar{A}_2A_2A_1$[#1013]
1026	A_1G_2	$\neq 2$	$\bar{A}_2G_2$[#979], A_1E_6[#981] $(p \neq 7)$
1027	A_1A_2	$\neq 2$	$\bar{A}_2A_2$[#980], A_1E_6[#981] $(p \geq 5)$, A_1G_2[#1026] $(p=3)$
1028	$B_2\bar{A}_4$	$\neq 2$	$\bar{A}_4^2$[#104]
1029	$A_1\bar{A}_4$	≥ 5	$B_2\bar{A}_4$[#1028]
1030	A_2F_4	3	G_2F_4[#105]
1031	A_1F_4	≥ 7	G_2F_4[#105]
1032	G_2C_4	2	G_2F_4[#105]
1033	$G_2^2A_1$	$\neq 2$	G_2F_4[#105]
1034	G_2A_1	≥ 13	G_2F_4[#105]
1035	G_2G_2	7	G_2F_4[#105]
1036	$A_2A_1G_2$	3	A_2F_4[#1030], $G_2^2A_1$[#1033]
1037	$A_2^2A_1$	3	$A_2A_1G_2$[#1036]
1039	$A_1A_1G_2$	≥ 7	A_1F_4[#1031], $G_2^2A_1$[#1033]
1040	A_1A_1	≥ 13	A_1F_4[#1031], G_2A_1[#1034]
1041	A_1G_2	7	A_1F_4[#1031], G_2G_2[#1035]
1042	$A_1^2A_1$	≥ 7	$A_1A_1G_2$[#1039]
1046	G_2D_4	2	G_2C_4[#1032]
1049	F_4	3	E_8[#0]

11.1. Irreducible diagonal subgroups

In this section we give the tables of diagonal subgroups referred to in Tables 11.4 and 11.5. The first column gives the ID number, as in the previous tables, and the second column gives the embedding of the diagonal subgroups. To describe the embeddings we use a slightly modified notation, to shorten the tables. Specifically, we introduce a shorthand for diagonal subgroups of A_1^nZ, where Z has no simple factor of type A_1. For example, instead of writing $A_1^3B_2 \hookrightarrow A_1^4B_2$ via $(1_a^{[r]}, 1_a^{[s]}, 1_b, 1_c, 10)$ we just write $(a^{[r]}, a^{[s]}, b, c, 10)$; from any such vector it is easy to recover the isomorphism type of the diagonal subgroup. Similarly, in Tables 11.37, 11.46 and 11.47 the usual notation for diagonal subgroups is used but we again omit the isomorphism type of each diagonal subgroup as they too can be easily recovered from the listed embedding.

There are further tables which give the extra restrictions on the field twists in certain diagonal embeddings. These restrictions ensure there is no repetition of conjugacy classes and further, that each conjugacy class is G-irreducible. The restrictions are given in rows of the tables: the first column lists all permitted equalities amongst certain subsets of the field twists; the second column lists any further requirements. So an ordered set $\{0, r, \dots\}$ is permitted if it satisfies the conditions in the first and second column of a row of the table. We note that a set of field twists satisfies the conditions of at most one row. We emphasise that an ordered set may be excluded either because it yields a G-reducible subgroup, or because it yields a repeated diagonal subgroup.

We give an example to illustrate this. Let $X = A_1A_1 = E_8(\#165)$, so that X is a diagonal subgroup of $E_8(\#124) = \bar{A}_1^8 < D_8$ via $(1_a^{[r]}, 1_a^{[s]}, 1_a^{[t]}, 1_a^{[u]}, 1_b^{[v]}, 1_b^{[w]}, 1_b^{[x]}, 1_b^{[y]})$ with $rstu = vwxy = 0$. Table 11.24 gives the extra conditions that an ordered set $r, s, \dots, y$ needs to satisfy. The conditions in the first column only restrict the equalities allowed among elements from r, s, t, u and separately the equalities allowed among elements from v, w, x, y. So in every row of the table r is permitted to be equal to v, for example.

In particular, the ordered set $0, 1, 2, 3, 0, 3, 1, 2$ satisfies the conditions of the first row, as does $0, 1, 2, 3, 0, 1, 2, 3$. We also note that the ordered set $0, 0, 1, 2, 0, 1, 2, 3$ satisfies the conditions of the second row, whereas $2, 1, 0, 0, 0, 1, 2, 3$ and $0, 1, 2, 3, 0, 0, 1, 2$ do not satisfy the conditions of any of the rows. This is because these three ordered sets of field twists yield the same conjugacy class of G-irreducible subgroups and thus only one of them can be permitted.

11.1.1. Irreducible diagonal subgroups contained in E_7.

Table 11.6: Irreducible diagonal subgroups of $\bar{A}_1^7 = E_7(\#43)$.

ID	Embedding
49	$(a, a^{[r]}, b, c, d, e, f)$
50	$(a, a^{[r]}, a^{[s]}, b, c, d, e)$ $(r \leq s)$
51	$(a, a^{[r]}, b, a^{[s]}, c, d, e)$ $(r \leq s)$
52	$(a, a^{[r]}, b^{[s]}, b^{[t]}, c, d, e)$ $(st = 0)$
53	$(a^{[r]}, a^{[s]}, b, c, c^{[t]}, d, e)$ $(rs = 0)$
54	$(a, a^{[r]}, b, c, d, e, c^{[s]})$ $(r \leq s; s \neq 0)$
55	$(a^{[r]}, a^{[s]}, a^{[t]}, a^{[u]}, b, c, d)$ $(ru = 0; r \leq s \leq t)$
56	$(a, a^{[r]}, b, a^{[s]}, c, d, a^{[t]})$ ($r \leq s \leq t$; if two of $0, r, s, t$ are equal then the other two are not equal)
57	$(a^{[r]}, a^{[s]}, a^{[t]}, b, b^{[u]}, c, d)$ ($rs = 0; s \leq t$; if $u = 0$ then $s < t$)
58	$(a^{[r]}, a^{[s]}, b, a^{[t]}, b^{[u]}, c, d)$ ($rst = 0$; if $u = 0$ then $s < t$)
59	$(a^{[r]}, a^{[s]}, b^{[u]}, a^{[t]}, c, d, b^{[v]})$ $(rt = uv = 0; r \leq s)$
60	$(a, a^{[r]}, b^{[s]}, b^{[t]}, c^{[u]}, c^{[v]}, d)$ ($st = uv = 0$; if $r = 0$ then $u \neq v$)
61	$(a^{[r]}, a^{[s]}, b^{[t]}, b^{[u]}, c^{[v]}, d, c^{[w]})$ $(rs = tu = vw = 0)$
62	$(a^{[r]}, a^{[s]}, b, c, c^{[t]}, d^{[u]}, d^{[v]})$ ($rs = uv = 0$; $t \leq u + v$; if $t = 0$ then $u \neq v$)
63	$(a, a^{[r]}, b, c^{[s]}, d^{[u]}, d^{[v]}, c^{[t]})$ ($st = uv = 0$; $s + t \leq u + v$; if $r = 0$ then $s \neq t$ and $u \neq v$; if $s = t$ then $u \neq v$)

64	$(a^{[r]}, a^{[s]}, a^{[t]}, a^{[u]}, a^{[v]}, b, c)$ ($rs = 0$; $s \leq \min\{t, u\}$; $u \leq v$; if two of s, t, u, v are equal then the other two are not equal)
65	$(a^{[r]}, a^{[s]}, a^{[t]}, a^{[u]}, b, b^{[v]}, c)$ ($rstu = 0$; if $v = 0$ then $r \neq s$)
66	$(a, a^{[r]}, b, a^{[s]}, b^{[u]}, c, a^{[t]})$ (if $u = 0$ then $t \neq 0$ and $r \neq s$; if $t = 0$ then $r < s$; if $r = 0$ then $s < t$; if $s = 0$ then $r < t$)
67	$(a, a^{[r]}, a^{[s]}, b^{[t]}, b^{[u]}, b^{[v]}, c)$ ($tuv = 0$; $r \leq s$; if $r = 0$ then $u \neq v$; if $s = 0$ then $t \neq v$; if $r = s$ then $t \neq u$)
68	$(a^{[r]}, a^{[s]}, b^{[u]}, a^{[t]}, b^{[v]}, c, b^{[w]})$ ($rs = uvw = 0$; $s \leq t$; if $s = t$ then $u \neq v$)
69	$(a^{[r]}, a^{[s]}, a^{[t]}, b, b^{[u]}, c, c^{[v]})$ ($rst = 0; u \leq v$; if $u = 0$ then $v \neq 0$ and $s < t$; if $v = 0$ then $s < t$)
70	$(a^{[r]}, a^{[s]}, b, a^{[t]}, b^{[u]}, c^{[v]}, c^{[w]})$ ($rst = vw = 0$; if $u = 0$ then $s < t$)
71	$(a, a^{[r]}, a^{[s]}, a^{[t]}, a^{[u]}, a^{[v]}, b)$ (see Table 11.7)
72	$(a^{[r]}, a^{[s]}, a^{[t]}, a^{[u]}, a^{[v]}, b, b^{[w]})$ ($rs = 0$; if $w \neq 0$ then either $s < \min\{t, u, v\}$ or $s > \max\{t, u, v\}$; if $w = 0$ then $s < \min\{t, u\}$ and $u < v$)
73	$(a^{[r]}, a^{[s]}, a^{[t]}, a^{[u]}, b, b^{[v]}, b^{[w]})$ ($rstu = 0$ and see Table 11.8)
74	$(a, a^{[r]}, b^{[u]}, a^{[s]}, b^{[v]}, b^{[w]}, a^{[t]})$ ($uvw = 0$ and see Table 11.9)
12	$(a, a^{[r]}, a^{[s]}, a^{[t]}, a^{[u]}, a^{[v]}, a^{[w]})$ (see Table 11.10)

Table 11.7: Conditions on field twists for $E_7(\#71)$.

Equalities among $0, r, \ldots, v$	Further requirements on $0, r, \ldots, v$
none	$r < \min\{s, t, u\}$
$r = s$	$t < u$
$r = t$	$s < u$
$r = u$	$s < t$
$r = v$	none
$r = s = t$	none
$r = s = v$	$t < u$
$r = t = v$	$s < u$
$r = u = v$	$s < t$
$r = s = t = v$	none
$r = 0$	$u < v$
$r = 0$; $s = t$	$u < v$
$r = 0$; $s = u$	none
$r = 0$; $t = u$	none
$r = 0$; $s = t = u$	none
$v = 0$	$r < \min\{s, u\}$; $s < t$
$v = 0$; $r = s$	$t < u$
$r = s = 0$	$t < u < v$
$r = t = 0$	$s < u < v$
$r = u = 0$	$s < t$
$r = u = 0$; $s = v$	none

$r = s = t = 0$	$u < v$

Table 11.8: Conditions on field twists for $E_7(\#73)$.

Equalities among $0, v, w$	Further requirements on $0, r, \ldots, w$
none	$v < w$
$v = 0$	$r < s$
$v = w = 0$	$0 < r < s$

Table 11.9: Conditions on field twists for $E_7(\#74)$.

Equalities among $0, r, s, t$	Further requirements on $0, r, \ldots, w$
none	$0 < r < s < t$
$r = 0$	$s < t$; $v < w$
$r = s = 0$	$v < w$

Table 11.10: Conditions on field twists for $E_7(\#12)$ and $E_8(\#162)$.

Equalities among $0, r, \ldots, w$	Further requirements on $0, r, \ldots, w$
none	$r < \min\{s, t\}$; $t < \min\{u, v, w\}$
$r = s$	$t < \min\{u, v\}$; $v < w$
$r = t$	$s < v$
$r = s = t$	none
$r = t = w$	$s < u, v$
$r = s = t = v$	$u < w$
$r = t$; $u = v = w$	none
$r = 0$	$t < \min\{u, v\}$; $v < w$
$r = 0$; $s = t$	$u < v$
$r = 0$; $t = u$	$v < w$
$r = 0$; $s = t = u$	$v < w$
$r = 0$; $s = t$; $u = w$	none
$r = s = 0$	$t < u < v < w$
$r = t = 0$	$s < u < v$
$r = t = 0$; $s = w$	$u < v$
$r = s = t = 0$	$u < v < w$

Table 11.11: Diagonal irreducible subgroups of $\bar{A}_1^3 A_1 B_2 = E_7(\#45)$.

ID	Embedding
89	$(a^{[r]}, a^{[s]}, b, c, 10)$ $(rs = 0)$
90	$(a^{[r]}, b, c, a^{[s]}, 10)$ $(rs = 0)$
91	$(a, b, b^{[r]}, c, 10)$ (if $p = 2$ then $r \neq 0$)
92	$(a, b^{[r]}, c, b^{[s]}, 10)$ $(rs = 0)$
93	$(a^{[r]}, a^{[s]}, a^{[t]}, b, 10)$ $(rs = 0; s \leq t$; if $p = 2$ then $s < t)$
94	$(a, b^{[r]}, b^{[s]}, b^{[t]}, 10)$ $(rt = 0; r \leq s$; if $p = 2$ then $r < s$; if $r = s$ then $r < t)$
95	$(a^{[r]}, b, a^{[s]}, a^{[t]}, 10)$ $(rst = 0)$
96	$(a^{[r]}, a^{[s]}, b^{[t]}, b^{[u]}, 10)$ $(rs = tu = 0)$
97	$(a^{[r]}, b, b^{[t]}, a^{[s]}, 10)$ $(rs = 0$; if $p = 2$ then $t \neq 0)$
98	$(a^{[r]}, a^{[s]}, a^{[t]}, a^{[u]}, 10)$ $(rsu = 0; s \leq t$; if $p = 2$ then $s < t$; if $s = t$ then $s < u)$

Table 11.12: Diagonal irreducible subgroups of $\bar{A}_1^3 A_1 A_1 = E_7(\#87)$ $(p \geq 5)$.

ID	Embedding
99	$(a^{[r]}, a^{[s]}, b, c, d)$ $(rs = 0)$
100	$(a^{[r]}, b, c, a^{[s]}, d)$ $(rs = 0)$
101	$(a^{[r]}, b, c, d, a^{[s]})$ $(rs = 0)$
102	$(a, b, b^{[r]}, c, d)$
103	$(a, b^{[r]}, c, b^{[s]}, d)$ $(rs = 0)$
104	$(a, b^{[r]}, c, d, b^{[s]})$ $(rs = 0)$
105	$(a, b, c, d^{[r]}, d^{[s]})$ $(rs = 0)$
106	$(a^{[r]}, a^{[s]}, a^{[t]}, b, c)$ $(rst = 0; s \leq t)$
107	$(a^{[r]}, a^{[s]}, b, a^{[t]}, c)$ $(rst = 0)$
108	$(a^{[r]}, a^{[s]}, b, c, a^{[t]})$ $(rst = 0)$
109	$(a^{[r]}, b, c, a^{[s]}, a^{[t]})$ $(rst = 0)$
110	$(a, b^{[r]}, b^{[s]}, b^{[t]}, c)$ $(rt = 0; r \leq s$; if $r = s$ then $r < t)$
111	$(a, b^{[r]}, b^{[s]}, c, b^{[t]})$ $(rt = 0; r \leq s)$
112	$(a, b^{[r]}, c, b^{[s]}, b^{[t]})$ $(rst = 0; s \neq t)$
113	$(a^{[r]}, a^{[s]}, b^{[t]}, b^{[u]}, c)$ $(rs = tu = 0)$
114	$(a^{[r]}, a^{[s]}, b^{[t]}, c, b^{[u]})$ $(rs = tu = 0)$
115	$(a^{[r]}, a^{[s]}, b, c^{[t]}, c^{[u]})$ $(rs = tu = 0$; if $t = u$ then $r \leq s)$
116	$(a^{[r]}, b, b^{[t]}, a^{[s]}, c)$ $(rs = 0)$
117	$(a^{[r]}, b^{[t]}, c, a^{[s]}, b^{[u]})$ $(rs = tu = 0)$
118	$(a^{[r]}, b, b^{[t]}, c, a^{[s]})$ $(rs = 0)$
119	$(a^{[r]}, b^{[t]}, c, b^{[u]}, a^{[s]})$ $(rs = tu = 0)$
120	$(a, b, b^{[r]}, c^{[s]}, c^{[t]})$ $(st = 0; s \neq t)$
121	$(a, b^{[r]}, c^{[t]}, b^{[s]}, c^{[u]})$ $(rs = tu = 0)$
122	$(a^{[r]}, a^{[s]}, a^{[t]}, a^{[u]}, b)$ $(rsu = 0; s \leq t$; if $s = t$ then $s < u)$

123	$(a^{[r]}, a^{[s]}, a^{[t]}, b, a^{[u]})$ $(rsu = 0; s \leq t)$
124	$(a^{[r]}, b, a^{[s]}, a^{[t]}, a^{[u]})$ $(rstu = 0;$ if $t = u$ then $r < s$ or $r = s < t)$
125	$(a, b^{[r]}, b^{[s]}, b^{[t]}, b^{[u]})$ $(rtu = 0;\ r \leq s;\ t \neq u)$
126	$(a^{[r]}, a^{[s]}, b^{[t]}, b^{[u]}, b^{[v]})$ $(rs = tuv = 0;$ if $u = v$ then $r = 0)$
127	$(a^{[r]}, b^{[t]}, b^{[u]}, a^{[s]}, b^{[v]})$ $(rs = tv = 0; t \leq u)$
128	$(a^{[r]}, b^{[t]}, b^{[u]}, b^{[v]}, a^{[s]})$ $(rs = tv = 0; t \leq u;$ if $t = u$ then $t < v)$
129	$(a^{[r]}, b, b^{[u]}, a^{[s]}, a^{[t]})$ $(rst = 0; s \neq t)$
130	$(a^{[r]}, a^{[s]}, b^{[u]}, b^{[v]}, a^{[t]})$ $(rst = uv = 0)$
131	$(a^{[r]}, a^{[s]}, b^{[u]}, a^{[t]}, b^{[v]})$ $(rst = uv = 0)$
132	$(a^{[r]}, a^{[s]}, a^{[t]}, b^{[u]}, b^{[v]})$ $(rs = uv = 0;\ s \leq t;$ if $u = v$ then $r \leq s)$
8	$(a^{[r]}, a^{[s]}, a^{[t]}, a^{[u]}, a^{[v]})$ $(rsuv = 0; u \neq v; s \leq t;$ if $s = t$ then $s < u)$
9	$(a^{[r]}, a^{[s]}, a^{[t]}, a^{[u]}, a^{[u]})$ $(rsu = 0;$ either $r < t < u$ or $s = t$ and $s \neq u)$

Table 11.13: Diagonal irreducible subgroups of $\bar{A}_1^3 A_1^3 = E_7(\#88)$ $(p = 2)$.

ID	Embedding
133	$(a^{[r]}, a^{[s]}, b, c, d, e)$ $(rs = 0)$
134	$(a^{[r]}, b, c, a^{[s]}, d, e)$ $(rs = 0)$
135	$(a, b, b^{[r]}, c, d, e)$ $(r \neq 0)$
136	$(a, b^{[r]}, c, b^{[s]}, d, e)$ $(rs = 0)$
137	$(a, b, c, d, d^{[r]}, e)$ $(r \neq 0)$
138	$(a^{[r]}, a^{[s]}, a^{[t]}, b, c, d)$ $(rs = 0; s < t)$
139	$(a^{[r]}, a^{[s]}, b, a^{[t]}, c, d)$ $(rst = 0)$
140	$(a^{[r]}, b, c, a^{[s]}, a^{[t]}, d)$ $(rs = 0; s < t)$
141	$(a, b^{[r]}, b^{[s]}, b^{[t]}, c, d)$ $(rt = 0; r < s)$
142	$(a, b, c^{[r]}, c^{[s]}, c^{[t]}, d)$ $(rs = 0; s < t)$
143	$(a, b, c, d, d^{[r]}, d^{[s]})$ $(0 < r < s)$
144	$(a^{[r]}, a^{[s]}, b^{[t]}, b^{[u]}, c, d)$ $(rs = tu = 0)$
145	$(a^{[r]}, a^{[s]}, b, c, c^{[t]}, d)$ $(rs = 0 \neq t)$
146	$(a^{[r]}, b, b^{[t]}, a^{[s]}, c, d)$ $(rs = 0 \neq t)$
147	$(a^{[r]}, b^{[t]}, c, a^{[s]}, b^{[u]}, d)$ $(rs = tu = 0)$
148	$(a^{[r]}, b, c, a^{[s]}, d, d^{[t]})$ $(rs = 0 \neq t)$
149	$(a, b, b^{[r]}, c, c^{[s]}, d)$ $(rs \neq 0)$
150	$(a, b^{[r]}, c^{[t]}, b^{[s]}, c^{[u]}, d)$ $(rs = tu = 0;\ r + s \leq t + u)$
151	$(a, b, c^{[r]}, c^{[s]}, d, d^{[t]})$ $(rs = 0 \neq t)$
152	$(a^{[r]}, a^{[s]}, a^{[t]}, a^{[u]}, b, c)$ $(rsu = 0; s < t)$
153	$(a^{[r]}, a^{[s]}, b, a^{[t]}, a^{[u]}, c)$ $(rst = 0; t < u)$
154	$(a^{[r]}, b, c, a^{[s]}, a^{[t]}, a^{[u]})$ $(rs = 0; s < t < u)$
155	$(a, b^{[r]}, b^{[s]}, b^{[t]}, b^{[u]}, c)$ $(rt = 0; r < s; t < u)$
156	$(a, b, c^{[r]}, c^{[s]}, c^{[t]}, c^{[u]})$ $(rs = 0; s < t < u)$
157	$(a^{[r]}, a^{[s]}, a^{[t]}, b, b^{[u]}, c)$ $(rs = 0 \neq u; s < t)$

158	$(a^{[r]}, a^{[s]}, b^{[u]}, a^{[t]}, b^{[v]}, c)$ $(rst = uv = 0)$
159	$(a^{[r]}, a^{[s]}, b, a^{[t]}, c, c^{[u]})$ $(rst = 0 \neq u)$
160	$(a^{[r]}, b, b^{[u]}, a^{[s]}, a^{[t]}, c)$ $(rs = 0 \neq u; s < t)$
161	$(a^{[r]}, b^{[u]}, c, a^{[s]}, a^{[t]}, b^{[v]})$ $(rs = uv = 0; s < t)$
162	$(a^{[r]}, b^{[t]}, b^{[u]}, b^{[v]}, a^{[s]}, c)$ $(rs = tv = 0; t < u)$
163	$(a, b^{[r]}, b^{[s]}, b^{[t]}, c, c^{[u]})$ $(rt = 0 \neq u; r < s)$
164	$(a^{[r]}, a^{[s]}, b^{[t]}, b^{[u]}, b^{[v]}, c)$ $(rs = tu = 0; u < v)$
165	$(a^{[r]}, b, c^{[t]}, c^{[u]}, c^{[v]}, a^{[s]})$ $(rs = tu = 0, u < v)$
166	$(a, b^{[r]}, c^{[t]}, c^{[u]}, c^{[v]}, b^{[s]})$ $(rs = tu = 0; u < v)$
167	$(a^{[r]}, a^{[s]}, b, c, c^{[t]}, c^{[u]})$ $(rs = 0; 0 < t < u)$
168	$(a, b, b^{[r]}, c, c^{[s]}, c^{[t]})$ $(r \neq 0; 0 < s < t)$
169	$(a^{[r]}, a^{[s]}, b^{[t]}, b^{[u]}, c, c^{[v]})$ $(rs = tu = 0 \neq v)$
170	$(a^{[r]}, b, b^{[t]}, a^{[s]}, c, c^{[u]})$ $(rs = 0 \neq tu)$
171	$(a^{[r]}, b^{[t]}, c^{[v]}, a^{[s]}, b^{[u]}, c^{[w]})$ $(rs = tu = vw = 0;\ t + u \leq v + w)$
172	$(a^{[r]}, a^{[s]}, a^{[t]}, a^{[u]}, a^{[v]}, b)$ $(rsu = 0; s < t; u < v)$
173	$(a^{[r]}, a^{[s]}, b, a^{[t]}, a^{[u]}, a^{[v]})$ $(rst = 0; t < u < v)$
174	$(a, b^{[r]}, b^{[s]}, b^{[t]}, b^{[u]}, b^{[v]})$ $(rt = 0; r < s; t < u < v)$
175	$(a^{[r]}, a^{[s]}, a^{[t]}, a^{[u]}, b, b^{[v]})$ $(rsu = 0 \neq v; s < t)$
176	$(a^{[r]}, a^{[s]}, b^{[v]}, a^{[t]}, a^{[u]}, b^{[w]})$ $(rst = vw = 0; t < u)$
177	$(a^{[r]}, b, b^{[v]}, a^{[s]}, a^{[t]}, a^{[u]})$ $(rs = 0 \neq v; s < t < u)$
178	$(a^{[r]}, b^{[t]}, b^{[u]}, b^{[v]}, b^{[w]}, a^{[s]})$ $(rs = tv = 0;\ t < u;\ v < w)$
179	$(a^{[r]}, a^{[s]}, b^{[t]}, b^{[u]}, b^{[v]}, b^{[w]})$ $(rs = tu = 0;\ u < v < w)$
180	$(a^{[r]}, a^{[s]}, a^{[t]}, b, b^{[u]}, b^{[v]})$ $(rs = 0; s < t; 0 < u < v)$
181	$(a^{[r]}, a^{[s]}, b^{[u]}, a^{[t]}, b^{[v]}, b^{[w]})$ $(rst = uv = 0; v < w)$
182	$(a^{[r]}, b^{[u]}, b^{[v]}, a^{[s]}, a^{[t]}, b^{[w]})$ $(rs = 0; s < t; u < v)$
13	$(a^{[r]}, a^{[s]}, a^{[t]}, a^{[u]}, a^{[v]}, a^{[w]})$ $(rsu = 0; s < t; u < v < w)$

Table 11.14: Diagonal irreducible subgroups of $\bar{A}_1 A_1^4 = E_7(\#191)$ $(p \neq 2)$.

ID	Embedding
195	$(a^{[r]}, a^{[s]}, b, c, d)$ $(rs = 0)$
196	$(a, b, b^{[r]}, c, d)$ $(r \neq 0)$
197	$(a^{[r]}, a^{[s]}, a^{[t]}, b, c)$ $(rs = 0; s < t)$
198	$(a, b, b^{[r]}, b^{[s]}, c)$ $(0 < r < s)$
199	$(a^{[r]}, a^{[s]}, b, b^{[t]}, c)$ $(rs = 0 \neq t)$
200	$(a, b, b^{[r]}, c, c^{[s]})$ $(0 < r \leq s)$
201	$(a^{[r]}, a^{[s]}, a^{[t]}, a^{[u]}, b)$ $(rs = 0; s < t < u)$
202	$(a, b, b^{[r]}, b^{[s]}, b^{[t]})$ $(0 < r < s < t)$
203	$(a^{[r]}, a^{[s]}, a^{[t]}, b, b^{[u]})$ $(rs = 0 \neq u; s < t)$
204	$(a^{[r]}, a^{[s]}, b, b^{[t]}, b^{[u]})$ $(rs = 0; 0 < t < u)$
11	$(a^{[r]}, a^{[s]}, a^{[t]}, a^{[u]}, a^{[v]})$ $(rs = 0; s < t < u < v)$

Table 11.15: Diagonal irreducible subgroups of $\bar{A}_1A_1A_1^2 = E_7(\#186)$ $(p \neq 2)$.

ID	Embedding	ID	Embedding
205	$(a^{[r]}, a^{[s]}, b, c)$ $(rs = 0)$	210	$(a^{[r]}, b, a^{[s]}, a^{[t]})$ $(rst = 0; s < t)$
206	$(a^{[r]}, b, a^{[s]}, c)$ $(rs = 0)$	211	$(a, b^{[r]}, b^{[s]}, b^{[t]})$ $(rs = 0; s < t)$
207	$(a, b^{[r]}, b^{[s]}, c)$ $(rs = 0)$	212	$(a^{[r]}, a^{[s]}, b, b^{[t]})$ $(rs = 0 \neq t)$
208	$(a, b, c, c^{[r]})$ $(r \neq 0)$	213	$(a^{[r]}, b^{[t]}, a^{[s]}, b^{[u]})$ $(rs = tu = 0)$
115$^{\{\underline{0}\}}$	(a, b, c, c) $(p \geq 5)$	10	$(a^{[r]}, a^{[s]}, a^{[t]}, a^{[u]})$ $(rst = 0; t < u)$
209	$(a^{[r]}, a^{[s]}, a^{[t]}, b)$ $(rst = 0)$		

Table 11.16: Diagonal irreducible subgroups of $\bar{A}_1A_1^5 = E_7(\#223)$ $(p = 2)$.

ID	Embedding
230	$(a^{[r]}, a^{[s]}, b, c, d, e)$ $(rs = 0)$
238	$(a^{[r]}, a^{[s]}, a^{[t]}, b, b^{[u]}, c)$ $(rs = 0 \neq u; s < t)$
231	$(a, b, b^{[r]}, c, d, e)$ $(r \neq 0)$
239	$(a^{[t]}, b, b^{[r]}, b^{[s]}, a^{[u]}, c)$ $(0 < r < s; tu = 0)$
232	$(a^{[r]}, a^{[s]}, a^{[t]}, b, c, d)$ $(rs = 0; s < t)$
240	$(a, b, b^{[r]}, b^{[s]}, c, c^{[t]})$ $(0 < r < s; t \neq 0)$
233	$(a, b, b^{[r]}, b^{[s]}, c, d)$ $(0 < r < s)$
241	$(a^{[r]}, a^{[s]}, b, b^{[t]}, c, c^{[u]})$ $(rs = 0; 0 < t \leq u)$
234	$(a^{[r]}, a^{[s]}, b, b^{[t]}, c, d)$ $(rs = 0 \neq t)$
242	$(a^{[r]}, a^{[s]}, a^{[t]}, a^{[u]}, a^{[v]}, b)$ $(rs = 0; s < t < u < v)$
235	$(a, b, b^{[r]}, c, c^{[s]}, d)$ $(0 < r \leq s)$
243	$(a, b, b^{[r]}, b^{[s]}, b^{[t]}, b^{[u]})$ $(0 < r < s < t < u)$
236	$(a^{[r]}, a^{[s]}, a^{[t]}, a^{[u]}, b, c)$ $(rs = 0; s < t < u)$
237	$(a, b, b^{[r]}, b^{[s]}, b^{[t]}, c)$ $(0 < r < s < t)$
14	$(a^{[r]}, a^{[s]}, a^{[t]}, a^{[u]}, a^{[v]}, a^{[w]})$ $(rs = 0;\ s < t < u < v < w)$

Table 11.17: Diagonal irreducible subgroups of $\bar{A}_1A_1A_1A_1 = E_7(\#268)$ $(p \neq 2)$.

ID	Embedding
273	$(a^{[r]}, a^{[s]}, b, c)$ $(rs = 0)$
274	$(a^{[r]}, b, a^{[s]}, c)$ $(rs = 0)$
275	$(a^{[r]}, b, c, a^{[s]})$ $(rs = 0)$
276	$(a, b^{[r]}, b^{[s]}, c)$ $(rs = 0; r \neq s)$
119$^{\{\underline{0}\}}$	(a, b, b, c) $(p \geq 5)$
277	$(a, b^{[r]}, c, b^{[s]})$ $(rs = 0; r \neq s)$
207$^{\{\underline{0}\}}$	(a, b, c, b)
278	$(a, b, c^{[r]}, c^{[s]})$ $(rs = 0; r \neq s)$

$\mathit{117}^{\{\underline{0}\}}$	(a,b,c,c) $(p \geq 5)$
279	$(a^{[r]},a^{[s]},a^{[t]},b)$ $(rst=0; s \neq t)$
280	$(a^{[r]},a^{[s]},b,a^{[t]})$ $(rst=0; s \neq t)$
281	$(a^{[r]},b,a^{[s]},a^{[t]})$ $(rst=0; s \neq t)$
282	$(a,b^{[r]},b^{[s]},b^{[t]})$ $(rst=0;$ r,s,t distinct$)$
283	$(a^{[r]},a^{[s]},b^{[t]},b^{[u]})$ $(rs=tu=0; t \neq u)$
284	$(a^{[r]},b^{[t]},a^{[s]},b^{[u]})$ $(rs=tu=0; t \neq u)$
285	$(a^{[r]},b^{[t]},b^{[u]},a^{[s]})$ $(rs=tu=0; t \neq u)$
3	$(a^{[r]},a^{[s]},a^{[t]},a^{[u]})$ $(rstu=0; s,t,u$ distinct$)$

11.1.2. Irreducible diagonal subgroups contained in E_8.

Table 11.18: Irreducible diagonal subgroups of $\bar{A}_1^8 = E_8(\#124)$.

ID	Embedding
132	$(a,a^{[r]},b,c,d,e,f,g)$
133	$(a,a^{[r]},a^{[s]},b,c,d,e,f)$ $(0 \leq r \leq s)$
134	$(a,a^{[r]},b,b^{[s]},c,d,e,f)$ $(r \leq s;\ s \neq 0)$
135	$(a,a^{[r]},b,c,b^{[s]},d,e,f)$ $(r \leq s)$
136	$(a,a^{[r]},a^{[s]},a^{[t]},b,c,d,e)$ $(0 \leq r \leq s \leq t;$ if two of $0,r,s,t$ are equal then the other two are not equal$)$
137	$(a,a^{[r]},a^{[s]},b,a^{[t]},c,d,e)$ $(0 \leq r \leq s \leq t)$
138	$(a,a^{[r]},a^{[s]},b^{[t]},b^{[u]},c,d,e)$ $(0 \leq r \leq s; tu=0)$
139	$(a^{[r]},a^{[s]},a^{[t]},b,c,c^{[u]},d,e)$ $(rt=0; r \leq s;$ if $u=0$ then $r<s)$
140	$(a,a^{[r]},b,b^{[s]},c,c^{[t]},d,e)$ $(r \leq s \leq t;$ if $r=0$ then $s \neq 0)$
141	$(a,a^{[r]},b,c^{[t]},b^{[s]},c^{[u]},d,e)$ $(tu=0;\ s \leq t+u;$ if $s=0$ then $t<u)$
142	$(a^{[r]},a^{[s]},b^{[t]},c,b^{[u]},d,c^{[v]},e)$ $(rs=tu=0;$ if $v=0$ then either $r<t$ or $r=t$ and $s \leq u)$
143	$(a,a^{[r]},b,c,b^{[s]},d,e^{[t]},e^{[u]})$ $(tu=0;$ if $r=0$ then $t<u;$ if $t=0$ then $r \leq u;$ if $u=0$ then $r \leq t)$
144	$(a^{[r]},a^{[s]},a^{[t]},a^{[u]},a^{[v]},b,c,d)$ $(rv=0; r \leq s \leq t \leq u;$ if two of r,s,t,u are equal then the other two are not equal$)$
145	$(a,a^{[r]},a^{[s]},a^{[t]},b,b^{[u]},c,d)$ $(s \leq t;$ if $r=0$ then $s<t$ and $u \neq 0;$ if $s=t$ then $u \neq 0;$ $\{0,r\} \neq \{s,t\})$
146	$(a^{[r]},a^{[s]},a^{[t]},b,a^{[u]},b^{[v]},c,d)$ $(rtu=0; r \leq s;$ if $v=0$ then $t<u)$
147	$(a^{[r]},a^{[s]},a^{[t]},b^{[u]},b^{[v]},b^{[w]},c,d)$ $(rt=uv=0; r \leq s; v \leq w;$ if $r=s$ then $v<w;$ $r \leq v;$ if $r=v$ then either $s<w$ or $s=w$ and $t \leq u)$
148	$(a,a^{[r]},a^{[s]},b,c^{[t]},c^{[u]},c^{[v]},d)$ $(0 \leq r \leq s;\ tuv=0;$ if $r=0$ then $t<u;$ if $r=s$ then $u<v;$ if $i=0$ then either $r<\min\{j,k\}$ or $r=\min\{j,k\}$ and $s \leq \max\{j,k\}$ for distinct $i,j,k \in \{t,u,v\})$
149	$(a^{[r]},a^{[s]},a^{[t]},b^{[u]},b^{[v]},c,c^{[w]},d)$ $(rst=uv=0;$ if $w=0$ then $s<t)$
150	$(a^{[r]},a^{[s]},a^{[t]},b,c,c^{[u]},d,d^{[v]})$ $(rst=0;\ u \leq v;$ if $u=0$ then $v \neq 0$ and $r<s;$ if $v=0$ then $r<s)$
151	$(a,a^{[r]},b,b^{[s]},c,c^{[t]},d^{[u]},d^{[v]})$ $(uv=0;\ r \leq s \leq t;$ if $v=0$ then $t \leq u;$ if $r=0$ then $s \neq 0)$

152 $(a, a^{[r]}, b, c^{[t]}, b^{[s]}, c^{[u]}, d^{[v]}, d^{[w]})$ ($tu = vw = 0$; $r \le v + w$; $s \le t + u$; if $r = 0$ then $v < w$; if $s = 0$ then $t < u$)

153 $(a^{[r]}, a^{[s]}, b^{[t]}, c^{[v]}, b^{[u]}, d^{[x]}, c^{[w]}, d^{[y]})$ ($rs = tu = vw = xy = 0$; if $v = w$ and $x = y$ then either $r < t$ or $r = t$ and $s \le u$; if $r = s$ and $t = u$ then either $v < x$ or $v = x$ and $w \le y$)

154 $(a, a^{[r]}, a^{[s]}, a^{[t]}, a^{[u]}, a^{[v]}, b, c)$ (see Table 11.19)

155 $(a^{[r]}, a^{[s]}, a^{[t]}, a^{[u]}, a^{[v]}, b, b^{[w]}, c)$ ($rv = 0$; if $w \neq 0$ then either $r < \min\{s, t, u\}$ or $r > \max\{s, t, u\}$; if $w = 0$ then $r < \min\{s, u\}$ and $s < t$)

156 $(a^{[r]}, a^{[s]}, a^{[t]}, a^{[u]}, b^{[v]}, b^{[w]}, b^{[x]}, c)$ ($rstu = vwx = 0$ and see Table 11.20)

157 $(a^{[r]}, a^{[s]}, a^{[t]}, b^{[v]}, a^{[u]}, b^{[w]}, b^{[x]}, c)$ ($rstu = vwx = 0$ and see Table 11.21)

158 $(a, a^{[r]}, a^{[s]}, a^{[t]}, b, b^{[u]}, c, c^{[v]})$ ($u \le v$; if $u = 0$ then $v \neq 0$ and $s < t$; if $r = 0$ then $s < t$ and $u \neq 0$; if $s = t$ then $u \neq 0$; if $s = 0$ then $r < t$; if $t = 0$ then $r < s$)

159 $(a^{[r]}, a^{[s]}, a^{[t]}, b, a^{[u]}, b^{[v]}, c, c^{[w]})$ ($rstu = 0$; $v \le w$; if $v = 0$ then $t < u$; if $w = 0$ then $r < s$)

160 $(a^{[r]}, a^{[s]}, a^{[t]}, b^{[u]}, b^{[v]}, b^{[w]}, c, c^{[x]})$ ($rt = uvw = 0$; either $r < s$ or $r = s$ and $v < w$; if $x = 0$ then $r < s$, $v < w$ and $r \le u$; if $x = 0$ and $r = u$ then either $s < v$ or $s = v$ and $t < w$)

161 $(a, a^{[r]}, a^{[s]}, b^{[w]}, c^{[t]}, c^{[u]}, c^{[v]}, b^{[x]})$ ($tuv = wx = 0$; $0 \le r \le s$; if $r = 0$ then $t < u$; if $r = s$ then $u < v$; if $i = 0$ then either $r < \min\{j, k\}$ or $r = \min\{j, k\}$ and $s \le \max\{j, k\}$ for distinct $i, j, k \in \{t, u, v\}$; if $\{0, r, s\} = \{t, u, v\}$ then $w \le x$)

162 $(a, b, b^{[r]}, b^{[s]}, b^{[t]}, b^{[u]}, b^{[v]}, b^{[w]})$ (see Table 11.10)

163 $(a, a^{[r]}, a^{[s]}, a^{[t]}, a^{[u]}, a^{[v]}, b^{[w]}, b^{[x]})$ (see Table 11.22)

164 $(a^{[r]}, a^{[s]}, a^{[t]}, a^{[u]}, a^{[v]}, b^{[w]}, b^{[x]}, b^{[y]})$ ($rstuv = wxy = 0$ and see Table 11.23)

165 $(a^{[r]}, a^{[s]}, a^{[t]}, a^{[u]}, b^{[v]}, b^{[w]}, b^{[x]}, b^{[y]})$ ($rstu = vwxy = 0$ and see Table 11.24)

166 $(a^{[r]}, a^{[s]}, a^{[t]}, b^{[v]}, a^{[u]}, b^{[w]}, b^{[x]}, b^{[y]})$ ($rstu = vwxy = 0$ and see Table 11.25)

26 $(a, a^{[r]}, a^{[s]}, a^{[t]}, a^{[u]}, a^{[v]}, a^{[w]}, a^{[x]})$ (see Table 11.26)

Table 11.19: Conditions on field twists for $E_8(\#154)$.

Equalities among $0, r, \ldots, v$	Further requirements on $0, r, \ldots, v$
none	$s < \min\{t, u\}$; $u < v$
$r = s$	$u < v$
$s = t$	$u < v$
$s = u$	$t < v$
$r = s = t$	$u < v$
$r = s = u$	$t < v$
$s = t = u$	none
$r = s = t = u$	none
$r = 0$	$s < \min\{t, u\}$; $u < v$
$r = 0$; $s = u$	$t < v$
$s = 0$	$r < t$; $u < v$

$s = 0;\ r = u$	none
$s = 0;\ u = v$	$r < t$
$s = 0;\ r = u = v$	none
$r = s = 0$	$u < v$
$r = s = 0;\ t = u$	none
$s = t = 0$	$u < v$
$s = t = 0;\ r = u$	none
$s = u = 0$	$r < t < v$
$s = t = u = 0$	$r < v$

Table 11.20: Conditions on field twists for $E_8(\#156)$.

Equalities among r, s, t, u	Further requirements on $0, r, \ldots, w$
none	$r = 0;\ s < t < u$
$r = s$	$t < u;\ v < w$
$r = s = t$	$v < w < x$

Table 11.21: Conditions on field twists for $E_8(\#157)$.

Equalities among v, w, x	Further requirements on $r, \ldots, x$
none	$0 < v < w$
$v = w$	$t < u$
$v = w = x$	$s < t < u$

Table 11.22: Conditions on field twists for $E_8(\#163)$.

Equalities among $0, r, \ldots, x$	Further requirements on $0, r, \ldots, x$
none	$s < \min\{t, u\};\ u < v$
$r = s$	$u < v$
$s = t$	$u < v;\ w < x$
$s = u$	$t < v$
$r = s = t$	$u < v$
$r = s = u$	$t < v$
$s = t = u$	none
$r = s = t = u$	none
$r = 0$	$s < \min\{t, u\};\ u < v;\ w < x$
$r = 0;\ s = u$	$t < v;\ w < x$
$s = 0$	$r < t;\ u < v$
$s = 0;\ r = u$	none
$s = 0;\ u = v$	$r < t;\ w < x$
$s = 0;\ r = u = v$	$w < x$

$r = s = 0$	$u < v$; $w < x$
$r = s = 0$; $t = u$	$w < x$
$s = t = 0$	$u < v$; $w < x$
$s = t = 0$; $r = u$	$w < x$
$s = u = 0$	$r < t < v$
$s = t = u = 0$	$r < v$; $w < x$

Table 11.23: Conditions on field twists for $E_8(\#164)$.

Equalities among r, s, t, u, v	Further requirements on $r, \dots, y$
none	$r < s < t < u$
$r = s$	$t < u$; $x < y$
$r = v$	$s < t < u$
$r = s = t$	$w < x < y$
$r = s = v$	$t < u$; $x < y$
$r = s = t = v$	$w < x < y$

Table 11.24: Conditions on field twists for $E_8(\#165)$.

Equalities among r, s, t, u and equalities among v, w, x, y	Further requirements on $r, \dots, y$
none	$r = v = 0$; $s < t < u$; let $i < j < k$ represent the order of w, x, y: $s \leq i$; if $s = i$ then $t \leq j$; if $s = i$ and $t = j$ then $u \leq k$
$r = s$	$t < u$; $v = 0$; $x < y$
$r = s = t$	$v = 0$; $w < x < y$
$r = s$; $v = x$	$t < u$; $w < y$; $r \leq v$; if $r = v$ then either $t < w$ or $t = w$ and $u \leq y$

Table 11.25: Conditions on field twists for $E_8(\#166)$.

Equalities among r, s, t, u and equalities among v, w, x, y	Further requirements on $r, \dots, y$
none	$r = 0$; $s < t < u$; let $0 = i < j < k < l$ represent the order of v, w, x, y: $s \leq j$; if $s = j$ then $t \leq k$; if $s = j$ and $t = k$ then $u \leq l$
$r = s$	$t < u$; $x < y$
$r = s = t$	$w < x < y$
$r = s$; $v = w$	$t < u$; $x < y$; $r \leq v$; if $r = v$ then $t \leq x$; if $r = v$ and $t = x$ then $u \leq y$
$r = s$; $v = x$	$r \leq v$; if $r = v$ then $t \leq w$; if $r = v$ and $t = w$ then $u \leq y$
$r = s$; $v = w = x$	$t < u$

$r = s = t;\ v = w$	$x < y$

Table 11.26: Conditions on field twists for $E_8(\#26)$.

Equalities among $0, r, \ldots, w$	Further requirements on $0, r, \ldots, w$
none	$r < s;\ s < \min\{t, u\};\ u < \min\{v, w, x\}$
$r = s$	$u < v, w, x;\ w < x$
$r = s = t$	$u < v < w < x$
$r = s = u$	$t < v < w$
$r = s = t = u$	$v < w < x$
$r = s;\ t = u$	$v < w$
$r = s;\ u = v$	$w < x$
$r = s;\ t = u = v$	$w < x$
$r = s;\ t = u;\ v = w$	none
$r = 0$	$s < \min\{t, u\};\ u < \min\{v, w\};\ w < x$
$r = 0;\ s = u$	$t < v;\ w < x$
$r = 0;\ s = u = w$	$t < v < x$
$r = 0;\ s = u;\ t = w$	none
$r = 0;\ s = u;\ t = w;\ v = x$	none
$r = s = 0$	$u < v < w < x$
$r = s = 0;\ t = u$	$v < w < x$
$r = s = u = 0$	$t < v < w < x$

Table 11.27: Irreducible diagonal subgroups of $\bar{A}_1^4 A_1 = E_8(\#174)$ $(p \geq 7)$.

ID	Embedding
179	$(a, a^{[r]}, b, c, d)\ (r \neq 0)$
180	$(a^{[r]}, b, c, d, a^{[s]})\ (rs = 0)$
181	$(a, a^{[r]}, a^{[s]}, b, c)\ (0 < r < s)$
182	$(a^{[r]}, a^{[s]}, b, c, a^{[t]})\ (rt = 0; r < s)$
183	$(a, a^{[r]}, b, b^{[s]}, c)\ (0 < r \leq s)$
184	$(a, a^{[r]}, b^{[s]}, c, b^{[t]})\ (r \neq 0 = st)$
185	$(a, a^{[r]}, a^{[s]}, a^{[t]}, b)\ (0 < r < s < t)$
186	$(a^{[r]}, a^{[s]}, a^{[t]}, b, a^{[u]})\ (ru = 0; r < s < t)$
187	$(a, a^{[r]}, b^{[s]}, b^{[t]}, b^{[u]})\ (r \neq 0 = su; s < t)$
188	$(a, a^{[r]}, a^{[s]}, b^{[t]}, b^{[u]})\ (0 < r < s; tu = 0)$
18	$(a^{[r]}, a^{[s]}, a^{[t]}, a^{[u]}, a^{[v]})\ (rv = 0; r < s < t < u)$

Table 11.28: Irreducible diagonal subgroups of $\bar{A}_1^4 A_1 B_2 = E_8(\#126)$.

ID	Embedding
191	$(a, a^{[r]}, b, c, d, 10)$ (if $p = 2$ then $r \neq 0$)
192	$(a, b, a^{[r]}, c, d, 10)$
193	$(a^{[r]}, b, c, d, a^{[s]}, 10)$ $(rs = 0)$
194	$(a^{[r]}, a^{[s]}, a^{[t]}, b, c, 10)$ $(rt = 0; r \leq s$; if $p = 2$ then $r < s)$
195	$(a^{[r]}, a^{[s]}, b, c, a^{[t]}, 10)$ $(rt = 0; r \leq s$; if $r = s$ then $r < t$; if $p = 2$ then $r < s)$
196	$(a^{[r]}, b, a^{[s]}, c, a^{[t]}, 10)$ $(rt = 0; r \leq s)$
197	$(a, a^{[r]}, b, b^{[s]}, c, 10)$ $(r \leq s$; if $p = 2$ then $rs \neq 0)$
198	$(a, a^{[r]}, b^{[s]}, c, b^{[t]}, 10)$ $(st = 0$; if $p = 2$ then $r \neq 0)$
199	$(a, b^{[s]}, a^{[r]}, b^{[t]}, c, 10)$ $(st = 0$; $r \leq s + t$; if $r = 0$ then $t \neq 0)$
200	$(a^{[r]}, b^{[t]}, a^{[s]}, c, b^{[u]}, 10)$ $(rs = tu = 0)$
201	$(a, a^{[r]}, a^{[s]}, a^{[t]}, b, 10)$ $(s \leq t$; if $r = 0$ then $s < t$; if $s = 0$ then $r < t$; if $p = 2$ then $r \neq 0$ and $s < t)$
202	$(a^{[r]}, a^{[s]}, a^{[t]}, b, a^{[u]}, 10)$ $(rtu = 0$; $r \leq s$; if $r = s$ then $r < u$; if $p = 2$ then $r < s$)
203	$(a^{[r]}, a^{[s]}, a^{[t]}, b^{[u]}, b^{[v]}, 10)$ $(rt = uv = 0; r \leq s$; if $p = 2$ then $r < s)$
204	$(a^{[r]}, a^{[s]}, b, b^{[u]}, a^{[t]}, 10)$ $(rt = 0; r \leq s$; if $r = s$ then $r < t$; if $p = 2$ then $r < s$ and $u \neq 0)$
205	$(a^{[r]}, b^{[u]}, a^{[s]}, b^{[v]}, a^{[t]}, 10)$ $(rt = uv = 0; r \leq s$; if $r = s$ then $u \leq v)$
206	$(a^{[r]}, a^{[s]}, a^{[t]}, a^{[u]}, a^{[v]}, 10)$ $(rv = 0$; $r \leq \min\{s, t\}$; $t \leq u$; if $r = s$ then $t < u$ and $r < v$; if $r = t$ then $s < u$; if $t = u$ then $t < v$; if $p = 2$ then $r < s$ and $t < u)$

Table 11.29: Irreducible diagonal subgroups of $\bar{A}_1^4 A_1 A_1 = E_8(\#189)$ $(p \geq 5)$.

ID	Embedding
207	$(a, a^{[r]}, b, c, d, e)$
208	$(a^{[r]}, b, a^{[s]}, c, d, e)$ $(rs = 0)$
209	$(a^{[r]}, b, c, d, a^{[s]}, e)$ $(rs = 0)$
210	$(a^{[r]}, b, c, d, e, a^{[s]})$ $(rs = 0)$
211	$(a, b, c, d, e^{[r]}, e^{[s]})$ $(rs = 0)$
212	$(a^{[r]}, a^{[s]}, a^{[t]}, b, c, d)$ $(rt = 0; r \leq s)$
213	$(a^{[r]}, a^{[s]}, b, c, a^{[t]}, d)$ $(rt = 0; r \leq s$; if $r = s$ then $r < t)$
214	$(a^{[r]}, a^{[s]}, b, c, d, a^{[t]})$ $(rt = 0; r \leq s)$
215	$(a^{[r]}, b, a^{[s]}, c, a^{[t]}, d)$ $(rt = 0; r \leq s)$
216	$(a^{[r]}, b, a^{[s]}, c, d, a^{[t]})$ $(rt = 0; r \leq s)$
217	$(a^{[r]}, b, c, d, a^{[s]}, a^{[t]})$ $(rst = 0)$
218	$(a, a^{[r]}, b, b^{[s]}, c, d)$ $(r \leq s$; $s \neq 0)$
219	$(a, a^{[r]}, b^{[s]}, c, b^{[t]}, d)$ $(st = 0)$

220 $(a, a^{[r]}, b^{[s]}, c, d, b^{[t]})$ $(st = 0)$
221 $(a, a^{[r]}, b, c, d^{[s]}, d^{[t]})$ $(st = 0)$
222 $(a, b^{[s]}, a^{[r]}, b^{[t]}, c, d)$ $(st = 0;\ r \leq s + t;$ if $r = 0$ then $t \neq 0)$
223 $(a^{[r]}, b^{[t]}, a^{[s]}, c, b^{[u]}, d)$ $(rs = tu = 0)$
224 $(a^{[r]}, b^{[t]}, a^{[s]}, c, d, b^{[u]})$ $(rs = tu = 0)$
225 $(a, b, a^{[r]}, c, d^{[s]}, d^{[t]})$ $(st = 0;\ s \neq t)$
226 $(a^{[r]}, b^{[t]}, c, d, a^{[s]}, b^{[u]})$ $(rs = tu = 0)$
227 $(a^{[r]}, b, c^{[t]}, d, a^{[s]}, c^{[u]})$ $(rs = tu = 0)$
228 $(a, a^{[r]}, a^{[s]}, a^{[t]}, b, c)$ $(s \leq t;$ if $r = 0$ then $s < t;$ if $s = 0$ then $r < t)$
229 $(a^{[r]}, a^{[s]}, a^{[t]}, b, a^{[u]}, c)$ $(rtu = 0; r \leq s;$ if $r = s$ then $r < u)$
230 $(a^{[r]}, a^{[s]}, a^{[t]}, b, c, a^{[u]})$ $(rtu = 0; r \leq s)$
231 $(a^{[r]}, a^{[s]}, b, c, a^{[t]}, a^{[u]})$ $(rtu = 0; r \leq s)$
232 $(a^{[r]}, b, a^{[s]}, c, a^{[t]}, a^{[u]})$ $(rtu = 0; r \leq s; t \neq u)$
233 $(a^{[r]}, a^{[s]}, a^{[t]}, b^{[u]}, b^{[v]}, c)$ $(rt = uv = 0; r \leq s)$
234 $(a^{[r]}, a^{[s]}, a^{[t]}, b^{[u]}, c, b^{[v]})$ $(rt = uv = 0; r \leq s)$
235 $(a^{[r]}, a^{[s]}, a^{[t]}, b, c^{[u]}, c^{[v]})$ $(rt = uv = 0; r \leq s;$ if $u = v$ then $r \leq s \leq t)$
236 $(a^{[r]}, a^{[s]}, b, b^{[u]}, a^{[t]}, c)$ $(rt = 0; r \leq s;$ if $r = s$ then $r < t$ and $u \neq 0)$
237 $(a^{[r]}, a^{[s]}, b^{[u]}, c, a^{[t]}, b^{[v]})$ $(rt = uv = 0; r \leq s;$ if $r = s$ then $r < t)$
238 $(a^{[r]}, a^{[s]}, b, b^{[u]}, c, a^{[t]})$ $(rt = 0; r \leq s;$ if $r = s$ then $u \neq 0)$
239 $(a^{[r]}, a^{[s]}, b^{[u]}, c, b^{[v]}, a^{[t]})$ $(rt = uv = 0; r \leq s)$
240 $(a^{[r]}, b^{[u]}, a^{[s]}, b^{[v]}, a^{[t]}, c)$ $(rt = uv = 0; r \leq s;$ if $r = s$ then $u < v)$
241 $(a^{[r]}, b^{[u]}, a^{[s]}, c, a^{[t]}, b^{[v]})$ $(rst = uv = 0)$
242 $(a^{[r]}, b^{[u]}, a^{[s]}, b^{[v]}, c, a^{[t]})$ $(rt = uv = 0; r \leq s;$ if $r = s$ then $u < v)$
243 $(a^{[r]}, b^{[u]}, a^{[s]}, c, b^{[v]}, a^{[t]})$ $(rst = uv = 0)$
244 $(a^{[r]}, b^{[u]}, b^{[v]}, c, a^{[s]}, a^{[t]})$ $(rst = uv = 0;$ if $s = t$ then $u \leq v)$
245 $(a^{[r]}, b, c, c^{[u]}, a^{[s]}, a^{[t]})$ $(rst = 0; s \neq t)$
246 $(a, a^{[r]}, b, b^{[s]}, c^{[t]}, c^{[u]})$ $(tu = 0;\ r \leq s;$ if $r = 0$ then $s \neq 0)$
247 $(a, a^{[r]}, b^{[s]}, c^{[u]}, b^{[t]}, c^{[v]})$ $(st = uv = 0)$
248 $(a, b^{[s]}, a^{[r]}, b^{[t]}, c^{[u]}, c^{[v]})$ $(st = uv = 0;\ u \neq v;\ r \leq s + t;$ if $r = 0$ then $t \neq 0)$
249 $(a^{[r]}, b^{[t]}, a^{[s]}, c^{[v]}, b^{[u]}, c^{[w]})$ $(rs = tu = vw = 0)$
250 $(a^{[r]}, a^{[s]}, a^{[t]}, a^{[u]}, a^{[v]}, b)$ $(rv = 0;\ r \leq \min\{s, t\};\ t \leq u;$ if $r = s$ then $t < u$ and $r < v;$ if $r = t$ then $s < u;$ if $t = u$ then $t < v)$
251 $(a^{[r]}, a^{[s]}, a^{[t]}, a^{[u]}, b, a^{[v]})$ $(rv = 0;\ r \leq \min\{s, t\};\ t \leq u;$ if $r = s$ then $t < u;$ if $r = t$ then $s < u)$
252 $(a^{[r]}, a^{[s]}, a^{[t]}, b, a^{[u]}, a^{[v]})$ $(rtuv = 0; r \leq s;$ if $r = s$ then $r < u;$ if $u = v$ then either $s < t$ or $s = t$ and $s < u)$
253 $(a, a^{[r]}, a^{[s]}, a^{[t]}, b^{[u]}, b^{[v]})$ $(uv = 0;\ s \leq t;$ if $u = v$ then $r \leq s;$ if $r = 0$ then $s < t;$ if $s = 0$ then $r < t)$
254 $(a^{[r]}, a^{[s]}, a^{[t]}, b^{[v]}, a^{[u]}, b^{[w]})$ $(rtu = vw = 0; r \leq s)$
255 $(a^{[r]}, a^{[s]}, a^{[t]}, b^{[v]}, b^{[w]}, a^{[u]})$ $(rtu = vw = 0; r \leq s)$
256 $(a^{[r]}, a^{[s]}, b, b^{[v]}, a^{[t]}, a^{[u]})$ $(rtu = 0; r \leq s;$ if $r = s$ then $v \neq 0)$
257 $(a^{[r]}, b^{[v]}, a^{[s]}, b^{[w]}, a^{[t]}, a^{[u]})$ $(rtu = vw = 0; r \leq s; t \neq u;$ if $r = s$ then $v < w)$
258 $(a^{[r]}, a^{[s]}, a^{[t]}, b^{[u]}, b^{[v]}, b^{[w]})$ $(rt = uvw = 0;\ r \leq s;$ if $v = w$ then $s \leq t)$

259	$(a^{[r]}, a^{[s]}, b^{[u]}, b^{[v]}, a^{[t]}, b^{[w]})$ ($rt = uw = 0$; $r \leq s$; $u \leq v$; if $r = s$ then $r < t$ and $u < v$)
260	$(a^{[r]}, b^{[u]}, a^{[s]}, b^{[v]}, a^{[t]}, b^{[w]})$ ($rt = uvw = 0$; $r \leq s$; if $r = s$ then $u < v$)
23	$(a^{[r]}, a^{[s]}, a^{[t]}, a^{[u]}, a^{[v]}, a^{[w]})$ ($rstuvw = 0$; see Table 11.30)

Table 11.30: Conditions on field twists for $A_1 = E_8(\#23)$.

Equalities among r, s, t, u and equalities among v, w	Further requirements on $r, \ldots, w$
none	$r < \min\{s, t\}$; $t < u$
$r = s$	$r < v$; $t < u$
$r = t$	$s < u$
$v = w$	$r < s < t < u$
$r = s$; $v = w$	$r < v$; $t < u$
$r = s = t$	none
$r = s = t$; $v = w$	$r \neq v$

Table 11.31: Irreducible diagonal subgroups of $\bar{A}_1^4 A_1^3 = E_8(\#190)$ ($p = 2$).

ID	Embedding
261	$(a, a^{[r]}, b, c, d, e, f)$ ($r \neq 0$)
262	$(a, b, a^{[r]}, c, d, e, f)$
263	$(a^{[r]}, b, c, d, a^{[s]}, e, f)$ ($rs = 0$)
264	$(a, b, c, d, e, e^{[r]}, f)$ ($r \neq 0$)
265	$(a^{[r]}, a^{[s]}, a^{[t]}, b, c, d, e)$ ($rt = 0; r < s$)
266	$(a^{[r]}, a^{[s]}, b, c, a^{[t]}, d, e)$ ($rt = 0; r < s$)
267	$(a^{[r]}, b, a^{[s]}, c, a^{[t]}, d, e)$ ($rt = 0; r \leq s$)
268	$(a^{[r]}, b, c, d, a^{[s]}, a^{[t]}, e)$ ($rs = 0; s < t$)
269	$(a, b, c, d, e, e^{[r]}, e^{[s]})$ ($0 < r < s$)
270	$(a, a^{[r]}, b, b^{[s]}, c, d, e)$ ($0 < r \leq s$)
271	$(a, a^{[r]}, b^{[s]}, c, b^{[t]}, d, e)$ ($r \neq 0 = st$)
272	$(a, a^{[r]}, b, c, d, d^{[s]}, e)$ ($rs \neq 0$)
273	$(a, b^{[s]}, a^{[r]}, b^{[t]}, c, d, e)$ ($st = 0$; $r \leq s + t$; if $r = 0$ then $t \neq 0$)
274	$(a^{[r]}, b^{[t]}, a^{[s]}, c, b^{[u]}, d, e)$ ($rs = tu = 0$)
275	$(a, b, a^{[r]}, c, d, d^{[s]}, e)$ ($s \neq 0$)
276	$(a^{[r]}, b^{[t]}, c, d, a^{[s]}, b^{[u]}, e)$ ($rs = tu = 0; r \leq t$; if $r = t$ then $s \leq u$)
277	$(a^{[r]}, b, c^{[t]}, d, a^{[s]}, c^{[u]}, e)$ ($rs = tu = 0; r \leq t$; if $r = t$ then $s \leq u$)
278	$(a^{[r]}, b, c, d, a^{[s]}, e, e^{[t]})$ ($rs = 0 \neq t$)
279	$(a, a^{[r]}, a^{[s]}, a^{[t]}, b, c, d)$ ($r \neq 0$; $s < t$; if $s = 0$ then $r < t$)
280	$(a^{[r]}, a^{[s]}, a^{[t]}, b, a^{[u]}, c, d)$ ($rtu = 0; r < s$)
281	$(a^{[r]}, a^{[s]}, b, c, a^{[t]}, a^{[u]}, d)$ ($rt = 0; r < s; t < u$)

282 $(a^{[r]}, b, a^{[s]}, c, a^{[t]}, a^{[u]}, d)$ $(rt = 0; r \leq s; t < u)$

283 $(a^{[r]}, b, c, d, a^{[s]}, a^{[t]}, a^{[u]})$ $(rs = 0; s < t < u)$

284 $(a^{[r]}, a^{[s]}, a^{[t]}, b^{[u]}, b^{[v]}, c, d)$ $(rt = uv = 0; r < s)$

285 $(a^{[r]}, a^{[s]}, a^{[t]}, b, c, c^{[u]}, d)$ $(rt = 0; r < s)$

286 $(a^{[r]}, a^{[s]}, b, b^{[u]}, a^{[t]}, c, d)$ $(rt = 0 \neq u; r < s)$

287 $(a^{[r]}, a^{[s]}, b^{[u]}, c, a^{[t]}, b^{[v]}, d)$ $(rt = uv = 0; r < s)$

288 $(a^{[r]}, a^{[s]}, b, c, a^{[t]}, d, d^{[u]})$ $(rt = 0 \neq u; r < s)$

289 $(a^{[r]}, b, a^{[s]}, b^{[u]}, a^{[t]}, c, d)$ $(rst = uv = 0;$ if $u = 0$ then $r < s)$

290 $(a^{[r]}, b^{[u]}, a^{[s]}, c, a^{[t]}, b^{[v]}, d)$ $(rst = uv = 0)$

291 $(a^{[r]}, b, a^{[s]}, c, a^{[t]}, d, d^{[u]})$ $(rst = 0 \neq u)$

292 $(a^{[r]}, b^{[u]}, b^{[v]}, c, a^{[s]}, a^{[t]}, d)$ $(rs = uv = 0; s < t)$

293 $(a^{[r]}, b, c, c^{[u]}, a^{[s]}, a^{[t]}, d)$ $(rs = 0 \neq u; s < t)$

294 $(a^{[r]}, b^{[u]}, c, d, a^{[s]}, a^{[t]}, b^{[v]})$ $(rs = uv = 0; s < t)$

295 $(a^{[r]}, b, c^{[u]}, d, a^{[s]}, a^{[t]}, c^{[v]})$ $(rs = uv = 0; s < t)$

296 $(a, a^{[r]}, b, c, d, d^{[s]}, d^{[t]})$ $(r \neq 0; 0 < s < t)$

297 $(a, b, a^{[r]}, c, d, d^{[s]}, d^{[t]})$ $(0 < s < t)$

298 $(a, a^{[r]}, b, b^{[s]}, c, c^{[t]}, d)$ $(rt \neq 0; r \leq s)$

299 $(a, a^{[r]}, b^{[s]}, c^{[u]}, b^{[t]}, c^{[v]}, d)$ $(r \neq 0 = st = uv; s \leq u;$ if $s = u$ then $t \leq v)$

300 $(a, a^{[r]}, b^{[s]}, c, b^{[t]}, d, d^{[u]})$ $(st = 0 \neq ru)$

301 $(a, b^{[s]}, a^{[r]}, b^{[t]}, c, c^{[u]}, d)$ $(st = 0 \neq u;$ $r \leq s + t;$ if $r = 0$ then $t \neq 0)$

302 $(a^{[r]}, b^{[t]}, a^{[s]}, c^{[v]}, b^{[u]}, c^{[w]}, d)$ $(rs = tu = vw = 0; t \leq v;$ if $t = v$ then $u \leq w)$

303 $(a^{[r]}, b^{[t]}, a^{[s]}, c, b^{[u]}, d, d^{[v]})$ $(rs = tu = 0 \neq v)$

304 $(a^{[r]}, b^{[t]}, c^{[v]}, d, a^{[s]}, b^{[u]}, c^{[w]})$ $(rs = tu = vw = 0; r \leq t;$ if $r = t$ then $s \leq u)$

305 $(a^{[r]}, a^{[s]}, a^{[t]}, a^{[u]}, a^{[v]}, b, c)$ $(rv = 0; r < s; r \leq t;$ $t < u;$ if $r = t$ then $s < u)$

306 $(a^{[r]}, a^{[s]}, a^{[t]}, b, a^{[u]}, a^{[v]}, c)$ $(rtu = 0; r < s; u < v)$

307 $(a^{[r]}, a^{[s]}, b, c, a^{[t]}, a^{[u]}, a^{[v]})$ $(rt = 0; r < s; t < u < v)$

308 $(a^{[r]}, b, a^{[s]}, c, a^{[t]}, a^{[u]}, a^{[v]})$ $(rt = 0; r \leq s; t < u < v)$

309 $(a, a^{[r]}, a^{[s]}, a^{[t]}, b, b^{[u]}, c)$ $(ru \neq 0; s < t;$ if $s = 0$ then $r < t)$

310 $(a^{[r]}, a^{[s]}, a^{[t]}, b^{[v]}, a^{[u]}, b^{[w]}, c)$ $(rtu = vw = 0; r < s)$

311 $(a^{[r]}, a^{[s]}, a^{[t]}, b, a^{[u]}, c, c^{[v]})$ $(rtu = 0 \neq v; r < s)$

312 $(a^{[r]}, a^{[s]}, b, b^{[v]}, a^{[t]}, a^{[u]}, c)$ $(rt = 0 \neq v; r < s; t < u)$

313 $(a^{[r]}, a^{[s]}, b^{[v]}, c, a^{[t]}, a^{[u]}, b^{[w]})$ $(rt = vw = 0; r < s; t < u)$

314 $(a^{[r]}, b^{[v]}, a^{[s]}, b^{[w]}, a^{[t]}, a^{[u]}, c)$ $(rt = vw = 0; t < u; r \leq s;$ if $r = s$ then $v < w)$

315 $(a^{[r]}, b^{[v]}, a^{[s]}, c, a^{[t]}, a^{[u]}, b^{[w]})$ $(rst = vw = 0; t < u)$

316 $(a^{[r]}, b^{[v]}, b^{[w]}, c, a^{[s]}, a^{[t]}, a^{[u]})$ $(rs = vw = 0; s < t < u)$

317 $(a^{[r]}, b, c, c^{[v]}, a^{[s]}, a^{[t]}, a^{[u]})$ $(rs = 0 \neq v; s < t < u)$

318 $(a^{[r]}, a^{[s]}, a^{[t]}, b^{[u]}, b^{[v]}, b^{[w]}, c)$ $(rt = uv = 0; r < s; v < w)$

319 $(a^{[r]}, a^{[s]}, a^{[t]}, b, c, c^{[u]}, c^{[v]})$ $(rt = 0; 0 < u < v; r < s)$

320 $(a^{[r]}, a^{[s]}, b^{[u]}, b^{[v]}, a^{[t]}, b^{[w]}, c)$ $(rt = uw = 0; r < s; u < v; r \leq u;$ if $r = u$ then either $s < v$ or $s = v$ and $t \leq w)$

321 $(a^{[r]}, a^{[s]}, b^{[u]}, c, a^{[t]}, b^{[v]}, b^{[w]})$ $(rt = uv = 0; r < s; v < w)$

322 $(a^{[r]}, b^{[u]}, a^{[s]}, b^{[v]}, a^{[t]}, b^{[w]}, c)$ $(rt = uvw = 0; r \leq s;$ if $r = s$ then $u < v)$

323	$(a^{[r]}, b^{[u]}, a^{[s]}, c, a^{[t]}, b^{[v]}, b^{[w]})$ $(rst = uv = 0; v < w)$
324	$(a^{[r]}, b^{[u]}, b^{[v]}, c, a^{[s]}, a^{[t]}, b^{[w]})$ $(rs = uvw = 0; s < t)$
325	$(a^{[r]}, b, c^{[u]}, c^{[v]}, a^{[s]}, a^{[t]}, c^{[w]})$ $(rs = uw = 0; s < t; u < v)$
326	$(a^{[r]}, a^{[s]}, a^{[t]}, b^{[u]}, b^{[v]}, c, c^{[w]})$ $(rt = uv = 0 \neq w; r < s)$
327	$(a^{[r]}, a^{[s]}, b, b^{[u]}, a^{[t]}, c, c^{[w]})$ $(rt = 0 \neq uw; r < s)$
328	$(a^{[r]}, a^{[s]}, b^{[u]}, c^{[w]}, a^{[t]}, b^{[v]}, c^{[x]})$ $(rt = uv = wx = 0; r < s; u \leq w;$ if $u = w$ then $v \leq x)$
329	$(a^{[r]}, b^{[u]}, a^{[s]}, b^{[v]}, a^{[t]}, c, c^{[w]})$ $(rst = uv = 0 \neq w;$ if $u = 0$ then $r < s)$
330	$(a^{[r]}, b^{[u]}, a^{[s]}, c^{[w]}, a^{[t]}, b^{[v]}, c^{[x]})$ $(rst = uv = wx = 0; r \leq s;$ if $r = s$ then $u \leq w;$ if $r = s$ and $u = w$ then $v \leq x)$
331	$(a^{[r]}, b^{[u]}, b^{[v]}, c^{[w]}, a^{[s]}, a^{[t]}, c^{[x]})$ $(rs = uv = wx = 0; s < t)$
332	$(a^{[r]}, b^{[u]}, c, c^{[w]}, a^{[s]}, a^{[t]}, b^{[v]})$ $(rs = uv = 0 \neq w; s < t)$
333	$(a, a^{[r]}, b, b^{[s]}, c, c^{[t]}, c^{[u]})$ $(rs \neq 0; r \leq s; 0 < t < u)$
334	$(a, b^{[s]}, a^{[r]}, b^{[t]}, c, c^{[u]}, c^{[v]})$ $(st = 0; 0 < u < v;\ r \leq s + t;$ if $r = 0$ then $t \neq 0)$
335	$(a^{[r]}, a^{[s]}, a^{[t]}, a^{[u]}, a^{[v]}, a^{[w]}, b)$ $(rv = 0; r < s; t < u; v < w; r \leq t;$ if $r = t$ then $s < u)$
336	$(a^{[r]}, a^{[s]}, a^{[t]}, b, a^{[u]}, a^{[v]}, a^{[w]})$ $(rtu = 0; r < s; u < v < w)$
337	$(a^{[r]}, a^{[s]}, a^{[t]}, a^{[u]}, a^{[v]}, b, b^{[w]})$ $(rv = 0 \neq w; r < s; t < u; r \leq t;$ if $r = t$ then $s < u)$
338	$(a^{[r]}, a^{[s]}, a^{[t]}, b^{[w]}, a^{[u]}, a^{[v]}, b^{[x]})$ $(rtu = wx = 0; r < s; u < v)$
339	$(a^{[r]}, a^{[s]}, b, b^{[w]}, a^{[t]}, a^{[u]}, a^{[v]})$ $(rt = 0 \neq w; r < s; t < u < v)$
340	$(a^{[r]}, b^{[w]}, a^{[s]}, b^{[x]}, a^{[t]}, a^{[u]}, a^{[v]})$ $(rt = wx = 0; t < u < v; r \leq s;$ if $r = s$ then $w < x)$
341	$(a, a^{[r]}, a^{[s]}, a^{[t]}, b, b^{[u]}, b^{[v]})$ $(r \neq 0 < u < v; s < t;$ if $s = 0$ then $r < t)$
342	$(a^{[r]}, a^{[s]}, a^{[t]}, b^{[v]}, a^{[u]}, b^{[w]}, b^{[x]})$ $(rtu = vw = 0; r < s; w < x)$
343	$(a^{[r]}, a^{[s]}, b^{[v]}, b^{[w]}, a^{[t]}, a^{[u]}, b^{[x]})$ $(rt = vx = 0; r < s; v < w; t < u)$
344	$(a^{[r]}, b^{[v]}, a^{[s]}, b^{[w]}, a^{[t]}, a^{[u]}, b^{[x]})$ $(rt = vwx = 0; t < u; r \leq s;$ if $r = s$ then $v < w)$
345	$(a^{[r]}, b^{[v]}, b^{[w]}, b^{[x]}, a^{[s]}, a^{[t]}, a^{[u]})$ $(rs = vw = 0; s < t < u; w < x)$
27	$(a^{[r]}, a^{[s]}, a^{[t]}, a^{[u]}, a^{[v]}, a^{[w]}, a^{[x]})$ $(rsv = 0;\ v < w < x;\ t < u;$ if $r \neq t$ then $r < \min\{s, t\};$ if $r = t$ then $s < u)$

Table 11.32: Irreducible diagonal subgroups of $\bar{A}_1^2 A_1^4 = E_8(\#365)$ $(p \neq 2)$.

ID	Embedding
374	$(a, a^{[r]}, b, c, d, e)$
375	$(a^{[r]}, b, a^{[s]}, c, d, e)$ $(rs = 0)$
376	$(a, b, c, c^{[r]}, d, e)$ $(r \neq 0)$
377	$(a^{[r]}, a^{[s]}, a^{[t]}, b, c, d)$ $(rt = 0;\ r \leq s;$ if $r = s$ then $r < t)$
378	$(a^{[r]}, b, a^{[s]}, a^{[t]}, c, d)$ $(rs = 0;\ s < t)$
379	$(a, b, c, c^{[r]}, c^{[s]}, d)$ $(0 < r < s)$
380	$(a, a^{[r]}, b, b^{[s]}, c, d)$ $(s \neq 0)$

381	$(a^{[r]}, b^{[t]}, a^{[s]}, b^{[u]}, c, d)$ $(rs = tu = 0;\ r \leq t;$ if $r = t$ then $s \leq u)$
382	$(a^{[r]}, b, a^{[s]}, c, c^{[t]}, d)$ $(rs = 0 \neq t)$
383	$(a, b, c, c^{[r]}, d, d^{[s]})$ $(rs \neq 0; r \leq s)$
384	$(a^{[r]}, a^{[s]}, a^{[t]}, a^{[u]}, b, c)$ $(rt = 0;\ r \leq s;\ t < u;$ if $r = s$ then $r < t)$
385	$(a^{[r]}, b, a^{[s]}, a^{[t]}, a^{[u]}, c)$ $(rs = 0;\ s < t < u)$
386	$(a, b, c, c^{[r]}, c^{[s]}, c^{[t]})$ $(0 < r < s < t)$
387	$(a^{[r]}, a^{[s]}, a^{[t]}, b, b^{[u]}, c)$ $(rt = 0 \neq u;\ r \leq s;$ if $r = s$ then $r < t)$
388	$(a^{[r]}, b^{[u]}, a^{[s]}, a^{[t]}, b^{[v]}, c)$ $(rs = uv = 0;\ s < t)$
389	$(a^{[r]}, c, a^{[s]}, a^{[t]}, b, b^{[u]})$ $(rs = 0 \neq u;\ s < t)$
390	$(a, a^{[r]}, b, b^{[s]}, b^{[t]}, c)$ $(0 < s < t)$
391	$(a^{[r]}, b, c, c^{[t]}, c^{[u]}, a^{[s]})$ $(rs = 0; 0 < t < u)$
392	$(a, a^{[r]}, b, b^{[s]}, c, c^{[t]})$ $(st \neq 0; s \leq t)$
393	$(a^{[r]}, b^{[t]}, a^{[s]}, b^{[u]}, c, c^{[v]})$ $(rs = tu = 0 \neq v;\ r \leq t;$ if $r = t$ then $s \leq u)$
394	$(a^{[r]}, a^{[s]}, a^{[t]}, a^{[u]}, a^{[v]}, b)$ $(rt = 0;\ r \leq s;\ t < u < v;$ if $r = s$ then $r < t)$
395	$(a^{[r]}, b, a^{[s]}, a^{[t]}, a^{[u]}, a^{[v]})$ $(rs = 0;\ s < t < u < v)$
396	$(a^{[r]}, a^{[s]}, a^{[t]}, a^{[u]}, b, b^{[v]})$ $(rt = 0 \neq v;\ r \leq s;\ t < u;$ if $r = s$ then $r < t)$
397	$(a^{[r]}, b^{[v]}, a^{[s]}, a^{[t]}, a^{[u]}, b^{[w]})$ $(rs = vw = 0;\ s < t < u)$
398	$(a, a^{[r]}, b, b^{[s]}, b^{[t]}, b^{[u]})$ $(0 < s < t < u)$
399	$(a^{[r]}, a^{[s]}, a^{[t]}, b, b^{[u]}, b^{[v]})$ $(rt = 0 < u < v;\ r \leq s;$ if $r = s$ then $r < t)$
400	$(a^{[r]}, b^{[u]}, a^{[s]}, a^{[t]}, b^{[v]}, b^{[w]})$ $(rs = uv = 0;\ s < t;\ u < v;\ r \leq u;$ if $r = u$ then either $s < v$ or $s = v$ and $t \leq w)$
25	$(a^{[r]}, a^{[s]}, a^{[t]}, a^{[u]}, a^{[v]}, a^{[w]})$ $(rt = 0;\ r \leq s;\ t < u < v < w;$ if $r = s$ then $r < t)$

Table 11.33: Irreducible diagonal subgroups of $\bar{A}_1^2 A_1 A_1^2 = E_8(\#401)$ $(p \neq 2)$.

ID	Embedding
408	$(a, a^{[r]}, b, c, d)$
409	$(a^{[r]}, b, a^{[s]}, c, d)$ $(rs = 0)$
410	$(a^{[r]}, b, c, a^{[s]}, d)$ $(rs = 0)$
411	$(a, b, c^{[r]}, c^{[s]}, d)$ $(rs = 0)$
412	$(a, b, c, d, d^{[r]})$ $(r \neq 0)$
214$^{\{\underline{0}\}}$	(a, b, c, d, d)
413	$(a^{[r]}, a^{[s]}, a^{[t]}, b, c)$ $(rt = 0; r \leq s;$ if $r = s$ then $s < t)$
414	$(a^{[r]}, a^{[s]}, b, a^{[t]}, c)$ $(rt = 0; r \leq s)$
415	$(a^{[r]}, b, a^{[s]}, a^{[t]}, c)$ $(rst = 0)$
416	$(a^{[r]}, b, c, a^{[s]}, a^{[t]})$ $(rs = 0; s < t)$
417	$(a, b, c^{[r]}, c^{[s]}, c^{[t]})$ $(rs = 0; s < t)$
418	$(a, a^{[r]}, b^{[s]}, b^{[t]}, c)$ $(st = 0)$
419	$(a, a^{[r]}, b, c, c^{[s]})$ $(s \neq 0)$
420	$(a^{[r]}, b^{[t]}, a^{[s]}, b^{[u]}, c)$ $(rs = tu = 0)$
421	$(a^{[r]}, b, a^{[s]}, c, c^{[t]})$ $(rs = 0 \neq t)$

422	$(a^{[r]}, b, c^{[t]}, a^{[s]}, c^{[u]})$ $(rs = tu = 0)$
423	$(a^{[r]}, b^{[t]}, c, a^{[s]}, b^{[u]})$ $(rs = tu = 0;\ r \leq t;$ if $r = t$ then $s \leq u)$
424	$(a^{[r]}, a^{[s]}, a^{[t]}, a^{[u]}, b)$ $(rtu = 0; r \leq s;$ if $r = s$ then $s < t)$
425	$(a^{[r]}, a^{[s]}, b, a^{[t]}, a^{[u]})$ $(rt = 0; r \leq s; t < u)$
426	$(a^{[r]}, b, a^{[s]}, a^{[t]}, a^{[u]})$ $(rst = 0; t < u)$
427	$(a^{[r]}, a^{[s]}, a^{[t]}, b, b^{[u]})$ $(rt = 0 \neq u; r \leq s;$ if $r = s$ then $s < t)$
428	$(a^{[r]}, a^{[s]}, b^{[u]}, a^{[t]}, b^{[v]})$ $(rt = uv = 0; r \leq s)$
430	$(a^{[r]}, b^{[u]}, a^{[s]}, a^{[t]}, b^{[v]})$ $(rst = uv = 0)$
431	$(a^{[r]}, b^{[u]}, b^{[v]}, a^{[s]}, a^{[t]})$ $(rs = uv = 0; s < t)$
431	$(a, a^{[r]}, b^{[s]}, b^{[t]}, b^{[u]})$ $(st = 0; t < u)$
24	$(a^{[r]}, a^{[s]}, a^{[t]}, a^{[u]}, a^{[v]})$ $(rtu = 0; r \leq s;\ u < v;$ if $r = s$ then $s < t)$

Table 11.34: Irreducible diagonal subgroups of $\bar{A}_1^2 A_1 A_1 = E_8(\#402)$ $(p \geq 11)$.

ID	Embedding	ID	Embedding
432	$(a, a^{[r]}, b, c)$	437	$(a^{[r]}, a^{[s]}, b, a^{[t]})$ $(r \leq s; rt = 0)$
433	$(a^{[r]}, b, a^{[s]}, c)$ $(rs = 0)$	438	$(a^{[r]}, b, a^{[s]}, a^{[t]})$ $(rst = 0)$
434	$(a^{[r]}, b, c, a^{[s]})$ $(rs = 0)$	439	$(a, a^{[r]}, b^{[s]}, b^{[t]})$ $(st = 0)$
435	$(a, b, c^{[r]}, c^{[s]})$ $(rs = 0)$	440	$(a^{[r]}, b^{[t]}, a^{[s]}, b^{[u]})$ $(rs = tu = 0)$
436	$(a^{[r]}, a^{[s]}, a^{[t]}, b)$ $(r \leq s;\ rt = 0;$ if $r = s$ then $r < t)$	15	$(a^{[r]}, a^{[s]}, a^{[t]}, a^{[u]})$ $(rst = 0;\ r \leq s;$ if $r = s$ then $r < t)$

Table 11.35: Irreducible diagonal subgroups of $\bar{A}_1^2 A_1^3 B_2 = E_8(\#441)$ $(p = 2)$.

ID	Embedding
452	$(a, a^{[r]}, b, c, d, 10)$ $(r \neq 0)$
453	$(a^{[r]}, b, a^{[s]}, c, d, 10)$ $(rs = 0)$
454	$(a, b, c, c^{[r]}, d, 10)$ $(r \neq 0)$
455	$(a^{[r]}, a^{[s]}, a^{[t]}, b, c, 10)$ $(r < s; rt = 0)$
456	$(a^{[r]}, b, a^{[s]}, a^{[t]}, c, 10)$ $(rs = 0;\ s < t)$
457	$(a, b, c, c^{[r]}, c^{[s]}, 10)$ $(0 < r < s)$
458	$(a, a^{[r]}, b, b^{[s]}, c, 10)$ $(rs \neq 0)$
459	$(a^{[r]}, b^{[t]}, a^{[s]}, b^{[u]}, c, 10)$ $(rs = tu = 0;\ r \leq t;$ if $r = t$ then $s \leq u)$
460	$(a^{[r]}, b, a^{[s]}, c, c^{[t]}, 10)$ $(rs = 0 \neq t)$
461	$(a^{[r]}, a^{[s]}, a^{[t]}, a^{[u]}, b, 10)$ $(rt = 0; r < s; t < u)$
462	$(a^{[r]}, b, a^{[s]}, a^{[t]}, a^{[u]}, 10)$ $(rs = 0; s < t < u)$
463	$(a^{[r]}, a^{[s]}, a^{[t]}, b, b^{[u]}, 10)$ $(rt = 0 \neq u; r < s)$
464	$(a^{[r]}, b^{[u]}, a^{[s]}, a^{[t]}, b^{[v]}, 10)$ $(rs = uv = 0; s < t)$
465	$(a, a^{[r]}, b, b^{[s]}, b^{[t]}, 10)$ $(r \neq 0; 0 < s < t)$
466	$(a^{[r]}, a^{[s]}, a^{[t]}, a^{[u]}, a^{[v]}, 10)$ $(rt = 0; r < s; t < u < v)$

Table 11.36: Irreducible diagonal subgroups of $\bar{A}_1^2 A_1^5 = E_8(\#451)$ $(p = 2)$.

ID	Embedding
467	$(a, a^{[r]}, b, c, d, e, f)$ $(r \neq 0)$
468	$(a^{[r]}, b, a^{[s]}, c, d, e, f)$ $(rs = 0)$
469	$(a, b, c, c^{[r]}, d, e, f)$ $(r \neq 0)$
470	$(a^{[r]}, a^{[s]}, a^{[t]}, b, c, d, e)$ $(rt = 0; r < s)$
471	$(a^{[r]}, b, a^{[s]}, a^{[t]}, c, d, e)$ $(rs = 0; s < t)$
472	$(a, b, c, c^{[r]}, c^{[s]}, d, e)$ $(0 < r < s)$
473	$(a, a^{[r]}, b, b^{[s]}, c, d, e)$ $(rs \neq 0)$
474	$(a^{[r]}, b^{[t]}, a^{[s]}, b^{[u]}, c, d, e)$ $(rs = tu = 0; r \leq t;$ if $r = t$ then $s < u)$
475	$(a^{[r]}, b, a^{[s]}, c, c^{[t]}, d, e)$ $(rs = 0 \neq t)$
476	$(a, b, c, c^{[r]}, d, d^{[s]}, e)$ $(rs \neq 0; r \leq s)$
477	$(a^{[r]}, a^{[s]}, a^{[t]}, a^{[u]}, b, c, d)$ $(rt = 0; r < s; t < u)$
478	$(a^{[r]}, b, a^{[s]}, a^{[t]}, a^{[u]}, c, d)$ $(rs = 0; s < t < u)$
479	$(a, b, c, c^{[r]}, c^{[s]}, c^{[t]}, d)$ $(0 < r < s < t)$
480	$(a^{[r]}, a^{[s]}, a^{[t]}, b, b^{[u]}, c, d)$ $(rt = 0 \neq u; r < s)$
481	$(a^{[r]}, b^{[u]}, a^{[s]}, a^{[t]}, b^{[v]}, c, d)$ $(rs = uv = 0; s < t)$
482	$(a^{[r]}, b, a^{[s]}, a^{[t]}, c, c^{[u]}, d)$ $(rs = 0 \neq u; s < t)$
483	$(a, a^{[r]}, b, b^{[s]}, b^{[t]}, c, d)$ $(r \neq 0; 0 < s < t)$
484	$(a^{[r]}, b, c, c^{[t]}, c^{[u]}, a^{[s]}, d)$ $(rs = 0; 0 < t < u)$
485	$(a, b, c, c^{[r]}, c^{[s]}, d, d^{[t]})$ $(t \neq 0; 0 < r < s)$
486	$(a, a^{[r]}, b, b^{[s]}, c, c^{[t]}, d)$ $(rst \neq 0; s \leq t)$
487	$(a^{[r]}, b^{[t]}, a^{[s]}, b^{[u]}, c, c^{[v]}, d)$ $(rs = tu = 0 \neq v; r \leq t;$ if $r = t$ then $s < u)$
488	$(a^{[r]}, b, a^{[s]}, c, c^{[t]}, d, d^{[u]})$ $(rs = 0 \neq tu; t \leq u)$
489	$(a^{[r]}, a^{[s]}, a^{[t]}, a^{[u]}, a^{[v]}, b, c)$ $(rt = 0; r < s; t < u < v)$
490	$(a^{[r]}, b, a^{[s]}, a^{[t]}, a^{[u]}, a^{[v]}, c)$ $(rs = 0; s < t < u < v)$
491	$(a, b, c, c^{[r]}, c^{[s]}, c^{[t]}, c^{[u]})$ $(0 < r < s < t < u)$
492	$(a^{[r]}, a^{[s]}, a^{[t]}, a^{[u]}, b, b^{[v]}, c)$ $(rt = 0 \neq v; r < s; t < u)$
493	$(a^{[r]}, b^{[v]}, a^{[s]}, a^{[t]}, a^{[u]}, b^{[w]}, c)$ $(rs = vw = 0; s < t < u)$
494	$(a^{[r]}, b, a^{[s]}, a^{[t]}, a^{[u]}, c, c^{[v]})$ $(rs = 0 \neq v; s < t < u)$
495	$(a, a^{[r]}, b, b^{[s]}, b^{[t]}, b^{[u]}, c)$ $(r \neq 0; 0 < s < t < u)$
496	$(a^{[r]}, b, c, c^{[t]}, c^{[u]}, c^{[v]}, a^{[s]})$ $(rs = 0; 0 < t < u < v)$
497	$(a^{[r]}, a^{[s]}, a^{[t]}, b, b^{[u]}, b^{[v]}, c)$ $(rt = 0; 0 < u < v; r < s)$
498	$(a^{[r]}, b^{[u]}, a^{[s]}, a^{[t]}, b^{[v]}, b^{[w]}, c)$ $(rs = uv = 0; s < t; v < w; r \leq u;$ if $r = u$ then either $s < v$ or $s = v$ and $t \leq w)$
499	$(a^{[r]}, b, a^{[s]}, a^{[t]}, c, c^{[u]}, c^{[v]})$ $(rs = 0; 0 < u < v; s < t)$
500	$(a^{[r]}, a^{[s]}, a^{[t]}, b, b^{[u]}, c, c^{[v]})$ $(rt = 0 \neq uv; r < s; u \leq v)$
501	$(a^{[r]}, b^{[u]}, a^{[s]}, a^{[t]}, b^{[v]}, c, c^{[w]})$ $(rs = uv = 0 \neq w; s < t)$
502	$(a, a^{[r]}, b, b^{[s]}, b^{[t]}, c, c^{[u]})$ $(ru \neq 0; 0 < s < t)$
503	$(a^{[r]}, b^{[t]}, c, c^{[v]}, c^{[w]}, a^{[s]}, b^{[u]})$ $(rs = tu = 0 < v < w; r \leq t;$ if $r = t$ then $s \leq u)$
504	$(a^{[r]}, a^{[s]}, a^{[t]}, a^{[u]}, a^{[v]}, a^{[w]}, b)$ $(rt = 0; r < s; t < u < v < w)$

505	$(a^{[r]}, b, a^{[s]}, a^{[t]}, a^{[u]}, a^{[v]}, a^{[w]})$ $(rs = 0; s < t < u < v < w)$
506	$(a^{[r]}, a^{[s]}, a^{[t]}, a^{[u]}, a^{[v]}, b, b^{[w]})$ $(rt = 0 \neq w; r < s; t < u < v)$
507	$(a^{[r]}, b^{[w]}, a^{[s]}, a^{[t]}, a^{[u]}, a^{[v]}, b^{[x]})$ $(rs = wx = 0; s < t < u < v)$
508	$(a, a^{[r]}, b, b^{[s]}, b^{[t]}, b^{[u]}, b^{[v]})$ $(r \neq 0; 0 < s < t < u < v)$
509	$(a^{[r]}, a^{[s]}, a^{[t]}, a^{[u]}, b, b^{[v]}, b^{[w]})$ $(rt = 0; 0 < v < w; r < s; t < u)$
510	$(a^{[r]}, b^{[v]}, a^{[s]}, a^{[t]}, a^{[u]}, b^{[w]}, b^{[x]})$ $(rs = vw = 0; s < t < u; w < x)$
511	$(a^{[r]}, a^{[s]}, a^{[t]}, b, b^{[u]}, b^{[v]}, b^{[w]})$ $(rt = 0; 0 < u < v < w; r < s)$
29	$(a^{[r]}, a^{[s]}, a^{[t]}, a^{[u]}, a^{[v]}, a^{[w]}, a^{[x]})$ $(rt = 0; r < s; t < u < v < w < x)$

Table 11.37: Irreducible diagonal subgroups of $\bar{A}_1^2 A_1 B_2^2 = E_8(\#404)$ $(p = 2)$.

ID	Embedding	ID	Embedding
520	$(1_a, 1_a^{[r]}, 1_b, 10_c, 10_d)$ $(r \neq 0)$	526	$(1_a, 1_a^{[r]}, 1_b, 10_c, 02_c^{[s]})$ $(r \neq 0)$
521	$(1_a^{[r]}, 1_b, 1_a^{[s]}, 10_c, 10_d)$ $(rs = 0)$	527	$(1_a^{[r]}, 1_b, 1_a^{[s]}, 10_c, 10_c^{[t]})$ $(rs = 0 \neq t)$
522	$(1_a, 1_b, 1_c, 10_d, 10_d^{[r]})$ $(r \neq 0)$	528	$(1_a^{[r]}, 1_b, 1_a^{[s]}, 10_c, 02_c^{[t]})$ $(rs = 0)$
523	$(1_a, 1_b, 1_c, 10_d, 02_d^{[r]})$	529	$(1_a^{[r]}, 1_a^{[s]}, 1_a^{[t]}, 10_b, 10_b^{[u]})$ $(r < s; rt = 0 \neq u)$
524	$(1_a^{[r]}, 1_a^{[s]}, 1_a^{[t]}, 10_b, 10_c)$ $(rt = 0; r < s)$	530	$(1_a^{[r]}, 1_a^{[s]}, 1_a^{[t]}, 10_b, 02_b^{[u]})$ $(r < s; rt = 0)$
525	$(1_a, 1_a^{[r]}, 1_b, 10_c, 10_c^{[s]})$ $(rs \neq 0)$		

Table 11.38: Irreducible diagonal subgroups of $\bar{A}_1^2 A_1 A_1 = E_8(\#538)$ $(p \geq 7)$.

ID	Embedding	ID	Embedding
542	$(a, a^{[r]}, b, c)$	547	$(a^{[r]}, a^{[s]}, b, a^{[t]})$ $(r \leq s; rt = 0)$
543	$(a^{[r]}, b, a^{[s]}, c)$ $(rs = 0)$	548	$(a^{[r]}, b, a^{[s]}, a^{[t]})$ $(rst = 0)$
544	$(a^{[r]}, b, c, a^{[s]})$ $(rs = 0)$	549	$(a, a^{[r]}, b^{[s]}, b^{[t]})$ $(st = 0)$
545	$(a, b, c^{[r]}, c^{[s]})$ $(rs = 0)$	550	$(a^{[r]}, b^{[t]}, a^{[s]}, b^{[u]})$ $(rs = tu = 0)$
546	$(a^{[r]}, a^{[s]}, a^{[t]}, b)$ $(r \leq s;\ rt = 0)$	17	$(a^{[r]}, a^{[s]}, a^{[t]}, a^{[u]})$ $(rst = 0;\ r \leq s)$

Table 11.39: Irreducible diagonal subgroups of $\bar{A}_1^2 A_1^2 A_1 = E_8(\#558)$ $(p \neq 2)$.

ID	Embedding
566	$(a, a^{[r]}, b, c, d)$
567	$(a^{[r]}, b, a^{[s]}, c, d)$ $(rs = 0)$
568	$(a^{[r]}, b, c, a^{[s]}, d)$ $(rs = 0)$
569	$(a^{[r]}, b, c, d, a^{[s]})$ $(rs = 0)$
570	$(a, b, c, c^{[r]}, d)$ $(r \neq 0)$
411$^{\{0\}}$	(a, b, c, c, d)
571	$(a, b, c^{[r]}, d, c^{[s]})$ $(rs = 0; r \neq s)$

$227^{\{\underline{0}\}}$	(a, b, c, d, c)
572	$(a^{[r]}, a^{[s]}, a^{[t]}, b, c)$ $(rst = 0)$
573	$(a^{[r]}, a^{[s]}, b, c, a^{[t]})$ $(rt = 0; r \leq s)$
574	$(a^{[r]}, b, a^{[s]}, a^{[t]}, c)$ $(rst = 0; s \neq t)$
575	$(a^{[r]}, b, a^{[s]}, c, a^{[t]})$ $(rst = 0; s \neq t)$
576	$(a^{[r]}, b, c, a^{[s]}, a^{[t]})$ $(rst = 0; s \neq t)$
577	$(a, b, c^{[r]}, c^{[s]}, c^{[t]})$ $(rt = 0; r < s; t \notin \{r, s\})$
578	$(a, a^{[r]}, b^{[s]}, b^{[t]}, c)$ $(st = 0; s \neq t;$ if $r = 0$ then $s < t)$
579	$(a^{[r]}, a^{[s]}, b^{[t]}, c, b^{[u]})$ $(rs = tu = 0; t \neq u)$
580	$(a^{[r]}, b^{[t]}, a^{[s]}, b^{[u]}, c)$ $(rs = tu = 0; r \leq t;$ if $r = t$ then $s \leq u)$
581	$(a^{[r]}, b^{[t]}, a^{[s]}, c, b^{[u]})$ $(rs = tu = 0)$
582	$(a^{[r]}, b, a^{[s]}, c^{[t]}, c^{[u]})$ $(rs = tu = 0; t \neq u)$
583	$(a^{[r]}, b^{[t]}, b^{[u]}, a^{[s]}, c)$ $(rs = tu = 0; r \leq t;$ if $r = t$ then $s \leq u)$
584	$(a^{[r]}, b^{[t]}, c, a^{[s]}, b^{[u]})$ $(rs = tu = 0)$
585	$(a^{[r]}, b, c^{[t]}, a^{[s]}, c^{[u]})$ $(rs = tu = 0; t \neq u)$
586	$(a^{[r]}, b^{[t]}, b^{[u]}, c, a^{[s]})$ $(rs = tu = 0)$
587	$(a^{[r]}, b^{[t]}, c, b^{[u]}, a^{[s]})$ $(rs = tu = 0)$
588	$(a^{[r]}, b, c^{[t]}, c^{[u]}, a^{[s]})$ $(rs = tu = 0; t \neq u)$
589	$(a^{[r]}, a^{[s]}, a^{[t]}, a^{[u]}, b)$ $(rtu = 0; r \leq s; t \neq u;$ if $r = s$ then $t < u)$
590	$(a^{[r]}, a^{[s]}, a^{[t]}, b, a^{[u]})$ $(rstu = 0; t \neq u)$
591	$(a^{[r]}, b, a^{[s]}, a^{[t]}, a^{[u]})$ $(rstu = 0;$ s, t, u distinct$)$
592	$(a^{[r]}, a^{[s]}, a^{[t]}, b^{[u]}, b^{[v]})$ $(rst = uv = 0; u \neq v)$
593	$(a^{[r]}, a^{[s]}, b^{[u]}, b^{[v]}, a^{[t]})$ $(rt = uv = 0; u \neq v; r \leq s;$ if $r = s$ then $u < v)$
594	$(a^{[r]}, b^{[u]}, a^{[s]}, a^{[t]}, b^{[v]})$ $(rst = uv = 0; s \neq t)$
595	$(a^{[r]}, b^{[u]}, a^{[s]}, b^{[v]}, a^{[t]})$ $(rst = uv = 0; s \neq t)$
596	$(a^{[r]}, b^{[u]}, b^{[v]}, a^{[s]}, a^{[t]})$ $(rst = uv = 0; s \neq t)$
597	$(a, a^{[r]}, b^{[s]}, b^{[t]}, b^{[u]})$ $(stu = 0;$ s, t, u distinct; if $r = 0$ then $s < t)$
6	$(a^{[r]}, a^{[s]}, a^{[t]}, a^{[u]}, a^{[v]})$ $(rtuv = 0;$ $r \leq s;$ t, u, v distinct; if $r = s$ then $t < u)$

Table 11.40: Irreducible diagonal subgroups of $\bar{A}_1^2 A_1 A_1 = E_8(\#559)$ $(p \geq 7)$.

ID	Embedding	ID	Embedding
598	$(a^{[r]}, a^{[s]}, b, c)$ $(rs = 0)$	605	$(a^{[r]}, a^{[s]}, b, a^{[t]})$ $(rst = 0)$
599	$(a^{[r]}, b, a^{[s]}, c)$ $(rs = 0)$	606	$(a^{[r]}, b, a^{[s]}, a^{[t]})$ $(rst = 0; s \neq t)$
600	$(a^{[r]}, b, c, a^{[s]})$ $(rs = 0)$	607	$(a, b^{[r]}, b^{[s]}, b^{[t]})$ $(rst = 0; s \neq t)$
601	$(a, b^{[r]}, b^{[s]}, c)$ $(rs = 0)$	608	$(a^{[r]}, a^{[s]}, b^{[t]}, b^{[u]})$ $(rs = tu = 0; t \neq u)$
602	$(a, b^{[r]}, c, b^{[s]})$ $(rs = 0)$	609	$(a^{[r]}, b^{[t]}, a^{[s]}, b^{[u]})$ $(rs = tu = 0)$
603	$(a, b, c^{[r]}, c^{[s]})$ $(r \neq s)$	610	$(a^{[r]}, b^{[t]}, b^{[u]}, a^{[s]})$ $(rs = tu = 0)$
$545^{\{\underline{0}\}}$	(a, b, c, c)	5	$(a^{[r]}, a^{[s]}, a^{[t]}, a^{[u]})$ $(rstu = 0;$ $t \neq u)$
604	$(a^{[r]}, a^{[s]}, a^{[t]}, b)$ $(rst = 0)$		

Table 11.41: Irreducible diagonal subgroups of $A_1^6 = E_8(\#641)$ $(p = 2)$.

ID	Embedding
648	$(a, a^{[r]}, b, c, d, e)$ $(r \neq 0)$
649	$(a^{[r]}, b, c, a^{[s]}, d, e)$ $(rs = 0)$
650	$(a, a^{[r]}, a^{[s]}, b, c, d)$ $(0 < r < s)$
651	$(a^{[r]}, a^{[s]}, b, a^{[t]}, c, d)$ $(rt = 0; r < s)$
652	$(a, a^{[r]}, b^{[s]}, b^{[t]}, c, d)$ $(r \neq 0 = st)$
653	$(a, a^{[r]}, b, c, c^{[s]}, d)$ $(rs \neq 0; r \leq s)$
654	$(a, b^{[s]}, c, a^{[r]}, b^{[t]}, d)$ $(st = 0; r \leq s + t$; if $r = 0$ then $s \leq t)$
655	$(a^{[r]}, a^{[s]}, a^{[t]}, a^{[u]}, b, c)$ $(ru = 0; r < s < t)$
656	$(a^{[r]}, a^{[s]}, b, a^{[t]}, a^{[u]}, c)$ $(rt = 0; r < s; t < u; r \leq t$; if $r = t$ then $s \leq u)$
657	$(a, a^{[r]}, a^{[s]}, b, b^{[t]}, c)$ $(t \neq 0 < r < s)$
658	$(a^{[r]}, a^{[s]}, b^{[u]}, a^{[t]}, b^{[v]}, c)$ $(rt = uv = 0; r < s)$
659	$(a^{[r]}, a^{[s]}, b, a^{[t]}, c, c^{[u]})$ $(rt = 0 \neq u; r < s)$
660	$(a, a^{[r]}, b^{[s]}, b^{[t]}, c, c^{[u]})$ $(ru \neq 0 = st; r \leq u$; if $r = u$ then $s \leq t)$
661	$(a, b, c^{[t]}, a^{[r]}, b^{[s]}, c^{[u]})$ $(tu = 0; u \leq r \leq s$; if $t \neq 0$ and $0 \in \{r, s\}$ then $t \leq r + s)$
662	$(a^{[r]}, a^{[s]}, a^{[t]}, a^{[u]}, a^{[v]}, b)$ $(ru = 0; r < s < t; u < v)$
663	$(a^{[r]}, a^{[s]}, a^{[t]}, a^{[u]}, b, b^{[v]})$ $(ru = 0 \neq v; r < s < t)$
664	$(a^{[r]}, a^{[s]}, b^{[v]}, a^{[t]}, a^{[u]}, b^{[w]})$ $(rt = 0; r < s; t < u; r \leq t$; if $r = t$ then either $s < u$ or $s = u$ and $v \leq w)$
665	$(a, a^{[r]}, a^{[s]}, b, b^{[t]}, b^{[u]})$ $(0 < r < s; 0 < t < u;\ r \leq t$; if $r = t$ then $s \leq u)$
666	$(a^{[r]}, a^{[s]}, b^{[u]}, a^{[t]}, b^{[v]}, b^{[w]})$ $(rt = uv = 0; r < s; v < w;\ r \leq v$; if $r = v$ then either $s < w$ or $s = w$ and $t \leq u)$
28	$(a, a^{[r]}, a^{[s]}, a^{[t]}, a^{[u]}, a^{[v]})$ $(0 < r < s; t < u < v$; if $t = 0$ then either $r < u$ or $r = u$ and $s \leq v$ $)$

Table 11.42: Irreducible diagonal subgroups of $A_1^2 A_1^2 = E_8(\#683)$ $(p \geq 5)$.

ID	Embedding	ID	Embedding
687	$(a, a^{[r]}, b, c)$ $(r \neq 0)$	691	$(a^{[r]}, b, a^{[s]}, a^{[t]})$ $(rs = 0; s < t)$
688	$(a^{[r]}, b, a^{[s]}, c)$ $(rs = 0)$	692	$(a, a^{[r]}, b, b^{[s]})$ $(rs \neq 0)$
689	$(a, b, c, c^{[r]})$ $(r \neq 0)$	693	$(a^{[r]}, b^{[t]}, a^{[s]}, b^{[u]})$ $(rs = tu = 0; r \leq t$; if $r = t$ then $s \leq u)$
690	$(a^{[r]}, a^{[s]}, a^{[t]}, b)$ $(rt = 0; r < s)$	21	$(a^{[r]}, a^{[s]}, a^{[t]}, a^{[u]})$ $(rt = 0; r < s; t < u)$

Table 11.43: Irreducible diagonal subgroups of $A_1^2A_1^2 = E_8(\#694)$ $(p \geq 5)$.

ID	Embedding
699	$(a, a^{[r]}, b, c)$
700	$(a^{[r]}, b, a^{[s]}, c)$ $(rs = 0; r \neq s)$
688$^{\{\underline{0}\}}$	(a, b, a, c)
701	$(a^{[r]}, b, c, a^{[s]})$ $(rs = 0)$
702	$(a, b, c, c^{[r]})$
703	$(a^{[r]}, a^{[s]}, a^{[t]}, b)$ $(rst = 0; r \neq t)$
704	$(a^{[r]}, b, a^{[s]}, a^{[t]})$ $(rst = 0; r \neq s)$
705	$(a, a^{[r]}, b^{[s]}, b^{[t]})$ $(st = 0)$
706	$(a^{[r]}, b^{[t]}, a^{[s]}, b^{[u]})$ $(rs = tu = 0; r \neq s; t \neq u; r \leq t;$ if $r = t$ then $s \leq u)$
707	$(a^{[r]}, b^{[t]}, b^{[u]}, a^{[s]})$ $(rs = tu = 0;\ r \leq t;$ if $r = t$ then $s \leq u)$
22	$(a^{[r]}, a^{[s]}, a^{[t]}, a^{[u]})$ $(rtu = 0; r \neq t; s \neq u; r \leq s;$ if $r = s$ then $t \leq u)$

Table 11.44: Irreducible diagonal subgroups of $A_1^7 = E_8(\#791)$ $(p = 2)$.

ID	Embedding
798	$(a, a^{[r]}, b, c, d, e, f)$ $(r \neq 0)$
799	$(a, a^{[r]}, a^{[s]}, b, c, d, e)$ $(0 < r < s)$
800	$(a, a^{[r]}, b, b^{[s]}, c, d, e)$ $(0 < r \leq s)$
801	$(a, a^{[r]}, a^{[s]}, a^{[t]}, b, c, d)$ $(0 < r < s < t)$
802	$(a, a^{[r]}, a^{[s]}, b, b^{[t]}, c, d)$ $(t \neq 0; 0 < r < s)$
803	$(a, a^{[r]}, b, b^{[s]}, c, c^{[t]}, d)$ $(0 < r \leq s \leq t)$
804	$(a, a^{[r]}, a^{[s]}, a^{[t]}, a^{[u]}, b, c)$ $(0 < r < s < t < u)$
805	$(a, a^{[r]}, a^{[s]}, a^{[t]}, b, b^{[u]}, c)$ $(u \neq 0; 0 < r < s < t)$
806	$(a, a^{[r]}, a^{[s]}, b, b^{[t]}, b^{[u]}, c)$ $(0 < r < s; 0 < t < u; r \leq t;$ if $r = t$ then $s \leq u)$
807	$(a, a^{[r]}, a^{[s]}, b, b^{[t]}, c, c^{[u]})$ $(tu \neq 0; 0 < r < s)$
808	$(a, a^{[r]}, a^{[s]}, a^{[t]}, a^{[u]}, a^{[v]}, b)$ $(0 < r < s < t < u < v)$
809	$(a, a^{[r]}, a^{[s]}, a^{[t]}, a^{[u]}, b, b^{[v]})$ $(v \neq 0; 0 < r < s < t < u)$
810	$(a, a^{[r]}, a^{[s]}, a^{[t]}, b, b^{[u]}, b^{[v]})$ $(0 < r < s < t; 0 < u < v)$
30	$(a, a^{[r]}, a^{[s]}, a^{[t]}, a^{[u]}, a^{[v]}, a^{[w]})$ $(0 < r < s < t < u < v < w)$

Table 11.45: Irreducible diagonal subgroups of $\bar{A}_1A_1A_1A_1 = E_8(\#929)$ $(p \geq 7)$.

ID	Embedding	ID	Embedding
932	$(a^{[r]}, a^{[s]}, b, c)$ $(rs = 0)$	938	$(a^{[r]}, a^{[s]}, a^{[t]}, b)$ $(rst = 0; s \neq t)$
933	$(a^{[r]}, b, a^{[s]}, c)$ $(rs = 0)$	939	$(a^{[r]}, a^{[s]}, b, a^{[t]})$ $(rst = 0)$
934	$(a^{[r]}, b, c, a^{[s]})$ $(rs = 0)$	940	$(a^{[r]}, b, a^{[s]}, a^{[t]})$ $(rst = 0; s \neq t)$
935	$(a, b^{[r]}, b^{[s]}, c)$ $(rs = 0; r \neq s)$	941	$(a, b^{[r]}, b^{[s]}, b^{[t]})$ $(rst = 0; s \notin \{r, t\})$

$\mathit{600}^{\{\underline{0}\}}$	(a,b,b,c)	942	$(a^{[r]},a^{[s]},b^{[t]},b^{[u]})$ $(rs=tu=0; t\neq u)$
936	$(a,b^{[r]},c,b^{[s]})$ $(rs=0)$	943	$(a^{[r]},b^{[t]},a^{[s]},b^{[u]})$ $(rs=tu=0)$
937	$(a,b,c^{[r]},c^{[s]})$ $(rs=0)$	944	$(a^{[r]},b^{[t]},b^{[u]},a^{[s]})$ $(rs=tu; t\neq u)$
$\mathit{543}^{\{\underline{0}\}}$	(a,b,c,c)	32	$(a^{[r]},a^{[s]},a^{[t]},a^{[u]})$ $(rstu=0; t\notin\{s,u\})$

Table 11.46: Irreducible diagonal subgroups of $\bar{A}_2^4 = E_8(\#975)$.

ID	Embedding	ID	Embedding
983	$(10_a,10_b,10_b^{[r]},10_c)$	74	$(10,10^{[r]},10^{[r]},10^{[s]})$ $(0<r<s)$
984	$(10_a,10_b,10_b^{[r]},10_b^{[s]})$ $(r\leq s;$ if $r=0$ then $s\neq 0)$	75	$(10,10^{[r]},01^{[r]},10^{[s]})$ $(0<r<s)$
		76	$(10,01^{[r]},10^{[r]},10^{[s]})$ $(0<r<s)$
985	$(10_a,10_b,10_b^{[r]},01_b^{[s]})$	77	$(10,01^{[r]},01^{[r]},10^{[s]})$ $(0<r<s)$
986	$(10_a,10_b,01_b^{[r]},01_b^{[s]})$ $(0<r\leq s)$	78	$(10,10^{[r]},10^{[s]},10^{[s]})$ $(0<r<s)$
987	$(10_a,10_a^{[r]},10_b^{[s]},10_b^{[t]})$ $(st=0;\ r\leq s+t)$	79	$(10,10^{[r]},10^{[s]},01^{[s]})$ $(0<r<s)$
		80	$(10,10^{[r]},01^{[s]},10^{[s]})$ $(0<r<s)$
988	$(10_a,10_a^{[r]},10_b^{[s]},01_b^{[t]})$ $(st=0; r\neq 0)$	81	$(10,10^{[r]},01^{[s]},01^{[s]})$ $(0<r<s)$
		82	$(10,10^{[r]},10^{[s]},10^{[t]})$ $(0<r<s<t)$
66	$(10^{[r]},10,10,01)$ $(r\neq 0)$	83	$(10,10^{[r]},10^{[s]},01^{[t]})$ $(0<r<s<t)$
67	$(10,10^{[r]},10^{[r]},01^{[r]})$ $(r\neq 0)$	84	$(10,10^{[r]},01^{[s]},10^{[t]})$ $(0<r<s<t)$
68	$(10,10,10^{[r]},10^{[r]})$ $(r\neq 0)$	85	$(10,10^{[r]},01^{[s]},01^{[t]})$ $(0<r<s<t)$
69	$(10,10,10^{[r]},01^{[r]})$ $(r\neq 0)$	86	$(10,01^{[r]},10^{[s]},10^{[t]})$ $(0<r<s<t)$
70	$(10,10,10^{[r]},10^{[s]})$ $(0<r<s)$	87	$(10,01^{[r]},10^{[s]},01^{[t]})$ $(0<r<s<t)$
71	$(10,01,10^{[r]},10^{[s]})$ $(0<r<s)$	88	$(10,01^{[r]},01^{[s]},10^{[t]})$ $(0<r<s<t)$
72	$(01,10,10^{[r]},10^{[s]})$ $(0<r<s)$	89	$(10,01^{[r]},01^{[s]},01^{[t]})$ $(0<r<s<t)$
73	$(01,01,10^{[r]},10^{[s]})$ $(0<r<s)$		

Table 11.47: Irreducible diagonal subgroups of $\bar{A}_2A_2A_2 = E_8(\#1012)$ $(p=3)$.

ID	Embedding	ID	Embedding
1020	$(10_a^{[r]},10_a^{[s]},10_b)$ $(rs=0; r\neq s)$	94	$(10,10^{[r]},10)$ $(r\neq 0)$
1021	$(10_a^{[r]},01_a^{[s]},10_b)$ $(rs=0)$	95	$(10^{[r]},01^{[s]},10^{[t]})$ $(rst=0)$
1022	$(10_a^{[r]},10_b,10_a^{[s]})$ $(rs=0; r\neq s)$	96	$(10^{[r]},01^{[r]},10)$ $(r\neq 0)$
1023	$(10_a,10_b^{[r]},10_b^{[s]})$ $(rs=0)$	97	$(10,01,10^{[r]})$ $(r\neq 0)$
92	$(10^{[r]},10^{[s]},10^{[t]})$ $(rst=0)$	98	$(10^{[r]},01,10^{[r]})$ $(r\neq 0)$
93	$(10^{[r]},10,10^{[r]})$ $(r\neq 0)$	99	$(10,01^{[r]},10)$ $(r\neq 0)$

CHAPTER 12

Composition factors for G-irreducible subgroups

Let G be a simple exceptional algebraic group. In this chapter we give the composition factors of the action of each G-irreducible connected subgroup X on the minimal and adjoint modules for G. However, we do not explicitly list these for diagonal irreducible subgroups X of some semisimple irreducible subgroup Y as it would drastically lengthen the already large tables and, more importantly, because it is easy to recover them from the Y-composition factors of the G-modules in question. The notation used for describing the composition factors is given in Section 2.

The composition factors of each G-irreducible subgroup are found by repeated restrictions from a reductive, maximal connected overgroup M. When X is simple this has already been done in [**Tho16**, Tables 9–13], [**Tho15**, Tables 3–7], where the X-composition factors of the minimal and adjoint modules are given.

The first column of each table gives the identification number n of the irreducible subgroup $X = G(\#n)$. The second column gives the isomorphism type of X and the third gives any characteristic restrictions. In Tables 12.1–12.4, the fourth (resp. fifth) column gives the composition factors of the minimal module (resp. adjoint module) for G restricted to X. In Table 12.5 the fourth column gives the composition factors of $L(E_8) \downarrow X$.

Table 12.1: The composition factors of non-diagonal irreducible connected subgroups of G_2 acting on V_7 and $L(G_2)$.

ID	X	p	$V_7 \downarrow G_2(\#n)$	$L(G_2) \downarrow G_2(\#n)$
3	A_1	≥ 7	6	$W(10)/2$
4	A_2		$10/01/00$	$W(11)/10/01$
5	A_2	3	11	$30/11/03/00$
6	$\bar{A}_1\tilde{A}_1$		$(1,1)/(0,W(2))$	$(W(2),0)/(1,W(3))/(0,W(2))$

Table 12.2: The composition factors of non-diagonal irreducible connected subgroups of F_4 acting on V_{26} and $L(F_4)$.

ID	X	p	$V_{26} \downarrow X$	$L(F_4) \downarrow X$
5	G_2	7	20	$11/01$
7	A_1	≥ 11	$10/8/4/0$	$W(14)/10^2/6/4/2$
10	A_1	≥ 13	$W(16)/8$	$W(22)/W(14)/10/2$
12	B_4		$W(\lambda_1)/\lambda_4/0$	$W(\lambda_2)/\lambda_4$
13	$\bar{D}_4$		$\lambda_1/\lambda_3/\lambda_4/0$	$\lambda_1/W(\lambda_2)/\lambda_3/\lambda_4$
14	C_4	2	λ_2	$2\lambda_1/\lambda_2/\lambda_4/0^2$
15	$\tilde{D}_4$	2	λ_2	$2\lambda_1/\lambda_2/2\lambda_3/2\lambda_4/0^2$
16	G_2	7	20	$11/01$
24	$\bar{A}_1C_3$		$(1,100)/(0,W(010))$	$(W(2),000)/(1,W(001))/(0,W(200))$
25	A_1G_2	$\neq 2$	$(W(4),00)/(2,10)$	$(W(4),10)/(2,00)/(0,W(01))$
26	$\bar{A}_2\tilde{A}_2$		$(10,10)/(01,01)/(00,W(11))$	$(W(11),00)/(10,W(02))/(01,W(20))/(00,W(11))$
27	$\tilde{A}_1\bar{A}_3$		$(W(2),000)/(1,100)/(1,001)/(0,010)/(0,000)$	$(W(2),010)/(W(2),000)/(1,100)/(1,001)/(0,W(101))$
28	$\bar{A}_1^2B_2$		$(1,1,00)/(1,0,01)/(0,1,01)/(0,0,W(10))/$ $(0,0,00)$	$(W(2),0,00)/(1,1,W(10))/(1,0,01)/(0,W(2),00)/(0,1,01)/$ $(0,0,W(02))$
29	A_1^2	$\neq 2$	$(W(3),1)/(2,2)/(1,W(3))/(0,0)$	$(W(4),2)/(W(3),1)/(2,W(4))/(2,0)/(1,W(3))/(0,2)$

30	B_2^2	2	$(10,00)/(01,01)/(00,10)/(00,00)^2$	$(10,10)/(10,00)/(02,00)/(01,01)/(00,10)/(00,02)/(00,00)^4$
31	$\tilde{A}_1B_3$	2	$(2,000)/(1,001)/(0,100)/(0,000)$	$(2,100)/(2,000)/(1,001)/(0,100)/(0,010)/(0,000)^2$
32	$\bar{A}_1^4$		$(1,1,0,0)/(1,0,1,0)/(1,0,0,1)/(0,1,1,0)/$ $(0,1,0,1)/(0,0,1,1)/(0,0,0,0)^2$	$(W(2),0,0,0)/(1,1,1,1)/(1,1,0,0)/(1,0,1,0)/(1,0,0,1)/$ $(0,W(2),0,0)/(0,1,1,0)/(0,1,0,1)/(0,0,W(2),0)/(0,0,1,1)/$ $(0,0,0,W(2))$
36	$\tilde{A}_1^3$	$\neq 2$	$(2,0,0)/(1,1,1)^2/(0,2,0)/(0,0,2)/(0,0,0)$	$(2,2,0)/(2,0,2)/(2,0,0)/(1,1,1)^2/(0,2,2)/(0,2,0)/(0,0,2)$
38	$\bar{A}_1^2A_1$	≥ 5	$(1,1,0)/(1,0,3)/(0,1,3)/(0,0,4)/(0,0,0)$	$(2,0,0)/(1,1,4)/(1,0,3)/(0,2,0)/(0,1,3)/(0,0,W(6))/(0,0,2)$
39	$\bar{A}_1^2\tilde{A}_1^2$	2	$2\ (1,1,0,0)/(1,0,1,1)/(0,1,1,1)/(0,0,2,0)/$ $(0,0,0,2)/(0,0,0,0)^2$	$(2,0,0,0)/(1,1,2,0)/(1,1,0,2)/(1,1,0,0)/(1,0,1,1)/$ $(0,2,0,0)/(0,1,1,1)/(0,0,2,2)/(0,0,2,0)/(0,0,0,2)/(0,0,0,0)^4$
50	$\tilde{A}_1^2B_2$	2	$(2,0,00)/(1,1,01)/(0,2,00)/(0,0,10)/$ $(0,0,00)^2$	$(2,2,00)/(2,0,10)/(2,0,00)/(1,1,01)/(0,2,10)/(0,2,00)/$ $(0,0,10)/(0,0,02)/(0,0,00)^4$
51	$\tilde{A}_1^4$	2	$(2,0,0,0)/(1,1,1,1)/(0,2,0,0)/(0,0,2,0)/$ $(0,0,0,2)/(0,0,0,0)^2$	$(2,2,0,0)/(2,0,2,0)/(2,0,0,2)/(2,0,0,0)/(1,1,1,1)/(0,2,2,0)/$ $(0,2,0,2)/(0,2,0,0)/(0,0,2,2)/(0,0,2,0)/(0,0,0,2)/(0,0,0,0)^4$
56	$\tilde{A}_1G_2$	2	$(2,00)/(1,10)/(1,00)^2/(0,10)/(0,00)^2$	$(2,10)/(2,00)/(1,10)/(1,00)^2/(0,10)/(0,01)/(0,00)^2$
57	$\bar{A}_1A_1A_1$	$\neq 2$	$(1,2,1)/(0,W(4),0)/(0,2,2)$	$(2,0,0)/(1,W(4),1)/(1,0,W(3))/(0,W(4),2)/(0,2,0)/(0,0,2)$
58	$\bar{A}_1A_1$	≥ 7	$(1,5)/(0,W(8))/(0,4)$	$(2,0)/(1,W(9))/(1,3)/(0,W(10))/(0,6)/(0,2)$
59	$\bar{A}_1\tilde{A}_3$	2	$(1,010)/(0,101)$	$(2,000)/(1,200)/(1,010)/(1,002)/(0,101)/(0,020)/(0,000)^2$
60	$\bar{A}_1G_2$	2	$(1,10)/(0,01)$	$(2,00)/(1,20)/(1,10)/(0,20)/(0,01)/(0,00)^2$
64	$A_1\bar{A}_2$	$\neq 2$	$(W(4),00)/(2,10)/(2,01)/(2,00)$	$(W(4),10)/(W(4),01)/(W(4),00)/(2,00)/(0,W(11))/(0,10)/$ $(0,01)$
65	A_1A_2	3	$(4,00)/(2,11)/(0,00)$	$(4,11)/(2,00)/(0,30)/(0,11)^2/(0,03)/(0,00)$
66	A_1A_1	≥ 7	$(4,0)/(2,6)$	$(4,6)/(2,0)/(0,W(10))/(0,2)$
67	$A_1\tilde{A}_2$	$\neq 2$	$(2,10)/(2,01)/(0,W(11))$	$(W(4),00)/(2,20)/(2,02)/(2,00)/(0,W(11))$

Table 12.3: The composition factors of non-diagonal irreducible connected subgroups of E_6 acting on V_{27} and $L(E_6)$.

ID	X	p	$V_{27} \downarrow X$	$L(E_6) \downarrow X$
5	A_1	≥ 13	$W(16)/8/0$	$W(22)/W(16)/W(14)/10/8/2$
6	A_1	≥ 11	$W(12)/8/4$	$W(16)/W(14)/10^2/8/6/4/2$
7	F_4		$W(\lambda_4)/0$	$W(\lambda_1)/W(\lambda_4)$
8	C_4		$W(\lambda_2)$	$W(2\lambda_1)/W(\lambda_4)$
9	D_4	2	$\lambda_2/0$	$2\lambda_1/\lambda_2^2/2\lambda_3/2\lambda_4/0^2$
10	G_2		$W(20)$	$W(11)/W(01)$
11	A_2	$\neq 2$	$W(22)$	$W(41)/W(14)/W(11)$
24	$\bar{A}_1 A_5$		$(1,\lambda_1)/(0,\lambda_4)$	$(W(2),0)/(1,\lambda_3)/(0,W(\lambda_1+\lambda_5))$
25	$\bar{A}_2^3$		$(10,01,00)/(01,00,10)/$ $(00,10,01)$	$(W(11),00,00)/(10,10,10)/(01,01,01)/(00,W(11),00)/(00,00,W(11))$
26	$A_2 G_2$		$(10,W(10))/(W(02),00)$	$(W(11),W(10))/(W(11),00)/(00,W(01))$
27	$\bar{A}_1 C_3$		$(1,100)/(0,W(010))/(0,000)$	$(W(2),000)/(1,100)/(1,W(001))/(0,W(200))/(0,W(010))$
28	$\bar{A}_1 A_1 A_2$		$(1,1,10)/(0,W(2),10)/$ $(0,0,W(02))$	$(W(2),0,00)/(1,W(3),00)/(1,1,W(11))/(0,W(2),W(11))/(0,W(2),00)/$ $(0,0,W(11))$
29	$\bar{A}_1 A_3$		$(1,010)/(0,W(101))$	$(W(2),000)/(1,W(200))/(1,W(002))/(0,W(101))/(0,W(020))$
30	$\bar{A}_1 A_2$	$\neq 2$	$(1,20)/(0,12)$	$(2,00)/(1,W(30))/(1,W(03))/(0,W(22))/(0,W(11))$
31	$\bar{A}_1 A_1 A_1$	$\neq 2$	$(1,2,1)/(0,W(4),0)/(0,2,2)/$ $(0,0,0)$	$(2,0,0)/(1,W(3),0)/(1,1,W(4))/(1,1,2)/(0,2,W(4))/(0,2,2)/(0,2,0)/$ $(0,0,W(4))/(0,0,2)$
32	$\bar{A}_1 A_1$	≥ 7	$(1,5)/(0,W(8))/(0,4)/(0,0)$	$(2,0)/(1,W(9))/(1,5)/(1,3)/(0,W(10))/(0,W(8))/(0,6)/(0,4)/(0,2)$
33	$\bar{A}_1 G_2$	2	$(1,10)/(0,01)/(0,00)$	$(2,00)/(1,10)^3/(1,00)^2/(0,20)/(0,01)^2/(0,00)^2$
38	$A_1 \bar{A}_2^2$	$\neq 2$	$(2,01,00)/(2,00,10)/(0,10,01)$	$(W(4),00,00)/(2,10,10)/(2,01,01)/(2,00,00)/(0,W(11),00)/(0,00,W(11))$
41	$A_1^2 \bar{A}_2$	$\neq 2$	$(2,2,00)/(2,0,10)/(0,2,01)$	$(W(4),0,00)/(2,2,10)/(2,2,01)/(2,0,00)/(0,W(4),00)/(0,2,00)/$ $(0,0,W(11))$
44	A_1^3	$\neq 2$	$(2,2,0)/(2,0,2)/(0,2,2)$	$(W(4),0,0)/(2,2,2)^2/(2,0,0)/(0,W(4),0)/(0,2,0)/(0,0,W(4))/(0,0,2)$

47	A_1G_2	$\neq 2$	$(W(4),00)/(2,10)/(0,00)$	$(W(4),10)/(W(4),00)/(2,10)/(2,00)/(0,W(01))$
48	A_2A_2	3	$(10,11)/(02,00)$	$(11,11)/(11,00)/(00,30)/(00,11)^2/(00,03)/(00,00)^2$
49	A_2A_1	≥ 7	$(10,6)/(02,0)$	$(11,6)/(11,0)/(00,W(10))/(00,2)$
50	A_1A_2	3	$(4,00)/(2,11)/(0,00)^2$	$(4,11)/(4,00)/(2,11)/(2,00)/(0,30)/(0,11)^2/(0,03)/(0,00)^2$
51	A_1A_1	≥ 7	$(4,0)/(2,6)/(0,0)$	$(4,6)/(4,0)/(2,6)/(2,0)/(0,W(10))/(0,2)$

Table 12.4: The composition factors of non-diagonal irreducible connected subgroups of E_7 acting on V_{56} and $L(E_7)$.

ID	X	p	$V_{56} \downarrow X$	$L(E_7) \downarrow X$
20	A_1	≥ 17	$W(21)/15/11/5$	$W(26)/W(22)/W(18)/16/14/10^2/6/2$
21	A_1	≥ 19	$W(27)/17/9$	$W(34)/W(26)/W(22)/18/14/10/2$
22	A_7		λ_2/λ_6	$W(\lambda_1+\lambda_7)/\lambda_4$
23	D_4	$\neq 2$	λ_2^2	$2\lambda_1/\lambda_2/2\lambda_3/2\lambda_4$
29	A_2	≥ 5	$W(60)/W(06)$	$W(44)/11$
30	$\bar{A}_1D_6$		$(1,\lambda_1)/(0,\lambda_5)$	$(W(2),0)/(1,\lambda_6)/(0,W(\lambda_2))$
31	$\bar{A}_2A_5$		$(10,\lambda_1)/(01,\lambda_5)/(00,\lambda_3)$	$(W(11),0)/(10,\lambda_4)/(01,\lambda_2)/(00,W(\lambda_1+\lambda_5))$
32	G_2C_3		$(W(10),100)/(00,W(001))$	$(W(10),W(010))/(W(01),000)/(00,W(200))$
33	A_1F_4		$(W(3),0)/(1,W(\lambda_4))$	$(W(2),W(\lambda_4))/(W(2),0)/(0,W(\lambda_1))$
34	A_1G_2	$\neq 2$	$(W(3),10)/(1,W(01))$	$(W(4),10)/(2,W(20))/(2,00)/(0,W(01))$
35	A_1A_1	≥ 5	$(W(5),2)/(3,W(6))/(1,4)$	$(W(6),4)/(4,W(6))/(4,2)/(2,W(8))/(2,4)/(2,0)/(0,2)$
36	$\bar{A}_1^3\bar{D}_4$		$(1,1,1,0)/(1,0,0,\lambda_1)/(0,1,0,\lambda_3)/$ $(0,0,1,\lambda_4)$	$(W(2),0,0,0)/(1,1,0,\lambda_4)/(1,0,1,\lambda_3)/(0,W(2),0,0)/(0,1,1,\lambda_1)/$ $(0,0,W(2),0)/(0,0,0,W(\lambda_2))$
37	$\bar{A}_1A_1B_4$		$(1,W(2),0)/(1,0,W(\lambda_1))/(0,1,\lambda_4)$	$(W(2),0,0)/(1,1,\lambda_4)/(0,W(2),W(\lambda_1))/(0,W(2),0)/(0,0,W(\lambda_2))$
38	$\bar{A}_1B_2B_3$		$(1,W(10),000)/(1,00,W(100))/$ $(0,01,001)$	$(W(2),00,000)/(1,01,001)/(0,W(10),W(100))/(0,W(02),000)/$ $(0,00,W(010))$

39	$\bar{A}_1\bar{A}_3^2$		$(1,010,000)/(1,000,010)/(0,100,100)/(0,001,001)$	$(W(2),000,000)/(1,100,001)/(1,001,100)/(0,W(101),000)/(0,010,010)/(0,000,W(101))$
40	$\bar{A}_1B_5$		$(1,W(\lambda_1))/(1,0)/(0,\lambda_5)$	$(W(2),0)/(1,\lambda_5)/(0,W(\lambda_1))/(0,W(\lambda_2))$
41	$\bar{A}_1A_1C_3$		$(1,1,100)/(0,W(3),000)/(0,1,W(010))$	$(W(2),0,000)/(1,W(2),100)/(1,0,W(001))/(0,W(2),W(010))/(0,W(2),000)/(0,0,W(200))$
42	$\bar{A}_1A_1C_3$		$(1,1,100)/(0,W(2),100)/(0,0,W(001))$	$(W(2),0,000)/(1,W(3),000)/(1,1,W(010))/(0,W(2),W(010))/(0,W(2),000)/(0,0,W(200))$
43	$\bar{A}_1^7$		$(1,1,1,0,0,0,0)/(1,0,0,1,1,0,0)/(1,0,0,0,0,1,1)/(0,1,0,1,0,1,0)/(0,1,0,0,1,0,1)/(0,0,1,1,0,0,1)/(0,0,1,0,1,1,0)$	$(W(2),0,0,0,0,0,0)/(1,1,0,1,0,0,1)/(1,1,0,0,1,1,0)/(1,0,1,1,0,1,0)/(1,0,1,0,1,0,1)/(0,W(2),0,0,0,0,0)/(0,1,1,1,1,0,0)/(0,1,1,0,0,1,1)/(0,0,W(2),0,0,0,0)/(0,0,0,1,1,1,1)/(0,0,0,W(2),0,0,0)/(0,0,0,0,W(2),0,0)/(0,0,0,0,0,W(2),0)/(0,0,0,0,0,0,W(2))$
44	$\bar{A}_1^3B_3$		$(1,1,1,000)/(1,0,0,W(100))/(1,0,0,000)/(0,1,0,001)/(0,0,1,001)$	$(W(2),0,0,000)/(1,1,0,001)/(1,0,1,001)/(0,W(2),0,000)/(0,1,1,W(100))/(0,1,1,000)/(0,0,W(2),000)/(0,0,0,W(100))/(0,0,0,W(010))$
45	$\bar{A}_1^3A_1B_2$		$(1,1,1,0,00)/(1,0,0,W(2),00)/(1,0,0,0,W(10))/(0,1,0,1,01)/(0,0,1,1,01)$	$(W(2),0,0,0,00)/(1,1,0,1,01)/(1,0,1,1,01)/(0,W(2),0,0,00)/(0,1,1,W(2),00)/(0,1,1,0,W(10))/(0,0,W(2),0,00)/(0,0,0,W(2),W(10))/(0,0,0,W(2),00)/(0,0,0,0,W(02))$
46	$\bar{A}_1^3A_2$		$(1,1,1,00)/(1,0,0,W(11))/(0,1,0,W(11))/(0,0,1,W(11))$	$(W(2),0,0,00)/(1,1,0,11)/(1,0,1,11)/(0,W(2),0,00)/(0,1,1,11)/(0,0,W(2),00)/(0,0,0,W(30))/(0,0,0,W(11))/(0,0,0,W(03))$
75	$\bar{A}_1^3G_2$		$(1,1,1,00)/(1,0,0,W(10))/(1,0,0,00)/(0,1,0,W(10))/(0,1,0,00)/(0,0,1,W(10))/(0,0,1,00)$	$(W(2),0,0,00)/(1,1,0,W(10))/(1,1,0,00)/(1,0,1,W(10))/(1,0,1,00)/(0,W(2),0,00)/(0,1,1,W(10))/(0,1,1,00)/(0,0,W(2),00)/(0,0,0,W(10))^2/(0,0,0,W(01))$
79	$\bar{A}_1^3A_1$	≥ 7	$(1,1,1,0)/(1,0,0,6))/(1,0,0,0)/(0,1,0,6)/(0,1,0,0)/(0,0,1,6)/(0,0,1,0)$	$(2,0,0,0)/(1,1,0,6)/(1,1,0,0)/(1,0,1,6)/(1,0,1,0)/(0,2,0,0)/(0,1,1,6)/(0,1,1,0)/(0,0,2,0)/(0,0,0,W(10))/(0,0,0,6)^2/(0,0,0,2)$

87	$\bar{A}_1^3A_1A_1$	≥ 5	$(1,1,1,0,0)/(1,0,0,2,0)/$ $(1,0,0,0,4)/(0,1,0,1,3)/(0,0,1,1,3)$	$(2,0,0,0,0)/(1,1,0,1,3)/(1,0,1,1,3)/(0,2,0,0,0)/(0,1,1,2,0)/$ $(0,1,1,0,4)/(0,0,2,0,0)/(0,0,0,2,4)/(0,0,0,2,0)/$ $(0,0,0,0,W(6))/(0,0,0,0,2)$
88	$\bar{A}_1^3A_1^3$	2	$(1,1,1,0,0,0)/(1,0,0,2,0,0)/$ $(1,0,0,0,2,0)/(1,0,0,0,0,2)/$ $(0,1,0,1,1,1)/(0,0,1,1,1,1)$	$(2,0,0,0,0,0)/(1,1,0,1,1,1)/(1,0,1,1,1,1)/(0,2,0,0,0,0)/$ $(0,1,1,2,0,0)/(0,1,1,0,2,0)/(0,1,1,0,0,2)/(0,0,2,0,0,0)/$ $(0,0,0,2,2,0)/(0,0,0,2,0,2)/(0,0,0,2,0,0)^2/(0,0,0,0,2,2)/$ $(0,0,0,0,2,0)^2/(0,0,0,0,0,2)^2/(0,0,0,0,0,0)^7$
185	$\bar{A}_1A_1^2\bar{A}_3$		$(1,W(2),0,000)/(1,0,W(2),000)/$ $(1,0,0,010)/(0,1,1,100)/(0,1,1,001)$	$(W(2),0,0,000)/(1,1,1,100)/(1,1,1,001)/(0,W(2),W(2),000)/$ $(0,W(2),0,010)/(0,W(2),0,000)/(0,0,W(2),010)/$ $(0,0,W(2),000)/(0,0,0,W(101))$
186	$\bar{A}_1A_1A_1^2$	$\neq 2$	$(1,2,0,0)/(1,0,2,2)/(0,1,W(3),1)/$ $(0,1,1,W(3))$	$(2,0,0,0)/(1,1,W(3),1)/(1,1,1,W(3))/(0,2,2,2)/(0,2,0,0)/$ $(0,0,W(4),2)/(0,0,2,W(4))/(0,0,2,0)/(0,0,0,2)$
187	$\bar{A}_1A_1A_1$	≥ 11	$(1,2,0)/(1,0,8)/(0,1,W(10))/(0,1,4)$	$(2,0,0)/(1,1,W(10))/(1,1,4)/(0,2,8)/(0,2,0)/(0,0,W(14))/$ $(0,0,W(10))/(0,0,6)/(0,0,2)$
188	$\bar{A}_1A_1B_2^2$	2	$(1,2,00,00)/(1,0,10,00)/$ $(1,0,00,10)/(1,0,00,00)^2/$ $(0,1,01,01)$	$(2,0,00,00)/(1,1,01,01)/(0,2,10,00)/(0,2,00,10)/(0,2,00,00)^2/$ $(0,0,10,10)/(0,0,10,00)^2/(0,0,02,00)/(0,0,00,10)^2/(0,0,00,02)/$ $(0,0,00,00)^7$
189	$\bar{A}_1A_1^2B_3$	2	$(1,2,0,000)/(1,0,2,000)/$ $(1,0,0,100)/\ (1,0,0,000)^2/$ $(0,1,1,001)$	$(2,0,0,000)/(1,1,1,001)/(0,2,2,000)/(0,2,0,100)/(0,2,0,000)^2/$ $(0,0,2,100)/(0,0,2,000)^2/(0,0,0,100)^2/(0,0,0,010)/\ (0,0,0,000)^5$
191	$\bar{A}_1A_1^4$	$\neq 2$	$(1,2,0,0,0)/(1,0,2,0,0)/$ $(1,0,0,2,0)/(1,0,0,0,2)/$ $(0,1,1,1,1)^2$	$(2,0,0,0,0)/(1,1,1,1,1)^2/(0,2,2,0,0)/(0,2,0,2,0)/(0,2,0,0,2)/$ $(0,2,0,0,0)/(0,0,2,2,0)/(0,0,2,0,2)/(0,0,2,0,0)/(0,0,0,2,2)/$ $(0,0,0,2,0)/(0,0,0,0,2)$
217	$\bar{A}_1A_1^3B_2$	2	$(1,2,0,0,00)/(1,0,2,0,00)/$ $(1,0,0,2,00)/(1,0,0,0,10)/$ $(1,0,0,0,00)^2/(0,1,1,1,01)$	$(2,0,0,0,00)/(1,1,1,1,01)/(0,2,2,0,00)/(0,2,0,2,00)/$ $(0,2,0,0,10)/(0,2,0,0,00)^2/(0,0,2,2,00)/(0,0,2,0,10)/$ $(0,0,2,0,00)^2/(0,0,0,2,10)/(0,0,0,2,00)^2/(0,0,0,0,10)^2/$ $(0,0,0,0,02)/(0,0,0,0,00)^7$

223	$\bar{A}_1A_1^5$	2	$(1,2,0,0,0,0)/(1,0,2,0,0,0)/(1,0,0,2,0,0)/(1,0,0,0,2,0)/(1,0,0,0,0,2)/(0,1,1,1,1,1)$	$(2,0,0,0,0,0)/(1,1,1,1,1,1)/(0,2,2,0,0,0)/(0,2,0,2,0,0)/(0,2,0,0,2,0)/(0,2,0,0,0,2)/(0,2,0,0,0,0)^2/(0,0,2,2,0,0)/(0,0,2,0,2,0)/(0,0,2,0,0,2)/(0,0,2,0,0,0)^2/(0,0,0,2,2,0)/(0,0,0,2,0,2)/(0,0,0,2,0,0)/(0,0,0,0,2,2)/(0,0,0,0,2,0)/(0,0,0,0,0,2)/(0,0,0,0,0,0)^7$
244	$\bar{A}_1A_1^2G_2$	2	$(1,2,0,00)/(1,0,2,00)/(1,0,0,10)/(1,0,0,00)^2/(0,1,1,10)/(0,1,1,00)^2$	$(2,0,0,00)/(1,1,1,10)/(1,1,1,00)^2/(0,2,2,00)/(0,2,0,10)/(0,2,0,00)^2/(0,0,2,10)/(0,0,2,00)^2/(0,0,0,10)^2/(0,0,0,01)/(0,0,0,00)^5$
251	$\bar{A}_1A_1B_3$	≥ 5	$(1,4,000)/(1,0,100)/(0,3,001)$	$(2,0,000)/(1,3,001)/(0,W(6),000)/(0,4,W(100))/(0,2,000)/(0,0,W(010))$
252	$\bar{A}_1B_2\bar{A}_3$		$(1,W(10),000)/(1,00,010)/(0,00,000)/(0,01,100)/(0,01,001)$	$(W(2),00,000)/(1,01,100)/(1,01,001)/(0,W(10),010)/(0,W(10),000)/(0,W(02),000)/(0,00,W(101))/(0,00,010)$
253	$\bar{A}_1B_2G_2$		$(1,W(10),00)/(1,00,W(10))/(0,01,W(10))/(0,01,00)$	$(W(2),00,00)/(1,01,W(10))/(1,01,00)/(0,W(10),W(10))/(0,W(02),00)/(0,00,W(10))/(0,00,W(01))$
254	$\bar{A}_1A_1\bar{A}_3$	≥ 5	$(1,4,000)/(1,0,000)/(1,0,010)/(0,3,100)/(0,3,001)$	$(2,0,000)/(1,3,100)/(1,3,001)/(0,W(6),000)/(0,4,000)/(0,3,010)/(0,2,000)/(0,0,W(101))$
255	$\bar{A}_1A_1G_2$	≥ 5	$(1,4,00)/(1,0,10)/(0,3,10)/(0,3,00)$	$(2,0,00)/(1,3,10)/(1,3,00)/(0,W(6),00)/(0,4,10)/(0,2,00)/(0,0,10)/(0,0,01)$
258	$\bar{A}_1A_1A_1$	≥ 7	$(1,4,0)/(1,0,6)/(0,3,6)/(0,3,0)$	$(2,0,0)/(1,3,6)/(1,3,0)/(0,6,0)/(0,4,6)/(0,2,0)/(0,0,W(10))/(0,0,6)/(0,0,2)$
263	$\bar{A}_1B_2A_2$	3	$(1,10,00)/(1,00,11)/(0,01,11)/(0,01,00)$	$(2,00,00)/(1,01,11)/(1,01,00)/(0,10,11)/(0,02,00)/(0,00,30)/(0,00,11)^2/(0,00,03)/(0,00,00)$
264	$\bar{A}_1B_2A_1$	≥ 7	$(1,10,0)/(1,00,6)/(0,01,6)/(0,01,0)$	$(2,00,0)/(1,01,6)/(1,01,0)/(0,10,6)/(0,02,0)/(0,00,W(10))/(0,00,6)/(0,00,2)$
267	$\bar{A}_1A_1$	≥ 11	$(1,10)/(1,0)/(0,W(15))/(0,9)/(0,5)$	$(2,0)/(1,W(15))/(1,9)/(1,5)/(0,W(18))/(0,W(14))/(0,10)^2/(0,6)/(0,2)$
268	$\bar{A}_1A_1A_1A_1$	$\neq 2$	$(1,1,2,1)/(0,W(3),0,0)/(0,1,W(4),0)/(0,1,2,2)$	$(2,0,0,0)/(1,2,2,1)/(1,0,W(4),1)/(1,0,0,W(3))/(0,2,W(4),0)/(0,2,2,2)/(0,2,0,0)/(0,0,W(4),2)/(0,0,2,0)/(0,0,0,2)$

269	$\bar{A}_1A_1A_1$	≥ 7	$(1,1,5)/(0,3,0)/(0,1,W(8))/(0,1,4)$	$(2,0,0)/(1,2,5)/(1,0,W(9))/(1,0,3)/(0,2,W(8))/(0,2,4)/(0,2,0)/$ $(0,0,W(10))/(0,0,6)/(0,0,2)$
270	$\bar{A}_1A_1A_3$	2	$(1,1,010)/(0,3,000)/(0,1,101)$	$(2,0,000)/(1,2,010)/(1,0,200)/(1,0,010)^2/(1,0,002)/(0,2,101)/$ $(0,2,000)/(0,0,101)^2/(0,0,020)/(0,0,000)^3$
271	$\bar{A}_1A_1G_2$	2	$(1,1,10)/(0,3,00)/(0,1,01)$	$(2,0,00)/(1,2,10)/(1,0,20)/(1,0,10)^2/(1,0,00)^2/(0,2,01)/$ $(0,2,00)/(0,0,20)/(0,0,01)^2/(0,0,00)^3$
291	$\bar{A}_1A_1A_1$	≥ 7	$(1,1,5)/(0,2,5)/(0,0,W(9))/(0,0,3)$	$(2,0,0)/(1,3,0)/(1,1,W(8))/(1,1,4)/(0,2,W(8))/(0,2,4)/(0,2,0)/$ $(0,0,W(10))/(0,0,6)/(0,0,2)$
292	$\bar{A}_1A_1A_3$	2	$(1,1,010)/(0,2,010)/(0,0,200)/$ $(0,0,010)^2/(0,0,002)$	$(2,0,000)/(1,3,000)/(1,1,101)/(0,2,101)/(0,2,000)/(0,0,101)^2/$ $(0,0,020)/(0,0,000)^3$
293	$\bar{A}_1A_1G_2$	2	$(1,1,10)/(0,2,10)/(0,0,20)/$ $(0,0,10)^2/(0,0,00)^2$	$(2,0,00)/(1,3,00)/(1,1,01)/(0,2,01)/(0,2,00)/(0,0,20)/$ $(0,0,01)^2/(0,0,00)^3$
300	$\bar{A}_2A_2A_1$		$(10,10,1)/(01,01,1)/(00,W(11),1)/$ $(00,00,W(3))$	$(W(11),00,0)/(10,10,W(2))/(10,W(02),0)/(01,W(20),0)/$ $(01,01,W(2))/(00,W(11),W(2))/(00,W(11),0)/(00,00,W(2))$
301	$\bar{A}_2C_3$		$(10,100)/(01,100)/(00,100)/$ $(00,W(001))$	$(W(11),000)/(10,W(010))/(10,000)/(01,W(010))/(01,000)/$ $(00,W(200))/(00,W(010))$
302	$\bar{A}_2A_3$		$(10,010)/(01,010)/(00,W(200))/$ $(00,W(002))$	$(W(11),000)/(10,W(101))/(01,W(101))/(00,W(101))/$ $(00,W(020))$
303	$\bar{A}_2A_2$	$\neq 2$	$(10,20)/(01,02)/(00,W(30))/$ $(00,W(03))$	$(W(11),00)/(10,12)/(01,21)/(00,W(22))/(00,W(11))$
304	A_1A_5	$\neq 2$	$(2,\lambda_1)/(2,\lambda_5)/(0,\lambda_3)$	$(W(4),0)/(2,\lambda_2)/(2,\lambda_4)/(2,0)/(0,W(\lambda_1+\lambda_5))$
305	$A_1A_2A_1$	$\neq 2$	$(2,10,1)/(2,01,1)/(0,W(11),1)/$ $(0,00,W(3))$	$(W(4),00,0)/(2,20,0)/(2,10,2)/(2,02,0)/(2,01,2)/(2,00,0)/$ $(0,W(11),2)/(0,W(11),0)/(0,00,W(2))$
306	$\bar{A}_2A_1A_1$	$\neq 2$	$(10,2,1)/(01,2,1)/(00,W(4),1)/$ $(00,2,1)/(00,0,W(3))$	$(W(11),0,0)/(10,W(4),0)/(10,2,2)/(10,0,0)/(01,W(4),0)/$ $(01,2,W(2))/(01,0,0)/(00,W(4),2)/(00,W(4),0)/(00,2,2)/$ $(00,2,0)/(00,0,2)$
311	$\bar{A}_2A_1$	≥ 7	$(10,5)/(01,5)/(00,W(9))/(00,5)/$ $(00,3)$	$(W(11),0)/(10,W(8))/(10,4)/(10,0)/(01,W(8))/(01,4)/(01,0)/$ $(00,W(10))/(00,W(8))/(00,6)/(00,4)/(00,2)$

312	$\bar{A}_2G_2$	2	$(10,10)/(01,10)/(00,20)/(00,10)^2/(00,00)^2$	$(W(11),00)/(10,01)/(10,00)/(01,01)/(01,00)/(00,20)/(00,01)^2$
313	A_1A_3	$\neq 2$	$(2,010)^2/(0,200)/\ (0,002)$	$(W(4),000)/(2,101)^2/(2,000)/(0,101)/(0,W(020))$
314	A_1A_2	$\neq 2$	$(2,20)/(2,02)/(0,W(30))/(0,W(03))$	$(W(4),00)/(2,21)/(2,12)/(2,00)/(0,W(22))/(0,W(11))$
315	A_2C_3	3	$(11,100)/(00,001)$	$(30,000)/\ (11,010)/(11,000)^2/(03,000)/(00,200)/(00,000)$
316	A_1C_3	≥ 7	$(6,100)/(0,001)$	$(W(10),000)/(6,010)/(2,000)/(0,200)$
317	$G_2A_1A_1$	$\neq 2$	$(10,2,1)/(00,W(4),1)/(00,0,W(3))$	$(10,W(4),0)/(10,2,2)/(01,0,0)/(00,W(4),2)/(00,2,0)/(00,0,2)$
318	G_2A_1	≥ 7	$(10,5)/(00,W(9))/(00,3)$	$(10,W(8))/(10,4)/(01,0)/(00,W(10))/(00,6)/(00,2)$
319	G_2A_3	2	$(10,010)/(00,200)/(00,010)^2/(00,002)$	$(10,101)/(01,000)/(00,101)^2/(00,020)/(00,000)$
320	G_2G_2	2	$(10,10)/(00,20)/(00,10)^2/(00,00)^2$	$(10,01)/(01,00)/(00,20)/(00,01)^2/(00,00)$
321	$A_2A_1A_1$	3	$(11,2,1)/(00,4,1)/\ (00,0,3)/(00,0,1)^2$	$(30,0,0)/(11,4,0)/(11,2,2)/(11,0,0)^2/(03,0,0)/(00,4,2)/(00,2,0)/(00,0,2)^2/(00,0,0)$
323	$A_1A_1A_1$	≥ 7	$(6,2,1)/(0,4,1)/(0,0,3)$	$(W(10),0,0)/(6,4,0)/(6,2,2)/(2,0,0)/(0,4,2)/(0,2,0)/(0,0,2)$
324	A_1A_1	≥ 7	$(6,5)/(0,W(9))/(0,3)$	$(W(10),0)/(6,W(8))/(6,4)/(2,0)/(0,W(10))/(0,6)/(0,2)$
329	A_1C_4	2	$(3,0)/(1,\lambda_2)$	$(2,\lambda_2)/(2,0)/(0,2\lambda_1)/(0,\lambda_2)^2/(0,\lambda_4)/(0,0)^3$
330	A_1G_2	7	$(3,00)/(1,20)$	$(2,20)/(2,00)/(0,11)/(0,01)$
331	A_1A_1	≥ 13	$(3,0)/(1,W(16))/(1,8)$	$(2,W(16))/(2,8)/(2,0)/(0,W(22))/(0,W(14))/(0,10)/(0,2)$
332	A_1D_4	2	$(3,0)/(1,\lambda_2)$	$(2,\lambda_2)/(2,0)/(0,\lambda_2)^2/(0,2\lambda_1)/(0,2\lambda_3)/(0,2\lambda_4)/(0,0)^3$
333	A_1A_2	3	$(3,11)/(1,30)/(1,11)^2/(1,03)/(1,00)$	$(4,11)/(2,W(22))/(2,00)/(0,30)/(0,11)^2/(0,03)/(0,00)$
334	A_1A_1	≥ 7	$(3,6)/(1,W(10))/(1,2)$	$(4,6)/(2,W(12))/(2,W(8)/(2,4)/(2,0)/(0,W(10))/(0,2)$

Table 12.5: The composition factors of non-diagonal irreducible connected subgroups of E_8 acting on $L(E_8)$.

ID	X	p	$L(E_8) \downarrow X$
10	A_1	≥ 17	$W(28)/W(26)/W(22)^2/W(18)^2/16/14^3/10^2/8/6/4/2$
40	A_1	≥ 23	$W(38)/W(34)/W(28)/W(26)/22^2/18/16/14/10/6/2$
41	A_1	≥ 29	$W(46)/W(38)/W(34)/28/26/22/18/14/10/2$
42	A_1	≥ 31	$W(58)/W(46)/W(38)/W(34)/26/22/14/2$
43	D_8		$W(\lambda_2)/\lambda_7$
44	B_7		$W(\lambda_1)/W(\lambda_2)/\lambda_7$
45	B_4		$W(\lambda_1+\lambda_4)/W(\lambda_2)/W(\lambda_3)$
46	B_4	2	$W(2\lambda_1)/\lambda_2/\lambda_3^2$
47	A_3	$\neq 2$	$W(111)^2/210/101^2/012$
62	A_8		$W(\lambda_1+\lambda_8)/\lambda_3/\lambda_5$
101	B_2	≥ 5	$W(32)/W(06)/02$
102	$\bar{A}_1E_7$		$(W(2),0)/(1,\lambda_7)/(0,W(\lambda_1))$
103	$\bar{A}_2E_6$		$(W(11),0)/(10,\lambda_6)/(01,\lambda_1)/(00,W(\lambda_2))$
104	$\bar{A}_4^2$		$(W(\lambda_1+\lambda_4),0)/(\lambda_1,\lambda_2)/(\lambda_2,\lambda_4)/(\lambda_3,\lambda_1)/(\lambda_4,\lambda_3)/(0,W(\lambda_1+\lambda_4))$
105	G_2F_4		$(W(10),W(\lambda_4))/(W(01),0)/(00,W(\lambda_1))$
106	A_1A_2	≥ 5	$(W(6),11)/(4,30)/(4,03)/(2,W(22))/(2,00)/(0,11)$
107	$\bar{A}_1^2D_6$		$(W(2),0,0)/(1,1,\lambda_1)/(1,0,\lambda_5)/(0,W(2),0)/(0,1,\lambda_6)/(0,0,W(\lambda_2))$
108	$\bar{D}_4^2$		$(\lambda_1,\lambda_1)/(W(\lambda_2),0)/(\lambda_3,\lambda_3)/(\lambda_4,\lambda_4)/(0,W(\lambda_2))$
109	$\bar{A}_3D_5$		$(W(101),0)/(100,\lambda_4)/(010,\lambda_1)/(001,\lambda_5)/(000,W(\lambda_2))$
110	A_1B_6		$(W(2),W(\lambda_1))/(W(2),0)/(1,\lambda_6)/(0,W(\lambda_2))$
111	B_2B_5		$(W(10),W(\lambda_1))/(W(02),0)/(01,\lambda_5)/(00,W(\lambda_2))$
112	B_3B_4		$(W(100),W(\lambda_1))/(W(010),0)/(001,\lambda_4)/(000,W(\lambda_2))$
113	B_2^2		$(W(11),01)/(W(10),W(02))/(W(02),W(10))/(W(02),00)/(01,W(11))/(00,W(02))$

114	B_2^2	$\neq 2$	$(W(20),00)/(10,02)^2/(02,10)^2/(02,00)/(00,W(20))/(00,02)$
115	A_1C_4		$(W(3),\lambda_1)/(W(2),W(\lambda_2))/(W(2),0)/(1,W(\lambda_3))/(0,W(2\lambda_1))$
116	A_1C_4	$\neq 2$	$(W(4),0)/(2,\lambda_2)^2/(2,0)/(0,2\lambda_1)/(0,W(\lambda_4))$
117	$\bar{A}_1^4\bar{D}_4$		$(W(2),0,0,0,0)/(1,1,1,1,0)/(1,1,0,0,\lambda_1)/(1,0,1,0,\lambda_3)/(1,0,0,1,\lambda_4)/(0,W(2),0,0,0)/(0,1,1,0,\lambda_4)/$ $(0,1,0,1,\lambda_3)/(0,0,W(2),0,0)/(0,0,1,1,\lambda_1)/(0,0,0,W(2),0)/(0,0,0,0,W(\lambda_2))$
118	$\bar{A}_1^2B_5$		$(W(2),0,0)/(1,1,W(\lambda_1))/(1,1,0)/(1,0,\lambda_5)/(0,W(2),0)/(0,1,\lambda_5)/(0,0,W(\lambda_1))/(0,0,W(\lambda_2))$
119	$\bar{A}_1^2\bar{A}_3^2$		$(W(2),0,000,000)/(1,1,010,000)/(1,1,000,010)/(1,0,100,100)/(1,0,001,001)/(0,W(2),000,000)/$ $(0,1,100,001)/(0,1,001,100)/(0,0,W(101),000)/(0,0,010,010)/(0,0,000,W(101))$
120	$\bar{A}_1^2A_1B_4$		$(W(2),0,0,0)/(1,1,W(2),0)/(1,1,0,W(\lambda_1))/(1,0,1,\lambda_4)/(0,W(2),0,0)/(0,1,1,\lambda_4)/(0,0,W(2),W(\lambda_1))/$ $(0,0,W(2),0)/(0,0,0,W(\lambda_2))$
121	$\bar{A}_1^2B_2B_3$	2	$(W(2),0,00,000)/(1,1,W(10),000)/(1,1,00,W(100))/(1,0,01,001)/(0,W(2),00,000)/(0,1,01,001)/$ $(0,0,W(10),W(100))/(0,0,W(02),000)/(0,0,00,W(010))$
122	$\bar{A}_1^2A_1C_3$		$(W(2),0,0,000)/(1,1,1,100)/(1,0,W(3),000)/(1,0,1,W(010))/(0,W(2),0,000)/(0,1,W(2),100)/$ $(0,1,0,W(001))/(0,0,W(2),W(010))/(0,0,W(2),000)/(0,0,0,W(200))$
124	$\bar{A}_1^8$		$(W(2),0,0,0,0,0,0,0)/(1,1,1,1,0,0,0,0)/(1,1,0,0,1,1,0,0)/(1,1,0,0,0,0,1,1)/(1,0,1,0,1,0,1,0)/$ $(1,0,1,0,0,1,0,1)/(1,0,0,1,1,0,0,1)/(1,0,0,1,0,1,1,0)/(0,W(2),0,0,0,0,0,0)/(0,1,1,0,1,0,0,1)/$ $(0,1,1,0,0,1,1,0)/(0,1,0,1,1,0,1,0)/(0,1,0,1,0,1,0,1)/(0,0,W(2),0,0,0,0,0)/(0,0,1,1,1,1,0,0)/$ $(0,0,1,1,0,0,1,1)/(0,0,0,W(2),0,0,0,0)/(0,0,0,0,W(2),0,0,0)/(0,0,0,0,1,1,1,1)/$ $(0,0,0,0,0,W(2),0,0)/(0,0,0,0,0,0,W(2),0)/(0,0,0,0,0,0,0,W(2))$
125	$\bar{A}_1^4B_3$		$(W(2),0,0,0,000)/(1,1,1,1,000)/(1,1,0,0,W(100))/(1,1,0,0,000)/(1,0,1,0,001)/(1,0,0,1,001)/$ $(0,W(2),0,0,000)/(0,1,1,0,001)/(0,1,0,1,001)/(0,0,W(2),0,000)/(0,0,1,1,W(100))/(0,0,1,1,000)/$ $(0,0,0,W(2),000)/(0,0,0,0,W(100))/(0,0,0,0,W(010))$
126	$\bar{A}_1^4A_1B_2$		$(W(2),0,0,0,0,00)/(1,1,1,1,0,00)/(1,1,0,0,W(2),00)/(1,1,0,0,0,W(10))/(1,0,1,0,1,01)/$ $(1,0,0,1,1,01)/(0,W(2),0,0,0,00)/(0,1,1,0,1,01)/(0,1,0,1,1,01)/(0,0,W(2),0,0,00)/$ $(0,0,1,1,W(2),00)/(0,0,1,1,0,W(10))/(0,0,0,W(2),0,00)/(0,0,0,0,W(2),W(10))/(0,0,0,0,W(2),00)/$ $(0,0,0,0,0,W(02))$

127	$\bar{A}_1^4 A_2$		$(W(2),0,0,0,00)/(1,1,1,1,00)/(1,1,0,0,W(11))/(1,0,1,0,W(11))/(1,0,0,1,W(11))/(0,W(2),0,0,00)/$ $(0,1,1,0,W(11))/(0,1,0,1,W(11))/(0,0,W(2),0,00)/(0,0,1,1,W(11))/(0,0,0,W(2),00)/$ $(0,0,0,0,W(30))/(0,0,0,0,W(11))/(0,0,0,0,W(03))$
167	$\bar{A}_1^4 G_2$		$(W(2),0,0,0,00)/(1,1,1,1,00)/(1,1,0,0,W(10))/(1,1,0,0,00)/(1,0,1,0,W(10))/(1,0,1,0,00)/$ $(1,0,0,1,W(10))/(1,0,0,1,00)/(0,W(2),0,0,00)/(0,1,1,0,W(10))/(0,1,1,0,00)/(0,1,0,1,W(10))/$ $(0,1,0,1,00)/(0,0,W(2),0,00)/(0,0,1,1,W(10))/(0,0,1,1,00)/(0,0,0,W(2),00)/(0,0,0,0,W(10))^2/$ $(0,0,0,0,01)$
174	$\bar{A}_1^4 A_1$	≥ 7	$(2,0,0,0,0)/(1,1,1,1,0)/(1,1,0,0,6)/(1,1,0,0,0)/(1,0,1,0,6)/(1,0,1,0,0)/(1,0,0,1,6)/(1,0,0,1,0)/$ $(0,2,0,0,0)/(0,1,1,0,6)/(0,1,1,0,0)/(0,1,0,1,6)/(0,1,0,1,0)/(0,0,2,0,0)/(0,0,1,1,6)/(0,0,1,1,0)/$ $(0,0,0,2,0)/(0,0,0,0,W(10))/(0,0,0,0,6)^2/(0,0,0,0,2)$
189	$\bar{A}_1^4 A_1 A_1$	≥ 5	$(2,0,0,0,0,0)/(1,1,1,1,0,0)/(1,1,0,0,2,0)/(1,1,0,0,0,4)/(1,0,1,0,1,3)/(1,0,0,1,1,3)/(0,2,0,0,0,0)/$ $(0,1,1,0,1,3)/(0,1,0,1,1,3)/(0,0,2,0,0,0)/(0,0,1,1,2,0)/(0,0,1,1,0,4)/(0,0,0,2,0,0)/(0,0,0,0,2,4)/$ $(0,0,0,0,2,0)/(0,0,0,0,0,W(6))/(0,0,0,0,0,2)$
190	$\bar{A}_1^4 A_1^3$	2	$(2,0,0,0,0,0,0)/(1,1,1,1,0,0,0)/(1,1,0,0,2,0,0)/(1,1,0,0,0,2,0)/(1,1,0,0,0,0,2)/(1,1,0,0,0,0,0)^2/$ $(1,0,1,0,1,1,1)/(1,0,0,1,1,1,1)/(0,2,0,0,0,0,0)/(0,1,1,0,1,1,1)/(0,1,0,1,1,1,1)/(0,0,2,0,0,0,0)/$ $(0,0,1,1,2,0,0)/(0,0,1,1,0,2,0)/(0,0,1,1,0,0,2)/(0,0,1,1,0,0,0)^2/(0,0,0,2,0,0,0)/(0,0,0,0,2,2,0)/$ $(0,0,0,0,2,0,2)/(0,0,0,0,2,0,0)^2/(0,0,0,0,0,2,2)/(0,0,0,0,0,2,0)^2/(0,0,0,0,0,0,2)^2/(0,0,0,0,0,0,0)^8$
350	$\bar{A}_1^2 B_2 \bar{A}_3$		$(W(2),0,00,000)/(1,1,W(10),000)/(1,1,00,010)/(1,1,00,000)/(1,0,01,100)/(1,0,01,001)/$ $(0,W(2),00,000)/(0,1,01,100)/(0,1,01,001)/(0,0,W(10),010)/(0,0,W(10),000)/(0,0,W(02),000)/$ $(0,0,00,W(101))/(0,0,00,010)$
351	$\bar{A}_1^2 A_1$	≥ 11	$(2,0,0)/(1,1,10)/(1,1,0)/(1,0,W(15))/(1,0,9)/(1,0,5)/(0,2,0)/(0,1,W(15))/(0,1,9)/(0,1,5)/$ $(0,0,W(18))/(0,0,W(14))/(0,0,10)^2/(0,0,6)/(0,0,2)$
355	$\bar{A}_1^2 A_1^2 \bar{A}_3$		$(W(2),0,0,0,000)/(1,1,W(2),0,000)/(1,1,0,W(2),000)/(1,1,0,0,010)/(1,0,1,1,100)/(1,0,1,1,001)/$ $(0,W(2),0,0,000)/(0,1,1,1,100)/(0,1,1,1,001)/(0,0,W(2),W(2),000)/(0,0,W(2),0,010)/$ $(0,0,W(2),0,000)/(0,0,0,W(2),010)/(0,0,0,W(2),000)/(0,0,0,0,W(101))$
360	$\bar{A}_1^2 A_1 \bar{A}_3$	≥ 5	$(2,0,0,000)/(1,1,4,000)/(1,1,0,010)/(1,1,0,000)/(1,0,3,100)/(1,0,3,001)/(0,2,0,000)/(0,1,3,100)/$ $(0,1,3,001)/(0,0,W(6),000)/(0,0,4,010)/(0,0,4,000)/(0,0,2,000)/(0,0,0,101)/(0,0,0,010)$

365	$\bar{A}_1^2 A_1^4$	$\neq 2$	$(2,0,0,0,0,0)/(1,1,2,0,0,0)/(1,1,0,2,0,0)/(1,1,0,0,2,0)/(1,1,0,0,0,2)/(1,0,1,1,1,1)^2/(0,2,0,0,0,0)/(0,1,1,1,1,1)^2/(0,0,2,0,2,0)/(0,0,2,0,0,2)/(0,0,0,2,2,0)/(0,0,0,2,0,2)/(0,0,2,2,0,0)/(0,0,2,0,0,0)/(0,0,0,2,0,0)/(0,0,0,0,2,2)/(0,0,0,0,2,0)/(0,0,0,0,0,2)$
401	$\bar{A}_1^2 A_1 A_1^2$	$\neq 2$	$(2,0,0,0,0)/(1,1,2,0,0)/(1,1,0,2,2)/(1,0,1,W(3),1)/(1,0,1,1,W(3))/(0,W(2),0,0,0)/(0,1,1,W(3),1)/(0,1,1,1,W(3))/(0,0,2,2,2)/(0,0,2,0,0)/(0,0,0,W(4),2)/(0,0,0,2,W(4))/(0,0,0,2,0)/(0,0,0,0,2)$
402	$\bar{A}_1^2 A_1 A_1$	≥ 11	$(2,0,0,0)/(1,1,2,0)/(1,1,0,8)/(1,0,1,10)/(1,0,1,4)/(0,2,0,0)/(0,1,1,10)/(0,1,1,4)/(0,0,2,8)/(0,0,2,0)/(0,0,0,W(14))/(0,0,0,10)/(0,0,0,6)/(0,0,0,2)$
403	$\bar{A}_1^2 A_1^2 B_3$	2	$(2,0,0,0,000)/(1,1,2,0,000)/(1,1,0,2,000)/(1,1,0,0,100)/(1,1,0,0,000)^2/(1,0,1,1,001)/(0,2,0,0,000)/(0,1,1,1,001)/(0,0,2,2,000)/(0,0,2,0,100)/(0,0,2,0,000)^2/(0,0,0,2,100)/(0,0,0,2,000)^2/(0,0,0,0,100)^2/(0,0,0,0,010)/(0,0,0,0,000)^6$
404	$\bar{A}_1^2 A_1 B_2^2$	2	$(2,0,0,00,00)/(1,1,2,00,00)/(1,1,0,10,00)/(1,1,0,00,10)/(1,1,0,00,00)^2/(1,0,1,01,01)/(0,2,0,00,00)/(0,1,1,01,01)/(0,0,2,10,00)/(0,0,2,00,10)/(0,0,2,00,00)^2/(0,0,0,10,10)/(0,0,0,10,00)^2/(0,0,0,02,00)/(0,0,0,00,10)^2/(0,0,0,00,02)/(0,0,0,00,00)^8$
441	$\bar{A}_1^2 A_1^3 B_2$	2	$(2,0,0,0,0,00)/(1,1,2,0,0,00)/(1,1,0,2,0,00)/(1,1,0,0,2,00)/(1,1,0,0,0,10)/(1,1,0,0,0,00)^2/(1,0,1,1,1,01)/(0,2,0,0,0,00)/(0,1,1,1,1,01)/(0,0,2,2,0,00)/(0,0,2,0,2,00)/(0,0,2,0,0,10)/(0,0,2,0,0,00)^2/(0,0,0,2,2,00)/(0,0,0,2,0,10)/(0,0,0,2,0,00)^2/(0,0,0,0,2,10)/(0,0,0,0,2,00)^2/(0,0,0,0,0,10)^2/(0,0,0,0,0,02)/(0,0,0,0,0,00)^8$
442	$\bar{A}_1^2 A_1^2 G_2$	2	$(2,0,0,0,00)/(1,1,2,0,00)/(1,1,0,2,00)/(1,1,0,0,10)/(1,1,0,0,00)^2/(1,0,1,1,10)/(1,0,1,1,00)^2/(0,2,0,0,00)/(0,1,1,1,10)/(0,1,1,1,00)^2/(0,0,2,2,00)/(0,0,2,0,10)/(0,0,2,0,00)^2/(0,0,0,2,10)/(0,0,0,2,00)^2/(0,0,0,0,10)^2/(0,0,0,0,01)/(0,0,0,0,00)^6$
451	$\bar{A}_1^2 A_1^5$	2	$(2,0,0,0,0,0,0)/(1,1,2,0,0,0,0)/(1,1,0,2,0,0,0)/(1,1,0,0,2,0,0)/(1,1,0,0,0,2,0)/(1,1,0,0,0,0,2)/(1,1,0,0,0,0,0)^2/(1,0,1,1,1,1,1)/(0,2,0,0,0,0,0)/(0,1,1,1,1,1,1)/(0,0,2,2,0,0,0)/(0,0,2,0,2,0,0)/(0,0,2,0,0,2,0)/(0,0,2,0,0,0,2)/(0,0,2,0,0,0,0)^2/(0,0,0,2,2,0,0)/(0,0,0,2,0,2,0)/(0,0,0,2,0,0,2)/(0,0,0,2,0,0,0)^2/(0,0,0,0,2,2,0)/(0,0,0,0,2,0,2)/(0,0,0,0,2,0,0)^2/(0,0,0,0,0,2,2)/(0,0,0,0,0,2,0)^2/(0,0,0,0,0,0,2)^2/(0,0,0,0,0,0,0)^8$
531	$\bar{A}_1^2 A_1 B_3$	≥ 5	$(2,0,0,000)/(1,1,4,000)/(1,1,0,100)/(1,0,3,001)/(0,2,0,000)/(0,1,3,001)/(0,0,W(6),000)/(0,0,4,100)/(0,0,2,000)/(0,0,0,010)$

532	$\bar{A}_1^2B_2G_2$		$(W(2),0,00,00)/(1,1,W(10),00)/(1,1,00,W(10))/(1,0,01,W(10))/(1,0,01,00)/(0,W(2),00,00)/$ $(0,1,01,W(10))/(0,1,01,00)/(0,0,W(10),W(10))/(0,0,W(02),00)/(0,0,00,W(10))/(0,0,00,W(01))$
534	$\bar{A}_1^2A_1G_2$	≥ 5	$(2,0,0,00)/(1,1,4,00)/(1,1,0,10)/(1,0,3,10)/(1,0,3,00)/(0,2,0,00)/(0,1,3,10)/(0,1,3,00)/$ $(0,0,W(6),00)/(0,0,4,10)/(0,0,2,00)/(0,0,0,10)/(0,0,0,01)$
538	$\bar{A}_1^2A_1A_1$	≥ 7	$(2,0,0,0)/(1,1,4,0)/(1,1,0,6)/(1,0,3,6)/(1,0,3,0)/(0,2,0,0)/(0,1,3,6)/(0,1,3,0)/(0,0,6,0)/(0,0,4,6)/$ $(0,0,2,0)/(0,0,0,W(10))/(0,0,0,6)/(0,0,0,2)$
551	$\bar{A}_1^2B_2A_2$	3	$(2,0,00,00)/(1,1,10,00)/(1,1,00,11)/(1,0,01,11)/(1,0,01,00)/(0,2,00,00)/(0,1,01,11)/(0,1,01,00)/$ $(0,0,10,11)/(0,0,10,00)/(0,0,02,00)/(0,0,00,30)/(0,0,00,11)^2/(0,0,00,03)/(0,0,00,00)$
552	$\bar{A}_1^2B_2A_1$	≥ 7	$(2,0,00,0)/(1,1,10,0)/(1,1,00,6)/(1,0,01,6)/(1,0,01,0)/(0,2,00,0)/(0,1,01,6)/(0,1,01,0)/(0,0,10,6)/$ $(0,0,02,0)/(0,0,00,W(10))/(0,0,00,6)/(0,0,00,2)$
558	$\bar{A}_1^2A_1^2A_1$	$\neq 2$	$(2,0,0,0,0)/(1,1,1,1,2)/(1,0,W(3),0,0)/(1,0,1,2,2)/(1,0,1,0,W(4))/(0,2,0,0,0)/(0,1,2,1,2)/$ $(0,1,0,W(3),0)/(0,1,0,1,W(4))/(0,0,2,2,2)/(0,0,2,0,W(4))/(0,0,2,0,0)/(0,0,0,2,W(4))/(0,0,0,2,0)/$ $(0,0,0,0,2)$
559	$\bar{A}_1^2A_1A_1$	≥ 7	$(2,0,0,0)/(1,1,1,5)/(1,0,3,0)/(1,0,1,W(8))/(1,0,1,4)/(0,2,0,0)/(0,1,2,6)/(0,1,0,W(9))/(0,1,0,3)/$ $(0,0,2,W(8))/(0,0,2,4)/(0,0,2,0)/(0,0,0,W(10))/(0,0,0,6)/(0,0,0,2)$
560	$\bar{A}_1^2A_1A_3$	2	$(2,0,0,000)/(1,1,1,010)/(1,0,3,000)/(1,0,1,101)/(0,2,0,000)/(0,1,2,010)/(0,1,0,010)^2/(0,1,0,200)/$ $(0,1,0,002)/(0,0,2,101)/(0,0,2,000)/(0,0,0,101)^2/(0,0,0,020)/(0,0,0,000)^4$
561	$\bar{A}_1^2A_1G_2$	2	$(2,0,0,00)/(1,1,1,10)/(1,0,3,00)/(1,0,1,01)/(0,2,0,00)/(0,1,2,10)/(0,1,0,10)^2/(0,1,0,20)/$ $(0,1,0,00)^2/(0,0,2,01)/(0,0,2,00)/(0,0,0,20)/(0,0,0,01)^2/(0,0,0,00)^4$
619	$B_3\bar{D}_4$		$(W(100),\lambda_1)/(W(100),0)/\ (W(010),0)/(001,\lambda_3)/(001,\lambda_4)/(000,\lambda_1)/(000,W(\lambda_2))$
620	$A_2\bar{D}_4$		$(W(30),0)/(W(11),\lambda_1)/(W(11),\lambda_3)/(W(11),\lambda_4)/(W(11),0)/(W(03),0)/(00,W(\lambda_2))$
621	$A_1B_2\bar{D}_4$		$(W(2),W(10),0)/(W(2),00,\lambda_1)/(W(2),00,0)/(1,01,\lambda_3)/(1,01,\lambda_4)/(0,W(10),\lambda_1)/(0,W(02),0)/$ $(0,00,W(\lambda_2))$
622	$G_2\bar{D}_4$		$(W(10),\lambda_1)/(W(10),\lambda_3)/(W(10),\lambda_4)/(W(10),0)^2/(W(01),0)/(00,\lambda_1)/(00,W(\lambda_2))/(00,\lambda_3)/(00,\lambda_4)$
623	B_3^2		$(W(100),000)/(W(100),001)/\ (W(010),000)/(001,W(100))/(001,001)/(001,000)/(000,W(100))/$ $(000,W(010))/(000,001)$
624	B_3A_2	$\neq 3$	$(W(100),11)/(W(100),00)/(W(010),00)/(001,11)^2/(000,W(30))/(000,11)^2/(000,W(03))$

625	$B_3A_1B_2$		$(W(100),1,01)/(W(100),0,00)/(W(010),0,00)/(001,W(2),00)/(001,1,01)/(001,0,W(10)/$ $(000,W(2),W(10))/(000,W(2),00)/(000,1,01)/(000,0,W(02))$
626	$A_1\bar{D}_4$	≥ 7	$(W(10),0)/(6,\lambda_1)/(6,\lambda_3)/(6,\lambda_4)/(6,0)^2/(2,0)/(0,\lambda_1)/(0,\lambda_2)/(0,\lambda_3)/(0,\lambda_4)$
627	G_2A_2	$\neq 3$	$(W(10),11)^3/(W(10),00)^2/(01,00)/(00,W(30))/(00,11)^4/(00,W(03))$
628	$A_1A_1B_2$	≥ 7	$(W(10),0,00)/(6,2,00)/(6,1,01)^2/(6,0,10)/(6,0,00)^2/(2,0,00)/(0,2,10)/(0,2,00)^2/(0,1,01)^2/(0,0,10)/$ $(0,0,02)$
629	A_1A_2	≥ 7	$(W(10),00)/(6,11)^3/(6,00)^2/(2,00)/(0,30)/(0,11)^4/(0,03)$
630	$A_1A_1A_1$	≥ 7	$(W(10),0,0)/(6,2,0)/(6,1,3)^2/(6,0,4)/(6,0,0)^2/(2,0,0)/(0,2,4)/(0,2,0)^2/(0,1,3)^2/(0,0,6)/(0,0,4)/$ $(0,0,2)$
635	$B_3A_1A_1$	≥ 5	$(100,1,3)/(100,0,0)/(010,0,0)/(001,2,0)/(001,1,3)/(001,0,4)/(000,2,4)/(000,2,0)/(000,1,3)/$ $(000,0,W(6))/(000,0,2)$
636	$B_3A_1^3$	2	$(100,1,1,1)/(100,0,0,0)^2/(010,0,0,0)/(001,2,0,0)/(001,1,1,1)/(001,0,2,0)/(001,0,0,2)/(001,0,0,0)^2/$ $(000,2,2,0)/(000,2,0,2)/(000,2,0,0)^2/(000,1,1,1)^2/(000,0,2,2)/(000,0,2,0)^2/(000,0,0,2)^2/$ $(000,0,0,0)^6$
638	$A_1^4B_2$	2	$(2,1,1,1,00)/(2,0,0,0,10)/(2,0,0,0,00)^2/(1,2,0,0,01)/(1,1,1,1,01)/(1,0,2,0,01)/(1,0,0,2,01)/$ $(1,0,0,0,01)^2/(0,2,2,0,00)/(0,2,0,2,00)/(0,2,0,0,00)^2/(0,1,1,1,10)/(0,1,1,1,00)^2/(0,0,2,2,00)/$ $(0,0,2,0,00)^2/(0,0,0,2,00)^2/(0,0,0,0,02)/(0,0,0,0,00)^8$
641	A_1^6	2	$(2,2,0,0,0,0)/(2,0,2,0,0,0)/(2,0,0,1,1,1)/(2,0,0,0,0,0)^2/(1,1,1,2,0,0)/(1,1,1,1,1,1)/(1,1,1,0,2,0)/$ $(1,1,1,0,0,2)/(1,1,1,0,0,0)^2/(0,2,2,0,0,0)/(0,2,0,1,1,1)/(0,2,0,0,0,0)^2/(0,0,2,1,1,1)/$ $(0,0,2,0,0,0)^2/(0,0,0,2,2,0)/(0,0,0,2,0,2)/(0,0,0,2,0,0)^2/(0,0,0,1,1,1)^2/(0,0,0,0,2,2)/$ $(0,0,0,0,2,0)^2/(0,0,0,0,0,2)^2/(0,0,0,0,0,0)^8$
667	$A_2A_1B_2$		$(W(30),0,00)/(W(11),W(2),00)/(W(11),1,01)^2/(W(11),0,W(10))/(W(11),0,00)/(W(03),0,00)/$ $(00,W(2),W(10))/(00,W(2),00)/(00,0,W(02))$
668	A_2^2	$\neq 3$	$(W(30),00)/(11,11)^3/(11,00)/(W(03),00)/(00,W(30))/(00,11)/(00,W(03))$
669	$A_2A_1A_1$	≥ 5	$(30,0,0)/(11,2,0)/(11,1,3)^2/(11,0,4)/(11,0,0)/(03,0,0)/(00,2,4)/(00,2,0)/(00,0,W(6))/(00,0,2)$
670	$A_2A_1^3$	2	$(30,0,0,0)/(11,2,0,0)/(11,1,1,1)^2/(11,0,2,0)/(11,0,0,2)/(11,0,0,0)^3/(03,0,0,0)/(00,2,2,0)/$ $(00,2,0,2)/(00,2,0,0)^2/(00,0,2,2)/(00,0,2,0)^2/(00,0,0,2)^2/(00,0,0,0)^6$

674	$A_1^3\bar{D}_4$	2	$(2,2,0,0)/(2,0,2,0)/(2,0,0,\lambda_1)/(2,0,0,0)^2/(1,1,1,\lambda_3)/(1,1,1,\lambda_4)/(0,2,2,0)/(0,2,0,\lambda_1)/(0,2,0,0)^2/$ $(0,0,2,\lambda_1)/(0,0,2,0)^2/(0,0,0,\lambda_1)^2/(0,0,0,\lambda_2)/(0,0,0,0)^6$
675	$A_1A_1\bar{D}_4$	≥ 5	$(2,4,0)/(2,0,\lambda_1)/(2,0,0)/(1,3,\lambda_3)/(1,3,\lambda_4)/(0,W(6),0)/(0,4,\lambda_1)/(0,2,0)/(0,0,\lambda_2)$
676	$A_1^2B_2^2$		$(W(2),1,00,01)/(W(2),0,W(10),00)/(W(2),0,00,00)/(1,W(2),01,00)/(1,1,01,01)/(1,0,01,W(10))/$ $(0,W(2),00,W(10))/(0,W(2),00,00)/(0,1,W(10),01)/(0,0,W(02),00)/(0,0,00,W(02))$
677	$A_1^2B_2^2$	$\neq 2$	$(2,2,00,00)/(2,0,10,00)/(2,0,00,10)/(2,0,00,00)/(1,1,01,01)^2/(0,2,10,00)/(0,2,00,10)/(0,2,00,00)/$ $(0,0,10,10)/(0,0,02,00)/(0,0,00,02)$
680	$A_1^2A_1B_2$	≥ 5	$(2,2,0,00)/(2,0,4,00)/(2,0,0,10)/(2,0,0,00)/(1,1,3,01)^2/(0,2,4,00)/(0,2,0,10)/(0,2,0,00)/$ $(0,0,W(6),00)/(0,0,4,10)/(0,0,2,00)/(0,0,0,02)$
681	$A_1^2A_1B_2$	≥ 5	$(2,1,0,01)/(2,0,4,00)/(2,0,0,00)/(1,2,3,00)/(1,1,3,01)/(1,0,3,10)/(0,2,0,10)/(0,2,0,00)/(0,1,4,01)/$ $(0,0,W(6),00)/(0,0,2,00)/(0,0,0,02)$
683	$A_1^2A_1^2$	≥ 5	$(2,2,0,0)/(2,0,4,0)/(2,0,0,4)/(2,0,0,0)/(1,1,3,3)^2/(0,2,4,0)/(0,2,0,4)/(0,2,0,0)/(0,0,W(6),0)/$ $(0,0,4,4)/(0,0,2,0)/(0,0,0,W(6))/(0,0,0,2)$
694	$A_1^2A_1^2$	≥ 5	$(2,1,0,3)/(2,0,4,0)/(2,0,0,0)/(1,2,3,0)/(1,1,3,3)/(1,0,3,4)/(0,2,0,4)/(0,2,0,0)/(0,1,4,3)/$ $(0,0,W(6),0)/(0,0,2,0)/(0,0,0,W(6))/(0,0,0,2)$
716	B_2D_5		$(W(10),\lambda_1)/(W(10),0)/(01,\lambda_4)/(01,\lambda_5)/(W(02),0)/(00,\lambda_1)/(00,W(\lambda_2))$
717	$A_1^2D_5$		$(W(2),W(2),0)/(W(2),0,\lambda_1)/(W(2),0,0)/(1,1,\lambda_4)/(1,1,\lambda_5)/(0,W(2),\lambda_1)/(0,W(2),0)/(0,0,W(\lambda_2))$
718	$\bar{A}_3B_4$		$(W(101),0)/(100,\lambda_4)/(010,\lambda_1)/(010,0)/(001,\lambda_4)/(000,W(\lambda_1))/(000,W(\lambda_2))$
719	$\bar{A}_3A_1B_3$		$(W(101),0,000)/(100,1,001)/(010,W(2),000)/(010,0,W(100))/(001,1,001)/(000,W(2),W(100))/$ $(000,W(2),000)/(000,0,W(010))$
720	$\bar{A}_3B_2^2$		$(W(101),00,00)/(100,01,01)/(010,W(10),00)/(010,00,W(10))/(001,01,01)/(000,W(10),W(10))/$ $(000,W(02),00)/(000,00,W(02))$
721	$\bar{A}_3B_2$	$\neq 2$	$(101,00)/(100,W(11))/(010,02)/(001,W(11))/(000,W(12))/(000,02)$
722	A_1D_5	≥ 5	$(W(6),0)/(4,\lambda_1)/(4,0)/(3,\lambda_4)/(3,\lambda_5)/(2,0)/(0,\lambda_1)/(0,W(\lambda_2))$
723	B_2^3	$\neq 2$	$(10,10,00)/(10,00,10)/(10,00,00)/(01,01,01)^2/(02,00,00)/(00,10,10)/(00,10,00)/(00,02,00)/$ $(00,00,10)/(00,00,02)$
724	B_2B_2	$\neq 2$	$(10,02)/(10,00)/(01,W(11))^2/(02,00)/(00,W(12))/(00,02)^2$

725	$A_1B_2^2$	≥ 5	$(W(6),00,00)/(4,10,00)/(4,00,10)/(4,00,00)/(3,01,01)^2/(2,00,00)/(0,10,10)/(0,10,00)/(0,02,00)/(0,00,10)/(0,00,02)$
726	A_1B_2	≥ 5	$(W(6),00)/(4,02)/(4,00)/(3,W(11))^2/(2,00)/(0,W(12))/(0,02)^2$
727	$A_1^2B_2$	≥ 5	$(W(6),0,00)/(4,4,00)/(4,0,10)/(4,0,00)/(3,3,01)^2/(2,0,00)/(0,W(6),00)/(0,4,10)/(0,4,00)/(0,2,00)/(0,0,10)/(0,0,02)$
729	A_1^3	≥ 5	$(W(6),0,0)/(4,4,0)/(4,0,4)/(4,0,0)/(3,3,3)^2/(2,0,0)/(0,W(6),0)/(0,4,4)/(0,4,0)/(0,2,0)/(0,0,W(6))/(0,0,4)/(0,0,2)$
733	$A_1^3B_3$	$\neq 2$	$(2,2,0,000)/(2,0,2,000)/(2,0,0,100)/(2,0,0,000)/(1,1,1,001)^2/(0,2,2,000)/(0,2,0,100)/(0,2,0,000)/(0,0,2,100)/(0,0,2,000)/(0,0,0,010)$
734	$A_1^2B_2$	$\neq 2$	$(2,2,00)/(2,0,02)/(2,0,00)/(1,1,W(11))^2/(0,2,02)/(0,2,00)/(0,0,W(12))/(0,0,02)$
736	$A_1^3G_2$	$\neq 2$	$(2,2,0,00)/(2,0,2,00)/(2,0,0,10)/(2,0,0,00)/(1,1,1,10)^2/(1,1,1,00)^2/(0,2,2,00)/(0,2,0,10)/(0,2,0,00)/(0,0,2,10)/(0,0,2,00)/(0,0,0,10)/(0,0,0,01)$
739	$A_1^3A_2$	3	$(2,2,0,00)/(2,0,2,00)/(2,0,0,11)/(2,0,0,00)/(1,1,1,11)^2/(1,1,1,00)^2/(0,2,2,00)/(0,2,0,11)/(0,2,0,00)/(0,0,2,11)/(0,0,2,00)/(0,0,0,30)/(0,0,0,11)^2/(0,0,0,03)/(0,0,0,00)$
740	$A_1^3A_1$	≥ 7	$(2,2,0,0)/(2,0,2,0)/(2,0,0,6)/(2,0,0,0)/(1,1,1,6)^2/(1,1,1,0)^2/(0,2,2,0)/(0,2,0,6)/(0,2,0,0)/(0,0,2,6)/(0,0,2,0)/(0,0,0,W(10))/(0,0,0,6)/(0,0,0,2)$
751	$\bar{A}_3A_1^2$	$\neq 2$	$(101,0,0)/(100,W(3),1)/(100,1,W(3))/(010,2,2)/(010,0,0)/(001,W(3),1)/(001,1,W(3))/(000,W(4),2)/(000,2,W(4))/(000,2,2)/(000,2,0)/(000,0,2)$
752	$\bar{A}_3A_1$	≥ 11	$(101,0)/(100,10)/(100,4)/(010,8)/(010,0)/(001,10)/(001,4)/(000,W(14))/(000,10)/(000,8)/(000,6)/(000,2)$
754	$\bar{A}_3A_1G_2$		$(W(101),0,00)/(100,1,W(10))/(100,1,00)/(010,W(2),00)/(010,0,W(10))/(001,1,W(10))/(001,1,00)/(000,W(2),W(10))/(000,W(2),00)/(000,0,W(10))/(000,0,W(01))$
755	$\bar{A}_3A_1A_2$	3	$(101,0,00)/(100,1,11)/(100,1,00)/(010,2,00)/(010,0,11)/(001,1,11)/(001,1,00)/(000,2,11)/(000,2,00)/(000,0,30)/(000,0,11)^2/(000,0,03)/(000,0,00)$
756	$\bar{A}_3A_1A_1$	≥ 7	$(101,0,0)/(100,1,6)/(100,1,0)/(010,2,0)/(010,0,6)/(001,1,6)/(001,1,0)/(000,2,6)/(000,2,0)/(000,0,W(10))/(000,0,6)/(000,0,2)$
758	$\bar{A}_3A_1B_2$	≥ 5	$(101,0,00)/(100,3,01)/(010,4,00)/(010,0,10)/(001,3,01)/(000,W(6),00)/(000,4,10)/(000,2,00)/(000,0,02)$

759	$\bar{A}_3A_1^2B_2$	2	$(101,0,0,00)/(100,1,1,01)/(010,2,0,00)/(010,0,2,00)/(010,0,0,10)/(010,0,0,00)^2/(001,1,1,01)/(000,2,2,00)/(000,2,0,10)/(000,2,0,00)^2/(000,0,2,10)/(000,0,2,00)^2/(000,0,0,10)^2/(000,0,0,02)/(000,0,0,00)^6$
762	$\bar{A}_3A_1^2$	≥ 5	$(101,0,0)/(100,3,3)/(010,4,0)/(010,0,4)/(001,3,3)/(000,W(6),0)/(000,4,4)/(000,2,0)/(000,0,W(6))/(000,0,2)$
764	$\bar{A}_3A_1^4$	2	$(101,0,0,0,0)/(100,1,1,1,1)/(010,2,0,0,0)/(010,0,2,0,0)/(010,0,0,2,0)/(010,0,0,0,2)/(010,0,0,0,0)^2/(001,1,1,1,1)/(000,2,2,0,0)/(000,2,0,2,0)/(000,2,0,0,2)/(000,2,0,0,0)^2/(000,0,2,2,0)/(000,0,2,0,2)/(000,0,2,0,0)^2/(000,0,0,2,2)/(000,0,0,2,0)^2/(000,0,0,0,2)^2/(000,0,0,0,0)^6$
770	A_1B_2	$\neq 2$	$(W(5),01)/(W(4),02)/(W(3),W(11))/(2,W(20))/(2,10)/(2,00)/(1,W(03))/(0,02)$
771	A_1A_1	≥ 5	$(W(5),4)/(4,W(6))/(4,2)/(3,W(7))/(3,W(5))/(3,1)/(2,W(8))/(2,4)^2/(2,0)/(1,W(9))/(1,W(5))/(1,3)/(0,W(6))/(0,2)$
772	A_1A_1	≥ 13	$(2,12)/(2,0)/(1,W(21))/(1,W(15))/(1,11)/(1,9)/(1,3)/(0,W(22))/(0,W(18))/(0,W(14))/(0,10)/(0,6)/(0,2)$
773	A_1B_2	5	$(2,20)/(2,00)/(1,13)/(1,11)/(0,22)/(0,02)$
774	A_1C_3	3	$(2,010)/(2,000)/(1,110)/(1,001)/(0,200)/(0,101)$
775	$A_1^2B_5$	2	$(2,2,0)/(2,0,\lambda_1)/(2,0,0)^2/(1,1,\lambda_5)/(0,2,\lambda_1)/(0,2,0)^2/(0,0,\lambda_1)^2/(0,0,\lambda_2)/(0,0,0)^4$
776	$A_1B_2B_4$	2	$(2,10,0)/(2,00,\lambda_1)/(2,00,0)^2/(1,01,\lambda_4)/(0,10,\lambda_1)/(0,10,0)^2/(0,02,0)/(0,00,\lambda_1)^2/(0,00,\lambda_2)/(0,00,0)^6$
777	$A_1B_3^2$	2	$(2,100,000)/(2,000,100)/(2,000,000)^2/(1,001,001)/(0,100,100)/(0,100,000)^2/(0,010,000)/(0,000,100)^2/(0,000,010)/(0,000,000)^4$
778	$A_1^3B_4$	2	$(2,2,0,0)/(2,0,2,0)/(2,0,0,\lambda_1)/(2,0,0,0)^2/(1,1,1,\lambda_4)/(0,2,2,0)/(0,2,0,\lambda_1)/(0,2,0,0)^2/(0,0,2,\lambda_1)/(0,0,2,0)^2/(0,0,0,\lambda_1)^2/(0,0,0,\lambda_2)/(0,0,0,0)^6$
779	$A_1^2B_2B_3$	2	$(2,2,00,000)/(2,0,10,000)/(2,0,00,100)/(2,0,00,000)^2/(1,1,01,001)/(0,2,10,000)/(0,2,00,100)/(0,2,00,000)^2/(0,0,10,100)/(0,0,10,000)^2/(0,0,02,000)/(0,0,00,100)^2/(0,0,00,010)/(0,0,00,000)^6$
781	$A_1^4B_3$	2	$(2,2,0,0,000)/(2,0,2,0,000)/(2,0,0,2,000)/(2,0,0,0,100)/(2,0,0,0,000)^2/(1,1,1,1,001)/(0,2,2,0,000)/(0,2,0,2,000)/(0,2,0,0,100)/(0,2,0,0,000)^2/(0,0,2,2,000)/(0,0,2,0,100)/(0,0,2,0,000)^2/(0,0,0,2,000)^2/(0,0,0,2,100)/(0,0,0,0,100)^2/(0,0,0,0,010)/(0,0,0,0,000)^6$

782	$A_1^3B_2^2$	2	$(2,2,0,00,00)/(2,0,2,00,00)/(2,0,0,10,00)/(2,0,0,00,10)/(2,0,0,00,00)^2/(1,1,1,01,01)/(0,2,2,00,00)/(0,2,0,10,00)/(0,2,0,00,10)/(0,2,0,00,00)^2/(0,0,2,10,00)/(0,0,2,00,10)/(0,0,2,00,00)^2/(0,0,0,10,10)/(0,0,0,10,00)^2/(0,0,0,02,00)/(0,0,0,00,10)^2/(0,0,0,00,02)/(0,0,0,00,00)^8$
785	$A_1^5B_2$	2	$(2,2,0,0,0,00)/(2,0,2,0,0,00)/(2,0,0,2,0,00)/(2,0,0,0,2,00)/(2,0,0,0,0,10)/(2,0,0,0,0,00)^2/(1,1,1,1,1,01)/(0,2,2,0,0,00)/(0,2,0,2,0,00)/(0,2,0,0,2,00)/(0,2,0,0,0,10)/(0,2,0,0,0,00)^2/(0,0,2,2,0,00)/(0,0,2,0,2,00)/(0,0,2,0,0,10)/(0,0,2,0,0,00)^2/(0,0,0,2,2,00)/(0,0,0,2,0,10)/(0,0,0,2,0,00)^2/(0,0,0,0,2,10)/(0,0,0,0,2,00)^2/(0,0,0,0,0,10)^2/(0,0,0,0,0,02)/(0,0,0,0,0,00)^8$
786	$A_1^4G_2$	2	$(2,2,0,0,00)/(2,0,2,0,00)/(2,0,0,2,00)/(2,0,0,0,10)/(2,0,0,0,00)^2/(1,1,1,1,10)/(1,1,1,1,00)^2/(0,2,2,0,00)/(0,2,0,2,00)/(0,2,0,0,10)/(0,2,0,0,00)^2/(0,0,2,2,00)/(0,0,2,0,10)/(0,0,2,0,00)^2/(0,0,0,2,00)^2/(0,0,0,2,10)/(0,0,0,0,10)^2/(0,0,0,0,01)/(0,0,0,0,00)^6$
791	A_1^7	2	$(2,2,0,0,0,0,0)/(2,0,2,0,0,0,0)/(2,0,0,2,0,0,0)/(2,0,0,0,2,0,0)/(2,0,0,0,0,2,0)/(2,0,0,0,0,0,2)/(2,0,0,0,0,0,0)^2/(1,1,1,1,1,1,1)/(0,2,2,0,0,0,0)/(0,2,0,2,0,0,0)/(0,2,0,0,2,0,0)/(0,2,0,0,0,2,0)/(0,2,0,0,0,0,2)/(0,2,0,0,0,0,0)^2/(0,0,2,2,0,0,0)/(0,0,2,0,2,0,0)/(0,0,2,0,0,2,0)/(0,0,2,0,0,0,2)/(0,0,2,0,0,0,0)^2/(0,0,0,2,0,0,0)^2/(0,0,0,2,2,0,0)/(0,0,0,2,0,2,0)/(0,0,0,2,0,0,2)/(0,0,0,0,2,2,0)/(0,0,0,0,2,0,2)/(0,0,0,0,2,0,0)^2/(0,0,0,0,0,2,2)/(0,0,0,0,0,2,0)^2/(0,0,0,0,0,0,2)^2/(0,0,0,0,0,0,0)^8$
823	$A_1^2B_2G_2$	2	$(2,2,00,00)/(2,0,10,00)/(2,0,00,10)/(2,0,00,00)^2/(1,1,01,10)/(1,1,01,00)^2/(0,2,10,00)/(0,2,00,10)/(0,2,00,00)^2/(0,0,10,10)/(0,0,10,00)^2/(0,0,02,00)/(0,0,00,10)^2/(0,0,00,01)/(0,0,00,00)^6$
826	$A_1B_2^3$	2	$(2,10,00,00)/(2,00,10,00)/(2,00,00,10)/(2,00,00,00)^2/(1,01,01,01)/(0,10,10,00)/(0,10,00,10)/(0,10,00,00)^2/(0,02,00,00)/(0,00,10,10)/(0,00,10,00)^2/(0,00,02,00)/(0,00,00,10)^2/(0,00,00,02)/(0,00,00,00)^8$
831	$A_1G_2B_3$	2	$(2,10,000)/(2,00,100)/(2,00,000)^2/(1,10,001)/(1,00,001)^2/(0,10,100)/(0,10,000)^2/(0,01,000)/(0,00,100)^2/(0,00,010)/(0,00,000)^4$
833	$A_1G_2^2$	2	$(2,10,00)/(2,00,10)/(2,00,00)^2/(1,10,10)/(1,10,00)^2/(1,00,10)^2/(1,00,00)^4/(0,10,10)/(0,10,00)^2/(0,01,00)/(0,00,10)^2/(0,00,01)/(0,00,00)^4$
835	A_1B_5	≥ 5	$(W(6),0)/(4,\lambda_1)/(3,\lambda_5)/(2,0)/(0,\lambda_2)$
836	B_2A_1	≥ 11	$(10,10))/(02,0)/(01,W(15))/(01,9)/(01,5)/(00,W(18))/(00,W(14))/(00,10)/(00,6)/(00,2)$
837	$B_2^2B_3$	2	$(10,10,000)/(10,00,100)/(10,00,000)^2/(02,00,000)/(01,01,001)/(00,10,100)/(00,10,000)^2/(00,02,000)/(00,00,100)^2/(00,00,010)/(00,00,000)^6$

838	A_1A_1	≥ 11	$(6,0)/(4,10)/(3,W(15))/(3,9)/(3,5)/(2,0)/(0,W(18))/(0,W(14))/(0,10)/(0,6)/(0,2)$
839	$B_2^2G_2$	2	$(10,10,00)/(10,00,10)/(10,00,00)^2/(02,00,00)/(01,01,10)/(01,01,00)^2/(00,10,10)/(00,10,00)^2/(00,02,00)/(00,00,10)^2/(00,00,01)/(00,00,00)^6$
844	G_2B_4		$(W(10),W(\lambda_1))/(W(10),\lambda_4)/(W(10),0)/(W(01),0)/(00,W(\lambda_2))/(00,\lambda_4)$
845	$B_3A_1^2$	$\neq 2$	$(100,2,2)/(010,0,0)/(001,W(3),1)/(001,1,W(3))/(000,W(4),2)/(000,2,W(4))/(000,2,0)/(000,0,2)$
846	B_3A_1	≥ 11	$(100,8))/(010,0)/(001,10)/(001,4)/(000,W(14))/(000,10)/(000,6)/(000,2)$
847	A_2B_4	3	$(30,0)/(11,\lambda_1)/(11,\lambda_4)/(11,0)^2/(03,0)/(00,W(\lambda_2))/(00,\lambda_4)/(00,0)$
848	A_1B_4	≥ 7	$(W(10),0)/(6,\lambda_1)/(6,\lambda_4)/(6,0)/(2,0)/(0,\lambda_2)/(0,\lambda_4)$
849	$G_2A_1^2$	$\neq 2$	$(10,W(3),1)/(10,2,2)/(10,1,W(3))/(10,0,0)/(01,0,0)/(00,W(4),2)/(00,W(3),1)/(00,2,W(4))/(00,2,0)/(00,1,W(3))/(00,0,2)$
850	G_2A_1	≥ 11	$(10,10)/(10,8)/(10,4)/(10,0)/(01,0)/(00,W(14))/(00,10)^2/(00,6)/(00,4)/(00,2)$
851	$A_2A_1^2$	3	$(30,0,0)/(11,3,1)/(11,2,2)/(11,1,3)/(11,1,1)^2/(11,0,0)^2/(03,0,0)/(00,4,2)/(00,3,1)/(00,2,4)/(00,2,0)^2/(00,1,3)/(00,1,1)^2/(00,0,2)^2/(00,0,0)$
853	$A_1A_1^2$	≥ 7	$(W(10),0,0)/(6,3,1)/(6,2,2)/(6,1,3)/(6,0,0)/(2,0,0)/(0,4,2)/(0,3,1)/(0,2,4)/(0,2,0)/(0,1,3)/(0,0,2)$
854	A_1A_1	≥ 11	$(W(10),0)/(6,10)/(6,8)/(6,4)/(6,0)/(2,0)/(0,W(14))/(0,10)^2/(0,6)/(0,4)/(0,2)$
859	$A_1^2B_2$	2	$(3,1,01)/(2,2,10)/(2,2,00)^2/(2,0,10)^2/(2,0,02)/(2,0,00)^4/(1,3,01)/(1,1,11)/(0,2,10)^2/(0,2,02)/(0,2,00)^4/(0,0,10)^3/(0,0,02)^2/(0,0,00)^8$
860	A_1B_2	≥ 5	$(W(7),01)/(W(6),10)/(W(6),00)/(5,01)/(4,02)/(3,W(11))/(2,10)/(2,00)/(1,01)/(0,02)$
862	A_1^2	≥ 5	$(W(7),3)/(W(6),4)/(W(6),0)/(5,3)/(4,W(6))/(4,2)/(3,W(7))/(3,5)/(3,1)/(2,4)/(2,0)/(1,3)/(0,W(6))/(0,2)$
863	A_1B_2	≥ 5	$(W(8),00)/(W(6),10)^2/(W(6),00)/(4,02)^2/(4,00)/(2,10)^2/(2,00)/(0,W(20))/(0,02)$
864	A_1^2	≥ 5	$(W(8),0)/(W(6),4)^2/(W(6),0)/(4,W(6))^2/(4,2)^2/(4,0)/(2,4)^2/(2,0)/(0,W(8))/(0,W(6))/(0,4)/(0,2)$
865	A_1^4		$(W(3),1,1,1)/(W(2),W(2),W(2),0)/(W(2),W(2),0,W(2))/(W(2),0,W(2),W(2))/(W(2),0,0,0)/(1,W(3),1,1)/(1,1,W(3),1)/(1,1,1,W(3))/(0,W(2),W(2),W(2))/(0,W(2),0,0)/(0,0,W(2),0)/(0,0,0,W(2))$
866	A_1A_1	≥ 11	$(3,7)/(2,W(12))/(2,8)/(2,4)/(2,0)/(1,W(15))/(1,W(11))/(1,9)/(1,5)/(1,3)/(0,W(14))/(0,10)/(0,6)/(0,2)$

867	A_1D_4	2	$(3,\lambda_1)/(2,\lambda_2)/(2,0)^2/(1,\lambda_3+\lambda_4)/(0,2\lambda_1)/(0,\lambda_2)^2/(0,0)^4$
871	A_1B_3	2	$(3,001)/(2,100)^2/(2,010)/(2,000)^2/(1,101)/(0,100)^4/(0,010)^2/(0,002)/(0,000)^4$
872	A_1A_2	2	$(3,11)/(2,30)/(2,11)/(2,03)/(2,00)^2/(1,30)^2/(1,22)/(1,03)^2/(1,00)^4/(0,30)^2/(0,22)/(0,11)^2/(0,03)^2/$ $(0,00)^4$
873	A_1^4	$\neq 2$	$(W(4),0,0,0)/(2,2,2,0)^2/(2,2,0,2)^2/(2,0,2,2)^2/(2,0,0,0)/(0,W(4),0,0)/(0,2,2,2)^2/(0,2,0,0)/$ $(0,0,W(4),0)/(0,0,2,0)/(0,0,0,W(4))/(0,0,0,2)$
874	A_1A_1	≥ 11	$(4,0)/(2,W(12))^2/(2,8)^2/(2,4)^2/(2,0)/(0,W(16))/(0,W(14))/(0,10)^2/(0,8)/(0,6)/(0,4)/(0,2)$
878	$\bar{A}_1\bar{A}_2A_5$		$(W(2),00,0)/(1,10,\lambda_1)/(1,01,\lambda_5)/(1,00,\lambda_3)/(0,W(11),0)/(0,10,\lambda_4)/(0,01,\lambda_2)/(0,00,W(\lambda_1+\lambda_5))$
879	$\bar{A}_1A_7$		$(W(2),0)/(1,\lambda_2)/(1,\lambda_6)/(0,W(\lambda_1+\lambda_7))/(0,\lambda_4)$
880	$\bar{A}_1G_2C_3$		$(W(2),00,000)/(1,W(10),100)/(1,00,W(001))/(0,W(10),W(010))/(0,W(01),000)/(0,00,W(200))$
881	$\bar{A}_1A_1F_4$		$(W(2),0,0)/(1,W(3),0)/(1,1,W(\lambda_4))/(0,W(2),W(\lambda_4))/(0,W(2),0)/(0,0,W(\lambda_1))$
882	$\bar{A}_1A_1G_2$	$\neq 2$	$(2,0,00)/(1,W(3),10)/(1,1,W(01))/(0,W(4),10)/(0,2,W(20))/(0,2,00)/(0,0,W(01))$
883	$\bar{A}_1A_1A_1$	≥ 5	$(2,0,0)/(1,W(6),3)/(1,4,1)/(1,2,W(5))/(0,W(6),4)/(0,4,W(6))/(0,4,2)/(0,2,W(8))/(0,2,4)/(0,2,0)/$ $(0,0,2)$
884	$\bar{A}_1A_2$	≥ 5	$(2,00)/(1,W(60))/(1,W(06))/(0,W(44))/(0,11)$
885	$\bar{A}_1A_1$	≥ 17	$(2,0)/(1,W(21))/(1,15)/(1,11)/(1,5)/(0,W(26))/(0,W(22))/(0,W(18))/(0,16)/(0,14)/(0,10)^2/(0,6)/$ $(0,2)$
886	$\bar{A}_1A_1$	≥ 19	$(2,0)/(1,W(27))/(1,17)/(1,9)/(0,W(34))/(0,W(26))/(0,W(22))/(0,18)/(0,14)/(0,10)/(0,2)$
887	$\bar{A}_1\bar{A}_2A_1A_2$		$(W(2),00,0,00)/(1,10,1,10)/(1,01,1,01)/(1,00,W(3),00)/(1,00,1,W(11))/(0,W(11),0,00)/$ $(0,10,W(2),10)/(0,10,0,W(02))/(0,01,W(2),01)/(0,01,0,W(20))/(0,00,W(2),W(11))/$ $(0,00,W(2),00)/(0,00,0,W(11))$
888	$\bar{A}_1\bar{A}_2C_3$		$(W(2),00,000)/(1,10,100)/(1,01,100)/(1,00,100)/(1,00,001)/(0,W(11),000)/(0,10,W(010))/$ $(0,10,000)/(0,01,W(010))/(0,01,000)/(0,00,W(200))/(0,00,W(010))$
889	$\bar{A}_1\bar{A}_2A_3$		$(W(2),00,000)/(1,10,010)/(1,01,010)/(1,00,W(200))/(1,00,W(002))/(0,W(11),000)/(0,10,W(101))/$ $(0,01,W(101))/(0,00,W(101))/(0,00,W(020))$
890	$\bar{A}_1\bar{A}_2A_2$	$\neq 2$	$(2,00,00)/(1,10,20)/(1,01,02)/(1,00,W(30))/(1,00,W(03))/(0,W(11),00)/(0,10,12)/(0,01,21)/$ $(0,00,W(22))/(0,00,W(11))$

891	$\bar{A}_1A_1A_5$	$\neq 2$	$(2,0,0)/(1,2,\lambda_1)/(1,2,\lambda_5)/(1,0,\lambda_3)/(0,W(4),0)/(0,2,\lambda_2)/(0,2,\lambda_4)/(0,2,0)/(0,0,W(\lambda_1+\lambda_5))$
892	$\bar{A}_1A_1A_1A_2$	$\neq 2$	$(2,0,0,00)/(1,2,1,10)/(1,2,1,01)/(1,0,W(3),00)/(1,0,1,W(11))/(0,W(4),0,00)/(0,2,2,10)/$ $(0,2,2,01)/(0,2,0,20)/(0,2,0,02)/(0,2,0,00)/(0,0,W(2),W(11))/(0,0,2,00)/(0,0,0,W(11))$
893	$\bar{A}_1\bar{A}_2A_1A_1$	$\neq 2$	$(2,00,0,0)/(1,10,1,2)/(1,01,1,2)/(1,00,W(3),0)/(1,00,1,W(4))/(1,00,1,2)/(0,W(11),0,0)/$ $(0,10,2,2)/(0,10,0,W(4))/(0,10,0,0)/(0,01,2,2)/(0,01,0,W(4))/(0,01,0,0)/(0,00,2,W(4))/$ $(0,00,2,0)^2/(0,00,0,W(4))/(0,00,0,2)$
907	$\bar{A}_1\bar{A}_2A_1$	≥ 7	$(2,00,0)/(1,10,5)/(1,01,5)/(1,00,W(9))/(1,00,5)/(1,00,3)/(0,W(11),0)/(0,10,W(8))/(0,10,4)/$ $(0,10,0)/(0,01,W(8))/(0,01,4)/(0,01,0)/(0,00,W(10))/(0,00,W(8))/(0,00,6)/(0,00,4)/(0,00,2)$
908	$\bar{A}_1\bar{A}_2G_2$	2	$(2,00,00)/(1,10,10)/(1,01,10)/(1,00,20)/(1,00,10)^2/(1,00,00)^2/(0,W(11),00)/(0,10,01)/(0,10,00)/$ $(0,01,01)/(0,01,00)/(0,00,20)/(0,00,01)^2$
910	$\bar{A}_1A_1A_3$	$\neq 2$	$(2,0,000)/(1,2,010)^2/(1,0,200)/(1,0,002)/(0,W(4),000)/(0,2,101)^2/(0,2,000)/(0,0,W(020))/(0,0,101)$
912	$\bar{A}_1A_1A_2$	$\neq 2$	$(2,0,00)/(1,2,20)/(1,2,02)/(1,0,W(30))/(1,0,W(03))/(0,W(4),00)/(0,2,21)/(0,2,12)/(0,2,00)/$ $(0,0,W(22))/(0,0,W(11))$
917	$\bar{A}_1D_4$	$\neq 2$	$(2,0)/(1,\lambda_2)^2/(0,2\lambda_1)/(0,\lambda_2)/(0,2\lambda_3)/(0,2\lambda_4)$
918	$\bar{A}_1A_2C_3$	3	$(2,00,000)/(1,11,100)/(1,00,001)/(0,30,000)/(0,11,010)/(0,11,000)^2/(0,03,000)/(0,00,200)/(0,00,000)$
919	$\bar{A}_1A_1C_3$	≥ 7	$(2,0,000)/(1,6,100)/(1,0,001)/(0,W(10),000)/(0,6,010)/(0,2,000)/(0,0,200)$
921	$\bar{A}_1G_2A_1A_1$	$\neq 2$	$(2,00,0,0)/(1,10,2,1)/(1,00,W(4),1)/(1,00,0,W(3))/(0,10,W(4),0)/(0,10,2,2)/(0,01,0,0)/$ $(0,00,W(4),2)/(0,00,2,0)/(0,00,0,2)$
921	$\bar{A}_1G_2A_1$	≥ 7	$(2,00,0)/(1,10,5)/(1,00,W(9))/(1,00,3)/(0,10,W(8))/(0,10,4)/(0,01,0)/(0,00,W(10))/(0,00,6)/$ $(0,00,2)$
922	$\bar{A}_1G_2A_3$	2	$(2,00,000)/(1,10,010)/(1,00,200)/(1,00,010)^2/(1,00,002)/(0,10,101)/(0,01,000)/(0,00,020)/$ $(0,00,101)^2/(0,00,000)^2$
923	$\bar{A}_1G_2G_2$	2	$(2,00,00)/(1,10,10)/(1,00,20)/(1,00,10)^2/(1,00,00)^2/(0,10,01)/(0,01,00)/(0,00,20)/(0,00,01)^2/$ $(0,00,00)^2$
924	$\bar{A}_1A_2A_1A_1$	3	$(2,00,0,0)/(1,11,2,1)/(1,00,4,1)/ (1,00,0,3)/(1,00,0,1)^2/(0,30,0,0)/(0,11,4,0)/(0,11,2,2)/$ $(0,11,0,0)^2/(0,03,0,0)/(0,00,4,2)/(0,00,2,0)/(0,00,0,2)^2/(0,00,0,0)$

929	$\bar{A}_1A_1A_1A_1$	≥ 7	$(2,0,0,0)/(1,6,2,1)/(1,0,4,1)/(1,0,0,3)/(0,W(10),0,0)/(0,6,4,0)/(0,6,2,2)/(0,2,0,0)/(0,0,4,2)/(0,0,2,0)/(0,0,0,2)$
930	$\bar{A}_1A_1A_1$	≥ 7	$(2,0,0)/(1,6,5)/(1,0,W(9))/(1,0,3)/(0,W(10),0)/(0,6,W(8))/(0,6,4)/(0,2,0)/(0,0,W(10))/(0,0,6)/(0,0,2)$
954	$\bar{A}_1A_1C_4$	2	$(2,0,0)/(1,3,0)/(1,1,\lambda_2)/(0,2,\lambda_2)/(0,2,0)/(0,0,2\lambda_1)/(0,0,\lambda_2)^2/(0,0,\lambda_4)/(0,0,0)^4$
955	$\bar{A}_1A_1G_2$	7	$(2,0,00)/(1,3,00)/(1,1,20)/(0,2,20)/(0,2,00)/(0,0,11)/(0,0,01)$
956	$\bar{A}_1A_1A_1$	≥ 13	$(2,0,0)/(1,3,0)/(1,1,W(16))/(1,1,8)/(0,2,W(16))/(0,2,8)/(0,2,0)/(0,0,W(22))/(0,0,W(14))/(0,0,10)/(0,0,2)$
958	$\bar{A}_1A_1D_4$	2	$(2,0,0)/(1,3,0)/(1,1,\lambda_2)/(0,2,\lambda_2)/(0,2,0)/(0,0,2\lambda_1)/(0,0,\lambda_2)^2/(0,0,2\lambda_3)/(0,0,2\lambda_4)/(0,0,0)^4$
965	$\bar{A}_1A_1A_2$	3	$(2,0,00)/(1,3,11)/(1,1,30)/(1,1,11)^2/(1,1,03)/(1,1,00)/(0,4,11)/(0,2,22)/(0,2,00)/(0,0,30)/(0,0,11)^2/(0,0,03)/(0,0,00)$
966	$\bar{A}_1A_1A_1$	≥ 7	$(2,0,0)/(1,3,6)/(1,1,W(10))/(1,1,2)/(0,4,6)/(0,2,W(12))/(0,2,W(8)/(0,2,4)/(0,2,0)/(0,0,W(10))/(0,0,2)$
975	$\bar{A}_2^4$		$(W(11),00,00,00)/(10,10,00,01)/(10,01,10,00)/(10,00,01,10)/(01,10,01,00)/(01,01,00,10)/(01,00,10,01)/(00,W(11),00,00)/(00,10,10,10)/(00,01,01,01)/(00,00,W(11),00)/(00,00,00,W(11))$
976	$\bar{A}_2F_4$		$(W(11),0)/(10,W(\lambda_4))/(10,0)/(01,W(\lambda_4))/(01,0)/(00,W(\lambda_1))/(00,W(\lambda_4))$
977	$\bar{A}_2C_4$		$(W(11),0)/(10,W(\lambda_2))/(01,W(\lambda_2))/(00,W(2\lambda_1))/(00,W(\lambda_4))$
978	$\bar{A}_2A_2G_2$		$(W(11),00,00)/(10,10,W(10))/(10,W(02),00)/(01,W(20),00)/(01,01,W(10))/(00,W(11),W(10))/(00,W(11),00)/(00,00,W(01))$
979	$\bar{A}_2G_2$		$(W(11),00)/(10,W(20))/(01,W(20))/(00,W(01))/(00,W(11))$
980	$\bar{A}_2A_2$	$\neq 2$	$(W(11),00)/(10,W(22))/(01,W(22))/(00,W(41))/(00,W(11))/(00,W(14))$
981	A_1E_6	$\neq 2$	$(W(4),0)/(2,\lambda_1)/(2,\lambda_6)/(2,0)/(0,W(\lambda_2))$
982	$A_1\bar{A}_2^3$	$\neq 2$	$(W(4),00,00,00)/(2,10,01,00)/(2,10,00,01)/(2,01,10,00)/(2,01,00,10)/(2,00,10,01)/(2,00,01,10)/(2,00,00,00)/(0,W(11),00,00)/(0,10,10,10)/(0,01,01,01)/(0,00,W(11),00)/(0,00,00,W(11))$
989	$A_1^2\bar{A}_2^2$	$\neq 2$	$(W(4),0,00,00)/(2,2,10,00)/(2,2,01,00)/(2,2,00,10)/(2,2,00,01)/(2,0,10,01)/(2,0,01,10)/(2,0,00,00)/(0,W(4),00,00)/(0,2,10,10)/(0,2,01,01)/(0,2,00,00)/(0,0,W(11),00)/(0,0,00,W(11))$

995	$A_1^3\bar{A}_2$	$\neq 2$	$(W(4),0,0,00)/(2,2,2,00)^4/(2,0,2,10)/(2,0,2,01)/(2,0,0,00)/(0,W(4),0,00)/(0,2,2,10)/(0,2,2,01)/(0,2,0,00)/(0,0,W(4),00)/(0,0,2,00)/(0,0,0,W(11))$
1002	$\bar{A}_2A_1G_2$	$\neq 2$	$(W(11),0,00)/(10,W(4),00)/(10,2,W(10))/(10,0,00)/(01,W(4),00)/(01,2,W(10))/(01,0,00)/(00,W(4),W(10))/(00,W(4),00)/(00,2,W(10))/(00,2,00)/(00,0,W(01))$
1003	$\bar{A}_2A_1$	≥ 13	$(W(11),0)/(10,W(16))/(10,8)/(10,0)/(01,W(16))/(01,8)/(01,0)/(00,W(22))/(00,W(16))/(00,W(14))/(00,10)/(00,8)/(00,2)$
1004	$\bar{A}_2A_1A_2$	3	$(11,0,00)/(10,4,00)/(10,2,11)/(10,0,00)^2/(01,4,00)/(01,2,11)/(01,0,00)^2/(00,4,11)/(00,4,00)/(00,2,11)/(00,2,00)/(00,0,30)/(00,0,11)^2/(00,0,03)/(00,0,00)^2$
1005	$\bar{A}_2A_1A_1$	≥ 7	$(11,0,0)/(10,4,0)/(10,2,6)/(10,0,0)/(01,W(4),0)/(01,2,6)/(01,0,0)/(00,4,6)/(00,4,0)/(00,2,6)/(00,2,0)/(00,0,W(10))/(00,0,2)$
1009	$\bar{A}_2A_1$	≥ 11	$(11,0)/(10,W(12))/(10,8)/(10,4)/(01,W(12))/(01,8)/(01,4)/(00,W(14))/(00,10)^2/(00,8)/(00,6)/(00,4)/(00,2)$
1010	$\bar{A}_2D_4$	2	$(11,0)/(10,\lambda_2)/(10,0)/(01,\lambda_2)/(01,0)/(00,2\lambda_1)/(00,\lambda_2)^2/(00,2\lambda_3)/(00,2\lambda_4)/(00,0)^2$
1011	$A_1A_2G_2$	$\neq 2$	$(W(4),00,00)/(2,W(20),00)/(2,10,W(10))/(2,W(02),00)/(2,01,W(10))/(2,00,00)/(0,W(11),W(10))/(0,W(11),00)/(0,00,W(01))$
1012	$\bar{A}_2A_2A_2$	3	$(11,00,00)/(10,10,11)/(10,02,00)/(01,20,00)/(01,01,11)/(00,11,11)/(00,11,00)/(00,00,30)/(00,00,11)^2/(00,00,03)/(00,00,00)^3$
1013	$\bar{A}_2A_2A_1$	≥ 7	$(11,00,0)/(10,10,6)/(10,02,0)/(01,01,6)/(01,20,0)/(00,11,6)/(00,11,0)/(00,00,W(10))/(00,00,2)$
1016	$A_1A_2A_2$	3	$(4,00,00)/(2,20,00)/(2,10,11)/(2,02,00)/(2,01,11)/(2,00,00)/(0,11,11)/(0,11,00)/(0,00,30)/(0,00,11)^2/(0,00,03)/(0,00,00)^3$
1017	$A_1A_2A_1$	≥ 7	$(4,00,0)/(2,20,0)/(2,10,6)/(2,02,0)/(2,01,6)/(2,00,0)/(0,11,6)/(0,11,0)/(0,00,W(10))/(0,00,2)$
1026	A_1G_2	$\neq 2$	$(W(4),00)/(2,W(20))^2/(2,00)/(0,W(11))/(0,W(01))$
1027	A_1A_2	$\neq 2$	$(W(4),00)/(2,W(22))^2/(2,00)/(0,W(41))/(0,W(11))/(0,W(14))$
1028	$B_2\bar{A}_4$	$\neq 2$	$(W(20),0)/(10,\lambda_2)/(10,\lambda_3)/(02,\lambda_1)/(02,\lambda_4)/(02,0)/(00,W(\lambda_1+\lambda_4))$
1029	$A_1\bar{A}_4$	≥ 5	$(W(8),0)/(W(6),\lambda_1)/(W(6),\lambda_4)/(W(6),0)/(4,\lambda_2)/(4,\lambda_3)/(4,0)/(2,\lambda_1)/(2,\lambda_4)/(2,0)/(0,W(\lambda_1+\lambda_4))$
1030	A_2F_4	3	$(30,0)/(11,\lambda_4)/(11,0)^2/(03,0)/(00,\lambda_1)/(00,0)$
1031	A_1F_4	≥ 7	$(W(10),0)/(6,\lambda_4)/(2,0)/(0,\lambda_1)$

1032	G_2C_4	2	$(10,\lambda_2)/(01,0)/(00,2\lambda_1)/(00,\lambda_2)^2/(00,\lambda_4)/(00,0)^2$
1033	$G_2^2A_1$	$\neq 2$	$(10,10,2)/(10,00,W(4))/(W(01),00,0)/(00,10,W(4))/(00,W(01),0)/(00,00,2)$
1034	G_2A_1	≥ 13	$(10,W(16))/(10,8)/(01,0)/(00,W(22))/(00,W(14))/(00,10)/(00,2)$
1035	G_2G_2	7	$(10,20)/(01,00)/(00,11)/(00,01)$
1036	$A_2A_1G_2$	3	$(30,0,00)/(11,W(4),00)/(11,2,10)/(11,0,00)/(03,0,00)/(00,W(4),10)/(00,2,00)/(00,0,10)/(00,0,01)/$ $(00,0,00)$
1037	$A_2^2A_1$	3	$(30,00,0)/(11,11,2)/(11,00,W(4))/(11,00,0)/(03,00,0)/(00,30,0)/(00,11,W(4))/(00,11,0)/(00,03,0)/$ $(00,00,2)/(00,00,0)^2$
1039	$A_1A_1G_2$	≥ 7	$(W(10),0,00)/(6,4,00)/(6,2,10)/(2,0,00)/(0,4,10)/(0,2,00)/(0,0,01)$
1040	A_1A_1	≥ 13	$(10,0)/(6,W(16))/(6,8)/(2,0)/(0,W(22))/(0,W(14))/(0,10)/(0,2)$
1041	A_1G_2	7	$(10,00)/(6,20)/(2,00)^2/(0,11)/(0,01)$
1042	$A_1^2A_1$	≥ 7	$(W(10),0,0)/(6,6,2)/(6,0,4)/(2,0,0)/(0,W(10),0)/(0,6,4)/(0,2,0)/(0,0,2)$
1046	G_2D_4	2	$(10,\lambda_2)/(01,0)/(00,2\lambda_1)/(00,\lambda_2)^2/(00,2\lambda_3)/(00,2\lambda_4)/(00,0)^2$
1049	F_4	3	1000/0010

CHAPTER 13

Composition factors for the action of Levi subgroups

Let G be a simple exceptional algebraic group. In this chapter we give the composition factors of the action of proper Levi subgroups L of G on the minimal and adjoint modules for G. If L' is simple and has rank at least 2 then these are found in [**LS96**, Tables 8.1–8.7], except for the case where $L < G = F_4$ acting on the minimal module. In all other cases the composition factors are deduced from those of a maximal subsystem subgroup containing L'.

We note that for $G = E_6, E_7$ or E_8 any simple factor of a Levi subgroup of type A_n, D_n or E_n is generated by long root subgroups of G and so we omit all bars in Tables 13.3–13.5.

Table 13.1: The composition factors for the action of Levi subgroups of G_2 on V_7 and $L(G_2)$.

L'	$V_7 \downarrow L'$	$L(G_2) \downarrow L'$
$\bar{A}_1$	$1^2/0^3$	$W(2)/1^4/0^3$
$\tilde{A}_1$	$W(2)/1^2$	$W(3)^2/W(2)/0^3$

Table 13.2: The composition factors for the action of Levi subgroups of F_4 on V_{26} and $L(F_4)$.

L'	$V_{26} \downarrow L'$	$L(F_4) \downarrow L'$
B_3	$W(100)/001^2/000^3$	$W(100)^2/W(010)/001^2/000$
B_2	$W(10)/01^4/00^5$	$W(10)^4/W(02)/01^4/00^6$
C_3	$100^2/W(010)$	$W(200)/W(001)^2/000^3$
$\bar{A}_2\tilde{A}_1$	$(10,1)/(10,0)/(01,1)/(01,0)/(00,W(2))/(00,1)^2/(00,0)$	$(W(11),0)/(10,W(2))/(10,1)/(10,0)/(01,W(2))/(01,1)/(01,0)/(00,W(2))/(00,1)^2/(00,0)$
$\tilde{A}_2\bar{A}_1$	$(10,1)/(10,0)/(01,1)/(01,0)/(W(11),0)$	$(W(20),1)/(W(20),0)/(W(11),0)/(W(02),1)/(W(02),0)/(00,W(2))/(00,1)^2/(00,0)$
$\bar{A}_2$	$10^3/01^3/00^8$	$W(11)/10^6/01^6/00^8$
$\tilde{A}_2$	$10^3/01^3/W(11)$	$W(20)^3/W(11)/W(02)^3/00^8$
$\bar{A}_1\tilde{A}_1$	$(1,1)^2/(1,0)^2/(0,W(2))/(0,1)^4/(0,0)^3$	$(W(2),0)/(1,W(2))^2/(1,1)^2/(1,0)^4/(0,W(2))^3/(0,1)^4/(0,0)^4$
$\bar{A}_1$	$1^6/0^{14}$	$W(2)/1^{14}/0^{21}$

$\tilde{A}_1$	$W(2)/1^8/0^7$	$W(2)^7/1^8/0^{15}$

Table 13.3: The composition factors for the action of Levi subgroups of E_6 on V_{27} and $L(E_6)$.

L'	$V_{27} \downarrow L'$	$L(E_6) \downarrow L'$
D_5	$\lambda_1/\lambda_4/0$	$W(\lambda_2)/\lambda_4/\lambda_5/0$
D_4	$\lambda_1/\lambda_3/\lambda_4/0^3$	$\lambda_1^2/W(\lambda_2)/\lambda_3^2/\lambda_4^2/0^3$
A_5	λ_1^2/λ_4	$W(\lambda_1+\lambda_5)/\lambda_3^2/0^3$
A_1A_4	$(1,\lambda_1)/(1,0)/(0,\lambda_3)/$ $(0,\lambda_4)$	$(W(2),0)/(1,\lambda_2)/(1,\lambda_3)/(0,W(\lambda_1+\lambda_4))/$ $(0,\lambda_1)/(0,\lambda_4)/(0,0)$
$A_1A_2^2$	$(1,01,00)/(1,00,10)/$ $(0,10,01)/(0,01,00)/$ $(0,00,10)$	$(W(2),00,00)/(1,10,10)/(1,01,01)/$ $(1,00,00)^2/(0,W(11),00)/(0,10,10)/$ $(0,01,01)/(0,00,W(11))/(0,00,00)$
A_4	$\lambda_1^2/\lambda_3/\lambda_4/0^2$	$W(\lambda_1+\lambda_4)/\lambda_1/\lambda_2^2/\lambda_3^2/\lambda_4/0^4$
A_1A_3	$(1,100)/(1,000)^2/$ $(0,010)/(0,001)^2/$ $(0,000)$	$(W(2),000)/(1,100)/(1,010)^2/(1,001)/$ $(0,W(101))/(0,100)^2/(0,001)^2/(0,000)^4$
A_2^2	$(10,01)/(01,00)^3/$ $(00,10)^3$	$(W(11),00)/(10,10)^3/(01,01)^3/$ $(00,W(11))/(00,00)^8$
$A_1^2A_2$	$(1,1,00)/(1,0,10)/$ $(1,0,00)/(0,1,01)/$ $(0,1,00)/(0,0,10)/$ $(0,0,01)/(0,0,00)$	$(W(2),0,00)/(1,1,10)/(1,1,01)/(1,0,10)/$ $(1,0,01)/(1,0,00)^2/(0,W(2),00)/$ $(0,1,10)/(0,1,01)/(0,1,00)^2/$ $(0,0,W(11))/(0,0,10)/(0,0,01)/(0,0,00)^2$
A_3	$100^2/010/001^2/000^5$	$W(101)/100^4/010^4/001^4/000^7$
A_1A_2	$(1,00)^3/(1,01)/(0,10)^3/$ $(0,01)/(0,00)^3$	$(W(2),00)/(1,10)^3/(1,01)^3/(1,00)^2/$ $(0,10)^3/(0,W(11))/(0,01)^3/(0,00)^9$
A_1^3	$(1,1,0)/(1,0,1)/$ $(1,0,0)^2/(0,1,1)/$ $(0,1,0)^2/(0,0,1)^2/$ $(0,0,0)^3$	$(W(2),0,0)/(1,1,1)^2/(1,1,0)^2/(1,0,1)^2/$ $(1,0,0)^4/(0,W(2),0)/(0,1,1)^2/(0,1,0)^4/$ $(0,0,W(2))/(0,0,1)^4/(0,0,0)^5$
A_2	$10^3/01^3/00^9$	$W(11)/10^9/01^9/00^{16}$
A_1^2	$(1,1)/(1,0)^4/(0,1)^4/$ $(0,0)^7$	$(W(2),0)/(1,1)^6/(1,0)^8/(0,W(2))/(0,1)^8/$ $(0,0)^{16}$
A_1	$1^6/0^{15}$	$W(2)/1^{20}/0^{35}$

Table 13.4: The composition factors for the action of Levi subgroups of E_7 on V_{56} and $L(E_7)$.

L'	$V_{56} \downarrow L'$	$L(E_7) \downarrow L'$
E_6	$\lambda_1/\lambda_6/0^2$	$\lambda_1/W(\lambda_2)/\lambda_6/0$
D_6	λ_1^2/λ_5	$W(\lambda_2)/\lambda_6^2/0^3$
A_1D_5	$(1,\lambda_1)/(1,0)^2/(0,\lambda_4)/$ $(0,\lambda_5)$	$(W(2),0)/(1,\lambda_4)/(1,\lambda_5)/(0,\lambda_1)^2/$ $(0,W(\lambda_2))/(0,0)$

D_5	$\lambda_1^2/\lambda_4/\lambda_5/0^4$	$\lambda_1^2/W(\lambda_2)/\lambda_4^2/\lambda_5^2/0^4$
A_1D_4	$(1,\lambda_1)/(1,0)^4/(0,\lambda_3)^2/$ $(0,\lambda_4)^2$	$(W(2),0)/(1,\lambda_3)^2/(1,\lambda_4)^2/(0,\lambda_1)^4/$ $(0,W(\lambda_2))/(0,0)^6$
D_4	$\lambda_1^2/\lambda_3^2/\lambda_4^2/0^8$	$\lambda_1^4/W(\lambda_2)/\lambda_3^4/\lambda_4^4/0^9$
A_6	$\lambda_1/\lambda_2/\lambda_5/\lambda_6$	$W(\lambda_1+\lambda_6)/\lambda_1/\lambda_3/\lambda_4/\lambda_6/0$
A_1A_5	$(1,\lambda_1)/(1,\lambda_5)/(0,\lambda_1)/$ $(0,\lambda_3)/(0,\lambda_5)$	$(W(2),0)/(1,\lambda_2)/(1,\lambda_4)/(1,0)^2/$ $(0,W(\lambda_1+\lambda_5))/(0,\lambda_2)/(0,\lambda_4)/(0,0)$
A_2A_4	$(10,\lambda_1)/(10,0)/(01,\lambda_4)/$ $(01,0)/(00,\lambda_2)/(00,\lambda_3)$	$(W(11),0)/(10,\lambda_3)/(10,\lambda_4)/(01,\lambda_1)/$ $(01,\lambda_2)/(00,W(\lambda_1+\lambda_4))/(00,\lambda_1)/$ $(00,\lambda_4)/(00,0)$
$A_1A_2A_3$	$(1,10,000)/(1,01,000)/$ $(1,00,010)/(0,10,100)/$ $(0,01,001)/(0,00,100)/$ $(0,00,001)$	$(W(2),00,000)/(1,10,001)/(1,01,100)/$ $(1,00,100)/(1,00,001)/(0,W(11),000)/$ $(0,10,010)/(0,10,000)/(0,01,010)/$ $(0,01,000)/(0,00,W(101))/(0,00,000)$
A_5	$\lambda_1^3/\lambda_3/\lambda_5^3$	$W(\lambda_1+\lambda_5)/\lambda_2^3/\lambda_4^3/0^8$
A_5'	$\lambda_1^2/\lambda_2/\lambda_4/\lambda_5^2/0^2$	$W(\lambda_1+\lambda_5)/\lambda_1^2/\lambda_2/\lambda_3^2/\lambda_4/\lambda_5^2/0^4$
A_1A_4	$(1,\lambda_1)/(1,\lambda_4)/(1,0)^2/$ $(0,\lambda_1)/(0,\lambda_2)/(0,\lambda_3)/$ $(0,\lambda_4)/(0,0)^2$	$(W(2),0)/(1,\lambda_1)/(1,\lambda_2)/(1,\lambda_3)/(1,\lambda_4)/$ $(1,0)^2/(0,W(\lambda_1+\lambda_4))/(0,\lambda_1)^2/(0,\lambda_2)/$ $(0,\lambda_3)/(0,\lambda_4)^2/(0,0)^2$
A_2A_3	$(10,100)/(10,000)^2/$ $(01,001)/(01,000)^2/$ $(00,100)/(00,010)^2/$ $(00,001)$	$(W(11),000)/(10,010)/(10,001)^2/$ $(10,000)/(01,100)^2/(01,010)/(01,000)/$ $(00,W(101))/(00,100)^2/(00,001)^2/$ $(00,000)^4$
$A_1^2A_3$	$(1,1,000)^2/(1,0,010)/$ $(1,0,000)^2/(0,1,100)/$ $(0,1,001)/(0,0,100)^2/$ $(0,0,001)^2$	$(W(2),0,000)/(1,1,100)/(1,1,001)/$ $(1,0,100)^2/(1,0,001)^2/(0,W(2),000)/$ $(0,1,010)^2/(0,1,000)^4/(0,0,W(101))/$ $(0,0,010)^2/(0,0,000)^4$
A_4	$\lambda_1^3/\lambda_2/\lambda_3/\lambda_4^3/0^6$	$W(\lambda_1+\lambda_4)/\lambda_1^4/\lambda_2^3/\lambda_3^3/\lambda_4^4/0^9$
A_1A_3	$(1,010)/(1,000)^6/$ $(0,100)^4/(0,001)^4$	$(W(2),000)/(1,100)^4/(1,001)^4/$ $(0,W(101))/(0,010)^6/(0,000)^{15}$
$(A_1A_3)'$	$(1,100)/(1,001)/$ $(1,000)^4/(0,100)^2/$ $(0,010)^2/(0,001)^2/$ $(0,000)^4$	$(W(2),000)/(1,100)^2/(1,010)^2/(1,001)^2/$ $(1,000)^4/(0,W(101))/(0,100)^4/(0,010)^2/$ $(0,001)^4/(0,000)^7$
A_2^2	$(10,10)/(10,00)^3/$ $(01,01)/(01,00)^3/$ $(00,10)^3/(00,01)^3/$ $(00,00)^2$	$(W(11),00)/(10,10)^3/(10,01)/(10,00)^3/$ $(01,10)/(01,01)^3/(01,00)^3/(00,W(11))/$ $(00,10)^3/(00,01)^3/(00,00)^9$
$A_1^2A_2$	$(1,1,00)^2/(1,0,10)/$ $(1,0,01)/(1,0,00)^2/$ $(0,1,10)/(0,1,01)/$ $(0,1,00)^2/(0,0,10)^2/$ $(0,0,01)^2/(0,0,00)^4$	$(W(2),0,00)/(1,1,10)/(1,1,01)/$ $(1,1,00)^2/(1,0,10)^2/(1,0,01)^2/(1,0,00)^4/$ $(0,W(2),00)/(0,1,10)^2/(0,1,01)^2/$ $(0,1,00)^4/(0,0,W(11))/(0,0,10)^3/$ $(0,0,01)^3/(0,0,00)^5$

A_1^4	$(1,1,1,0)/(1,0,0,1)^2/$ $(1,0,0,0)^4/(0,1,0,1)^2/$ $(0,1,0,0)^4/(0,0,1,1)^2/$ $(0,0,1,0)^4$	$(W(2),0,0,0)/(1,1,0,1)^2/(1,1,0,0)^4/$ $(1,0,1,1)^2/(1,0,1,0)^4/(0,W(2),0,0)/$ $(0,1,1,1)^2/(0,1,1,0)^4/(0,0,W(2),0)/$ $(0,0,0,W(2))/(0,0,0,1)^8/(0,0,0,0)^9$
A_3	$100^4/010^2/001^4/000^{12}$	$W(101)/100^8/010^6/001^8/000^{18}$
A_1A_2	$(1,10)/(1,01)/(1,00)^6/$ $(0,10)^4/(0,01)^4/(0,00)^8$	$(W(2),00)/(1,10)^4/(1,01)^4/(1,00)^8/$ $(0,W(11))/(0,10)^7/(0,01)^7/(0,00)^{16}$
A_1^3	$(1,1,0)^2/(1,0,1)^2/$ $(1,0,0)^4/(0,1,1)^2/$ $(0,1,0)^4/(0,0,1)^4/$ $(0,0,0)^8$	$(W(2),0,0)/(1,1,1)^2/(1,1,0)^4/(1,0,1)^4/$ $(1,0,0)^8/(0,W(2),0)/(0,1,1)^4/(0,1,0)^8/$ $(0,0,W(2))/(0,0,1)^8/(0,0,0)^{12}$
$(A_1^3)'$	$(1,1,1)/(1,0,0)^8/$ $(0,1,0)^8/(0,0,1)^8$	$(W(2),0,0)/(1,1,0)^8/(1,0,1)^8/$ $(0,W(2),0)/(0,1,1)^8/(0,0,W(2))/$ $(0,0,0)^{28}$
A_2	$10^6/01^6/00^{20}$	$W(11)/10^{15}/01^{15}/00^{35}$
A_1^2	$(1,1)^2/(1,0)^8/(0,1)^8/$ $(0,0)^{16}$	$(W(2),0)/(1,1)^7/(1,0)^{16}/(0,W(2))/$ $(0,1)^{16}/(0,0)^{31}$
A_1	$1^{12}/0^{32}$	$W(2)/1^{32}/0^{66}$

Table 13.5: The composition factors for the action of Levi subgroups of E_8 on $L(E_8)$.

L'	$L(E_8)\downarrow L'$
E_7	$W(\lambda_1)/\lambda_7^2/0^3$
A_1E_6	$(W(2),0)/(1,\lambda_1)/(1,\lambda_6)/(1,0)^2/(0,\lambda_1)/(0,W(\lambda_2))/(0,\lambda_6)/(0,0)$
E_6	$\lambda_1^3/W(\lambda_2)/\lambda_6^3/0^8$
D_7	$\lambda_1^2/W(\lambda_2)/\lambda_6/\lambda_7/0$
A_2D_5	$(W(11),0)/(10,\lambda_1)/(10,\lambda_4)/(10,0)/(01,\lambda_1)/(01,\lambda_5)/(01,0)/$ $(00,W(\lambda_2))/(00,\lambda_4)/(00,\lambda_5)/(00,0)$
D_6	$\lambda_1^4/W(\lambda_2)/\lambda_5^2/\lambda_6^2/0^6$
A_1D_5	$(W(2),0)/(1,\lambda_1)^2/(1,\lambda_4)/(1,\lambda_5)/(1,0)^4/(0,\lambda_1)^2/(0,W(\lambda_2))/(0,\lambda_4)^2/$ $(0,\lambda_5)^2/(0,0)^4$
A_2D_4	$(W(11),0)/(10,\lambda_1)/(10,\lambda_3)/(10,\lambda_4)/(10,0)^3/(01,\lambda_1)/(01,\lambda_3)/(01,\lambda_4)/$ $(01,0)^3/(00,\lambda_1)^2/(00,W(\lambda_2))/(00,\lambda_3)^2/(00,\lambda_4)^2/(00,0)^2$
D_5	$\lambda_1^6/W(\lambda_2)/\lambda_4^4/\lambda_5^4/0^{15}$
A_1D_4	$(W(2),0)/(1,\lambda_1)^2/(1,\lambda_3)^2/(1,\lambda_4)^2/(1,0)^8/(0,\lambda_1)^4/(0,W(\lambda_2))/(0,\lambda_3)^4/$ $(0,\lambda_4)^4/(0,0)^9$
$\bar{D}_4$	$\lambda_1^8/W(\lambda_2)/\lambda_3^8/\lambda_4^8/0^{28}$
A_7	$W(\lambda_1+\lambda_7)/\lambda_1/\lambda_2/\lambda_3/\lambda_5/\lambda_6/\lambda_7/0$
A_3A_4	$(W(101),0)/(100,\lambda_1)/(100,\lambda_3)/(100,0)/(010,\lambda_1)/(010,\lambda_4)/(001,\lambda_2)/$ $(001,\lambda_4)/(001,0)/(000,W(\lambda_1+\lambda_4))/(000,\lambda_2)/(000,\lambda_3)/(000,0)$
A_1A_6	$(W(2),0)/(1,\lambda_1)/(1,\lambda_2)/(1,\lambda_5)/(1,\lambda_6)/(0,W(\lambda_1+\lambda_6))/(0,\lambda_1)/(0,\lambda_3)/$ $(0,\lambda_4)/(0,\lambda_6)/(0,0)$

$A_1A_2A_4$ $(W(2),00,0)/(1,10,\lambda_4)/(1,10,0)/(1,01,\lambda_1)/(1,01,0)/(1,00,\lambda_2)/(1,00,\lambda_3)/(0,W(11),0)/(0,10,\lambda_1)/(0,10,\lambda_2)/(0,01,\lambda_3)/(0,01,\lambda_4)/(0,00,W(\lambda_1+\lambda_4))/(0,00,\lambda_1)/(0,00,\lambda_4)/(0,00,0)$

A_6 $W(\lambda_1+\lambda_6)/\lambda_1^3/\lambda_2^2/\lambda_3/\lambda_4/\lambda_5^2/\lambda_6^3/0^4$

A_1A_5 $(W(2),0)/(1,\lambda_1)^2/(1,\lambda_2)/(1,\lambda_4)/(1,\lambda_5)^2/(1,0)^2/(0,W(\lambda_1+\lambda_5))/(0,\lambda_1)^2/(0,\lambda_2)/(0,\lambda_3)^2/(0,\lambda_4)/(0,\lambda_5)^2/(0,0)^4$

A_2A_4 $(W(11),0)/(10,\lambda_1)/(10,\lambda_2)/(10,\lambda_4)^2/(10,0)^2/(01,\lambda_1)^2/(01,\lambda_3)/(01,\lambda_4)/(01,0)^2/(00,W(\lambda_1+\lambda_4))/(00,\lambda_1)/(00,\lambda_2)^2/(00,\lambda_3)^2/(00,\lambda_4)/(00,0)^4$

$A_1^2A_4$ $(W(2),0,0)/(1,1,\lambda_1)/(1,1,\lambda_4)/(1,1,0)^2/(1,0,\lambda_1)/(1,0,\lambda_2)/(1,0,\lambda_3)/(1,0,\lambda_4)/(1,0,0)^2/(0,W(2),0)/(0,1,\lambda_1)/(0,1,\lambda_2)/(0,1,\lambda_3)/(0,1,\lambda_4)/(0,1,0)^2/(0,0,W(\lambda_1+\lambda_4))/(0,0,\lambda_1)^2/(0,0,\lambda_2)/(0,0,\lambda_3)/(0,0,\lambda_4)^2/(0,0,0)^2$

A_3^2 $(W(101),000)/(100,100)/(100,010)/(100,001)/(100,000)^2/(010,100)/(010,001)/(010,000)^2/(001,100)/(001,010)/(001,001)/(001,000)^2/(000,W(101))/(000,100)^2/(000,010)^2/(000,001)^2/(000,000)^2$

$A_1A_2A_3$ $(W(2),00,000)/(1,10,001)/(1,10,000)^2/(1,01,100)/(1,01,000)^2/(1,00,100)/(1,00,010)^2/(1,00,001)/(0,W(11),000)/(0,10,100)^2/(0,10,010)/(0,10,000)/(0,01,010)/(0,01,001)^2/(0,01,000)/(0,00,W(101))/(0,00,100)^2/(0,00,001)^2/(0,00,000)^4$

$A_1^2A_2^2$ $(W(2),0,00,00)/(1,1,10,00)/(1,1,01,00)/(1,1,00,10)/(1,1,00,01)/(1,0,10,01)/(1,0,10,00)/(1,0,01,10)/(1,0,01,00)/(1,0,00,10)/(1,0,00,01)/(1,0,00,00)^2/(0,W(2),00,00)/(0,1,10,10)/(0,1,10,00)/(0,1,01,01)/(0,1,01,00)/(0,1,00,10)/(0,1,00,01)/(0,1,00,00)^2/(0,0,W(11),00)/(0,0,10,10)/(0,0,10,01)/(0,0,10,00)/(0,0,01,10)/(0,0,01,01)/(0,0,01,00)/(0,0,00,W(11))/(0,0,00,10)/(0,0,00,01)/(0,0,00,00)^2$

A_5 $W(\lambda_1+\lambda_5)/\lambda_1^6/\lambda_2^3/\lambda_3^2/\lambda_4^3/\lambda_5^6/0^{11}$

A_1A_4 $(W(2),0)/(1,\lambda_1)^3/(1,\lambda_2)/(1,\lambda_3)/(1,\lambda_4)^3/(1,0)^6/(0,W(\lambda_1+\lambda_4))/(0,\lambda_1)^4/(0,\lambda_2)^3/(0,\lambda_3)^3/(0,\lambda_4)^4/(0,0)^9$

A_2A_3 $(W(11),000)/(10,100)^2/(10,010)/(10,001)^2/(10,000)^5/(01,100)^2/(01,010)/(01,001)^2/(01,000)^5/(00,W(101))/(00,100)^4/(00,010)^4/(00,001)^4/(00,000)^7$

$A_1^2A_3$ $(W(2),0,000)/(1,1,100)/(1,1,001)/(1,1,000)^4/(1,0,100)^2/(1,0,010)^2/(1,0,001)^2/(1,0,000)^4/(0,W(2),000)/(0,1,100)^2/(0,1,010)^2/(0,1,001)^2/(0,1,000)^4/(0,0,W(101))/(0,0,100)^4/(0,0,010)^2/(0,0,001)^4/(0,0,000)^7$

$A_1A_2^2$ $(W(2),00,00)/(1,10,01)/(1,10,00)^3/(1,01,10)/(1,01,00)^3/(1,00,10)^3/(1,00,01)^3/(1,00,00)^2/(0,W(11),00)/(0,10,10)^3/(0,10,01)/(0,10,00)^3/(0,01,10)/(0,01,01)^3/(0,01,00)^3/(0,00,W(11))/(0,00,10)^3/(0,00,01)^3/(0,00,00)^9$

$A_1^3A_2$	$(W(2),0,0,00)/(1,1,1,00)^2/(1,1,0,10)/(1,1,0,01)/(1,1,0,00)^2/(1,0,1,10)/(1,0,1,01)/(1,0,1,00)^2/(1,0,0,10)^2/(1,0,0,01)^2/(1,0,0,00)^4/(0,W(2),0,00)/(0,1,1,10)/(0,1,1,01)/(0,1,1,00)^2/(0,1,0,10)^2/(0,1,0,01)^2/(0,1,0,00)^4/(0,0,W(2),00)/(0,0,1,10)^2/(0,0,1,01)^2/(0,0,1,00)^4/(0,0,0,W(11))/(0,0,0,10)^3/(0,0,0,01)^3/(0,0,0,00)^5$
A_4	$W(\lambda_1+\lambda_4)/\lambda_1^{10}/\lambda_2^5/\lambda_3^5/\lambda_4^{10}/0^{24}$
A_1A_3	$(W(2),000)/(1,100)^4/(1,010)^2/(1,001)^4/(1,000)^{12}/(0,W(101))/(0,100)^8/(0,010)^6/(0,001)^8/(0,000)^{18}$
A_2^2	$(W(11),00)/(10,10)^3/(10,01)^3/(10,00)^9/(01,10)^3/(01,01)^3/(01,00)^9/(00,W(11))/(00,10)^9/(00,01)^9/(00,00)^{16}$
$A_1^2A_2$	$(W(2),0,00)/(1,1,10)/(1,1,01)/(1,1,00)^6/(1,0,10)^4/(1,0,01)^4/(1,0,00)^8/(0,W(2),00)/(0,1,10)^4/(0,1,01)^4/(0,1,00)^8/(0,0,W(11))/(0,0,10)^7/(0,0,01)^7/(0,0,00)^{16}$
A_1^4	$(W(2),0,0,0)/(1,1,1,0)^2/(1,1,0,1)^2/(1,1,0,0)^4/(1,0,1,1)^2/(1,0,1,0)^4/(1,0,0,1)^4/(1,0,0,0)^8/(0,W(2),0,0)/(0,1,1,1)^2/(0,1,1,0)^4/(0,1,0,1)^4/(0,1,0,0)^8/(0,0,W(2),0)/(0,0,1,1)^4/(0,0,1,0)^8/(0,0,0,W(2))/(0,0,0,1)^8/(0,0,0,0)^{12}$
A_3	$W(101)/100^{16}/010^{10}/001^{16}/000^{45}$
A_1A_2	$(W(2),00)/(1,10)^6/(1,01)^6/(1,00)^{20}/(0,W(11))/(0,10)^{15}/(0,01)^{15}/(0,00)^{35}$
A_1^3	$(W(2),0,0)/(1,1,1)^2/(1,1,0)^8/(1,0,1)^8/(1,0,0)^{16}/(0,W(2),0)/(0,1,1)^8/(0,1,0)^{16}/(0,0,W(2))/(0,0,1)^{16}/(0,0,0)^{31}$
A_2	$W(11)/10^{27}/01^{27}/00^{78}$
A_1^2	$(W(2),0)/(1,1)^{12}/(1,0)^{32}/(0,W(2))/(0,1)^{32}/(0,0)^{66}$
A_1	$W(2)/1^{56}/0^{133}$

Bibliography

[ABS90] H. Azad, M. Barry, and G. Seitz, *On the structure of parabolic subgroups*, Comm. Algebra **18** (1990), no. 2, 551–562, DOI 10.1080/00927879008823931. MR1047327

[Ame05] B. Amende, *G-irreducible subgroups of type A1*, ProQuest LLC, Ann Arbor, MI, 2005. Thesis (Ph.D.)–University of Oregon. MR2707891

[BCP97] W. Bosma, J. Cannon, and C. Playoust, *The Magma algebra system. I. The user language*, J. Symbolic Comput. **24** (1997), no. 3-4, 235–265, DOI 10.1006/jsco.1996.0125. Computational algebra and number theory (London, 1993). MR1484478

[BGM19] M. Bate, H. Geranios, and B. Martin, *Orbit closures and invariants*, Math. Z. **293** (2019), no. 3-4, 1121–1159, DOI 10.1007/s00209-019-02228-6. MR4024580

[BMR05] M. Bate, B. Martin, and G. Röhrle, *A geometric approach to complete reducibility*, Invent. Math. **161** (2005), no. 1, 177–218, DOI 10.1007/s00222-004-0425-9. MR2178661

[BMRT11] M. Bate, B. Martin, G. Röhrle, and R. Tange, *Complete reducibility and conjugacy classes of tuples in algebraic groups and Lie algebras*, Math. Z. **269** (2011), no. 3-4, 809–832, DOI 10.1007/s00209-010-0763-9. MR2860266

[BMRT13] M. Bate, B. Martin, G. Röhrle, and R. Tange, *Closed orbits and uniform S-instability in geometric invariant theory*, Trans. Amer. Math. Soc. **365** (2013), no. 7, 3643–3673, DOI 10.1090/S0002-9947-2012-05739-4. MR3042598

[Bor91] A. Borel, *Linear algebraic groups*, 2nd ed., Graduate Texts in Mathematics, vol. 126, Springer-Verlag, New York, 1991. MR1102012

[Bou68] N. Bourbaki, *Éléments de mathématique. Fasc. XXXIV. Groupes et algébres de Lie. Chapitre IV: Groupes de Coxeter et systémes de Tits. Chapitre V: Groupes engendrés par des reflexions. Chapitre VI: systémes de racines.*, Hermann, Paris, 1968. Actualités Scientifiques et Industrielles, No. 1337. MR0240238

[BT71] A. Borel and J. Tits, *Éléments unipotents et sous-groupes paraboliques de groupes réductifs. I* (French), Invent. Math. **12** (1971), 95–104, DOI 10.1007/BF01404653. MR0294349

[Cap09] P.-E. Caprace, *"Abstract" homomorphisms of split Kac-Moody groups*, Mem. Amer. Math. Soc. **198** (2009), no. 924, xvi+84, DOI 10.1090/memo/0924. MR2499773

[Car72] R. W. Carter, *Conjugacy classes in the Weyl group*, Compositio Math. **25** (1972), 1–59. MR0318337

[CLSS92] A. M. Cohen, M. W. Liebeck, J. Saxl, and G. M. Seitz, *The local maximal subgroups of exceptional groups of Lie type, finite and algebraic*, Proc. London Math. Soc. (3) **64** (1992), no. 1, 21–48, DOI 10.1112/plms/s3-64.1.21. MR1132853

[CST] D. A. Craven, D. I. Stewart, and A. R. Thomas, *A New Maximal Subgroup of E_8 in Characteristic 3*, In preparation.

[Daw] D. K. Dawson, *Complete reducibility in Euclidean twin buildings*, ProQuest LLC, Ann Arbor, MI, 2011. Thesis (Ph.D.)–Cornell University. MR2982240

[GHT17] R. Guralnick, F. Herzig, and P. H. Tiep, *Adequate subgroups and indecomposable modules*, J. Eur. Math. Soc. (JEMS) **19** (2017), no. 4, 1231–1291, DOI 10.4171/JEMS/692. MR3626555

[GLS98] D. Gorenstein, R. Lyons, and R. Solomon, *The classification of the finite simple groups*, Mathematical Surveys and Monographs, vol. 40, American Mathematical Society, Providence, RI, 1994. MR1303592

[Gur99] R. M. Guralnick, *Small representations are completely reducible*, J. Algebra **220** (1999), no. 2, 531–541, DOI 10.1006/jabr.1999.7963. MR1717357

[KL90] P. Kleidman and M. Liebeck, *The subgroup structure of the finite classical groups*, London Mathematical Society Lecture Note Series, vol. 129, Cambridge University Press, Cambridge, 1990. MR1057341

[Kle87] P. B. Kleidman, *The maximal subgroups of the finite 8-dimensional orthogonal groups $P\Omega_8^+(q)$ and of their automorphism groups*, J. Algebra **110** (1987), no. 1, 173–242, DOI 10.1016/0021-8693(87)90042-1. MR904187

[Lit18] A. J. Litterick, *On non-generic finite subgroups of exceptional algebraic groups*, Mem. Amer. Math. Soc. **253** (2018), no. 1207, v+156, DOI 10.1090/memo/1207. MR3803556

[LS94] M. W. Liebeck and G. M. Seitz, *Subgroups generated by root elements in groups of Lie type*, Ann. of Math. (2) **139** (1994), no. 2, 293–361, DOI 10.2307/2946583. MR1274094

[LS96] M. W. Liebeck and G. M. Seitz, *Reductive subgroups of exceptional algebraic groups*, Mem. Amer. Math. Soc. **121** (1996), no. 580, vi+111, DOI 10.1090/memo/0580. MR1329942

[LS98] M. W. Liebeck and G. M. Seitz, *On the subgroup structure of classical groups*, Invent. Math. **134** (1998), no. 2, 427–453, DOI 10.1007/s002220050270. MR1650328

[LS99] M. W. Liebeck and G. M. Seitz, *On finite subgroups of exceptional algebraic groups*, J. Reine Angew. Math. **515** (1999), 25–72, DOI 10.1515/crll.1999.078. MR1717629

[LS03] M. W. Liebeck and G. M. Seitz, *Variations on a theme of Steinberg*, J. Algebra **260** (2003), no. 1, 261–297, DOI 10.1016/S0021-8693(02)00649-X. Special issue celebrating the 80th birthday of Robert Steinberg. MR1973585

[LS04] M. W. Liebeck and G. M. Seitz, *The maximal subgroups of positive dimension in exceptional algebraic groups*, Mem. Amer. Math. Soc. **169** (2004), no. 802, vi+227, DOI 10.1090/memo/0802. MR2044850

[LS12] M. W. Liebeck and G. M. Seitz, *Unipotent and nilpotent classes in simple algebraic groups and Lie algebras*, Mathematical Surveys and Monographs, vol. 180, American Mathematical Society, Providence, RI, 2012. MR2883501

[LST15] M. W. Liebeck, G. M. Seitz, and D. M. Testerman, *Distinguished unipotent elements and multiplicity-free subgroups of simple algebraic groups*, Pacific J. Math. **279** (2015), no. 1-2, 357–382, DOI 10.2140/pjm.2015.279.357. MR3437782

[LT99] R. Lawther and D. M. Testerman, *A_1 subgroups of exceptional algebraic groups*, Mem. Amer. Math. Soc. **141** (1999), no. 674, viii+131, DOI 10.1090/memo/0674. MR1466951

[LT04] M. W. Liebeck and D. M. Testerman, *Irreducible subgroups of algebraic groups*, Q. J. Math. **55** (2004), no. 1, 47–55, DOI 10.1093/qjmath/55.1.47. MR2043006

[LT18] A. J. Litterick and A. R. Thomas, *Complete reducibility in good characteristic*, Trans. Amer. Math. Soc. **370** (2018), no. 8, 5279–5340, DOI 10.1090/tran/7085. MR3803140

[Lüb01] F. Lübeck, *Small degree representations of finite Chevalley groups in defining characteristic*, LMS J. Comput. Math. **4** (2001), 135–169, DOI 10.1112/S1461157000000838. MR1901354

[McN07] G. McNinch, *Completely reducible Lie subalgebras*, Transform. Groups **12** (2007), no. 1, 127–135, DOI 10.1007/s00031-005-1130-5. MR2308032

[Rot09] J. J. Rotman, *An introduction to homological algebra*, 2nd ed., Universitext, Springer, New York, 2009. MR2455920

[Sei91] G. M. Seitz, *Maximal subgroups of exceptional algebraic groups*, Mem. Amer. Math. Soc. **90** (1991), no. 441, iv+197, DOI 10.1090/memo/0441. MR1048074

[Ser05] J.-P. Serre, *Complète réductibilité* (French, with French summary), Astérisque **299** (2005), Exp. No. 932, viii, 195–217. Séminaire Bourbaki. Vol. 2003/2004. MR2167207

[Ste63] R. Steinberg, *Representations of algebraic groups*, Nagoya Math. J. **22** (1963), 33–56. MR0155937

[Ste10] D. I. Stewart, *The reductive subgroups of G_2*, J. Group Theory **13** (2010), no. 1, 117–130, DOI 10.1515/JGT.2009.035. MR2604850

[Ste13] D. I. Stewart, *The reductive subgroups of F_4*, Mem. Amer. Math. Soc. **223** (2013), no. 1049, vi+88, DOI 10.1090/S0065-9266-2012-00668-X. MR3075783

[ST18] D. I. Stewart and A. R. Thomas, *The Jacobson-Morozov theorem and complete reducibility of Lie subalgebras*, Proc. Lond. Math. Soc. (3) **116** (2018), no. 1, 68–100, DOI 10.1112/plms.12067. MR3747044

[Tho15] A. R. Thomas, *Simple irreducible subgroups of exceptional algebraic groups*, J. Algebra **423** (2015), 190–238, DOI 10.1016/j.jalgebra.2014.10.011. MR3283715

[Tho16] A. R. Thomas, *Irreducible A_1 subgroups of exceptional algebraic groups*, J. Algebra **447** (2016), 240–296, DOI 10.1016/j.jalgebra.2015.08.026. MR3427635

Editorial Information

To be published in the *Memoirs*, a paper must be correct, new, nontrivial, and significant. Further, it must be well written and of interest to a substantial number of mathematicians. Piecemeal results, such as an inconclusive step toward an unproved major theorem or a minor variation on a known result, are in general not acceptable for publication.

Papers appearing in *Memoirs* are generally at least 80 and not more than 200 published pages in length. Papers less than 80 or more than 200 published pages require the approval of the Managing Editor of the Transactions/Memoirs Editorial Board. Published pages are the same size as those generated in the style files provided for AMS-LaTeX or AMS-TeX.

Information on the backlog for this journal can be found on the AMS website starting from `http://www.ams.org/memo`.

A Consent to Publish is required before we can begin processing your paper. After a paper is accepted for publication, the Providence office will send a Consent to Publish and Copyright Agreement to all authors of the paper. By submitting a paper to the *Memoirs*, authors certify that the results have not been submitted to nor are they under consideration for publication by another journal, conference proceedings, or similar publication.

Information for Authors

Memoirs is an author-prepared publication. Once formatted for print and on-line publication, articles will be published as is with the addition of AMS-prepared frontmatter and backmatter. Articles are not copyedited; however, confirmation copy will be sent to the authors.

Initial submission. The AMS uses Centralized Manuscript Processing for initial submissions. Authors should submit a PDF file using the Initial Manuscript Submission form found at `www.ams.org/submission/memo`, or send one copy of the manuscript to the following address: Centralized Manuscript Processing, MEMOIRS OF THE AMS, 201 Charles Street, Providence, RI 02904-2294 USA. If a paper copy is being forwarded to the AMS, indicate that it is for *Memoirs* and include the name of the corresponding author, contact information such as email address or mailing address, and the name of an appropriate Editor to review the paper (see the list of Editors below).

The paper must contain a *descriptive title* and an *abstract* that summarizes the article in language suitable for workers in the general field (algebra, analysis, etc.). The *descriptive title* should be short, but informative; useless or vague phrases such as "some remarks about" or "concerning" should be avoided. The *abstract* should be at least one complete sentence, and at most 300 words. Included with the footnotes to the paper should be the 2010 *Mathematics Subject Classification* representing the primary and secondary subjects of the article. The classifications are accessible from `www.ams.org/msc/`. The Mathematics Subject Classification footnote may be followed by a list of *key words and phrases* describing the subject matter of the article and taken from it. Journal abbreviations used in bibliographies are listed in the latest *Mathematical Reviews* annual index. The series abbreviations are also accessible from `www.ams.org/msnhtml/serials.pdf`. To help in preparing and verifying references, the AMS offers MR Lookup, a Reference Tool for Linking, at `www.ams.org/mrlookup/`.

Electronically prepared manuscripts. The AMS encourages electronically prepared manuscripts, with a strong preference for AMS-LaTeX. To this end, the Society has prepared AMS-LaTeX author packages for each AMS publication. Author packages include instructions for preparing electronic manuscripts, samples, and a style file that generates the particular design specifications of that publication series. Though AMS-LaTeX is the highly preferred format of TeX, author packages are also available in AMS-TeX.

Authors may retrieve an author package for *Memoirs of the AMS* from `www.ams.org/journals/memo/memoauthorpac.html`. The *AMS Author Handbook* is available in PDF format from the author package link. The author package can also be obtained free of charge by sending email to tech-support@ams.org or from the Publication Division, American Mathematical Society, 201 Charles St., Providence, RI 02904-2294, USA. When requesting an author package, please specify $\mathcal{A}\mathcal{M}\mathcal{S}$-LaTeX or $\mathcal{A}\mathcal{M}\mathcal{S}$-TeX and the publication in which your paper will appear. Please be sure to include your complete mailing address.

After acceptance. The source files for the final version of the electronic manuscript should be sent to the Providence office immediately after the paper has been accepted for publication. The author should also submit a PDF of the final version of the paper to the editor, who will forward a copy to the Providence office.

Accepted electronically prepared files can be submitted via the web at www.ams.org/submit-book-journal/, sent via FTP, or sent on CD to the Electronic Prepress Department, American Mathematical Society, 201 Charles Street, Providence, RI 02904-2294 USA. TeX source files and graphic files can be transferred over the Internet by FTP to the Internet node ftp.ams.org (130.44.1.100). When sending a manuscript electronically via CD, please be sure to include a message indicating that the paper is for the *Memoirs*.

Electronic graphics. Comprehensive instructions on preparing graphics are available at www.ams.org/authors/journals.html. A few of the major requirements are given here.

Submit files for graphics as EPS (Encapsulated PostScript) files. This includes graphics originated via a graphics application as well as scanned photographs or other computer-generated images. If this is not possible, TIFF files are acceptable as long as they can be opened in Adobe Photoshop or Illustrator.

Authors using graphics packages for the creation of electronic art should also avoid the use of any lines thinner than 0.5 points in width. Many graphics packages allow the user to specify a "hairline" for a very thin line. Hairlines often look acceptable when proofed on a typical laser printer. However, when produced on a high-resolution laser imagesetter, hairlines become nearly invisible and will be lost entirely in the final printing process.

Screens should be set to values between 15% and 85%. Screens which fall outside of this range are too light or too dark to print correctly. Variations of screens within a graphic should be no less than 10%.

Any graphics created in color will be rendered in grayscale for the printed version unless color printing is authorized by the Managing Editor and the Publisher. In general, color graphics will appear in color in the online version.

Inquiries. Any inquiries concerning a paper that has been accepted for publication should be sent to memo-query@ams.org or directly to the Electronic Prepress Department, American Mathematical Society, 201 Charles St., Providence, RI 02904-2294 USA.

Editors

This journal is designed particularly for long research papers, normally at least 80 pages in length, and groups of cognate papers in pure and applied mathematics. Papers intended for publication in the *Memoirs* should be addressed to one of the following editors. The AMS uses Centralized Manuscript Processing for initial submissions to AMS journals. Authors should follow instructions listed on the Initial Submission page found at www.ams.org/memo/memosubmit.html.

Managing Editor: Henri Darmon, Department of Mathematics, McGill University, Montreal, Quebec H3A 0G4, Canada; e-mail: darmon@math.mcgill.ca

1. GEOMETRY, TOPOLOGY & LOGIC
Coordinating Editor: Richard Canary, Department of Mathematics, University of Michigan, Ann Arbor, MI 48109-1043 USA; e-mail: canary@umich.edu

Algebraic topology, Michael Hill, Department of Mathematics, University of California Los Angeles, Los Angeles, CA 90095 USA; e-mail: mikehill@math.ucla.edu

Logic, Mariya Ivanova Soskova, Department of Mathematics, University of Wisconsin–Madison, Madison, WI 53706 USA; e-mail: msoskova@math.wisc.edu

Low-dimensional topology and geometric structures, Richard Canary

2. ALGEBRA AND NUMBER THEORY
Coordinating Editor: Henri Darmon, Department of Mathematics, McGill University, Montreal, Quebec H3A 0G4, Canada; e-mail: darmon@math.mcgill.ca

Algebra, Radha Kessar, Department of Mathematics, City, University of London, London EC1V 0HB, United Kingdom; e-mail: radha.kessar.1@city.ac.uk

Algebraic geometry, Lucia Caporaso, Department of Mathematics and Physics, Roma Tre University, Largo San Leonardo Murialdo, I-00146 Rome, Italy; e-mail: LCedit@mat.uniroma3.it

Analytic number theory, Lillian B. Pierce, Department of Mathematics, Duke University, 120 Science Drive Box 90320, Durham, NC 27708 USA; e-mail: pierce@math.duke.edu

Arithmetic geometry, Ted C. Chinburg, Department of Mathematics, University of Pennsylvania, Philadelphia, PA 19104-6395 USA; e-mail: ted@math.upenn.edu

Commutative algebra, Irena Peeva, Department of Mathematics, Cornell University, Ithaca, NY 14853 USA; e-mail: irena@math.cornell.edu

Number theory, Henri Darmon

3. GEOMETRIC ANALYSIS & PDE
Coordinating Editor: Alexander A. Kiselev, Department of Mathematics, Duke University, 120 Science Drive, Rm 117 Physics Bldg, Durham, NC 27708 USA; e-mail: kiselev@math.duke.edu

Differential geometry and geometric analysis, Ailana M. Fraser, Department of Mathematics, University of British Columbia, 1984 Mathematics Road, Room 121, Vancouver BC V6T 1Z2, Canada; e-mail: afraser@math.ubc.ca

Harmonic analysis and partial differential equations, Monica Visan, Department of Mathematics, University of California Los Angeles, 520 Portola Plaza, Los Angeles, CA 90095 USA; e-mail: visan@math.ucla.edu

Partial differential equations and functional analysis, Alexander A. Kiselev

Real analysis and partial differential equations, Joachim Krieger, Bâtiment de Mathématiques, École Polytechnique Fédérale de Lausanne, Station 8, 1015 Lausanne Vaud, Switzerland; e-mail: joachim.krieger@epfl.ch

4. ERGODIC THEORY, DYNAMICAL SYSTEMS & COMBINATORICS
Coordinating Editor: Vitaly Bergelson, Department of Mathematics, Ohio State University, 231 W. 18th Avenue, Columbus, OH 43210 USA; e-mail: vitaly@math.ohio-state.edu

Algebraic and enumerative combinatorics, Jim Haglund, Department of Mathematics, University of Pennsylvania, Philadelphia, PA 19104 USA; e-mail: jhaglund@math.upenn.edu

Probability theory, Robin Pemantle, Department of Mathematics, University of Pennsylvania, 209 S. 33rd Street, Philadelphia, PA 19104 USA; e-mail: pemantle@math.upenn.edu

Dynamical systems and ergodic theory, Ian Melbourne, Mathematics Institute, University of Warwick, Coventry CV4 7AL, United Kingdom; e-mail: I.Melbourne@warwick.ac.uk

Ergodic theory and combinatorics, Vitaly Bergelson

5. ANALYSIS, LIE THEORY & PROBABILITY
Coordinating Editor: Stefaan Vaes, Department of Mathematics, Katholieke Universiteit Leuven, Celestijnenlaan 200B, B-3001 Leuven, Belgium; e-mail: stefaan.vaes@wis.kuleuven.be

Functional analysis and operator algebras, Stefaan Vaes

Harmonic analysis, PDEs, and geometric measure theory, Svitlana Mayboroda, School of Mathematics, University of Minnesota, 206 Church Street SE, 127 Vincent Hall, Minneapolis, MN 55455 USA; e-mail: svitlana@math.umn.edu

Probability theory and stochastic analysis, Davar Khoshnevisan, Department of Mathematics, The University of Utah, Salt Lake City, UT 84112 USA; e-mail: davar@math.utah.edu

SELECTED PUBLISHED TITLES IN THIS SERIES

1296 **Christophe Cornut,** Filtrations and Buildings, 2020

1295 **Lisa Berger, Chris Hall, René Pannekoek, Jennifer Park, Rachel Pries, Shahed Sharif, Alice Silverberg, and Douglas Ulmer,** Explicit Arithmetic of Jacobians of Generalized Legendre Curves Over Global Function Fields, 2020

1294 **Jacob Bedrossian, Pierre Germain, and Nader Masmoudi,** Dynamics Near the Subcritical Transition of the 3D Couette Flow I: Below Threshold Case, 2020

1293 **Benjamin Jaye, Fedor Nazarov, Maria Carmen Reguera, and Xavier Tolsa,** The Riesz Transform of Codimension Smaller Than One and the Wolff Energy, 2020

1292 **Angel Castro, Diego Córdoba, and Javier Gómez-Serrano,** Global Smooth Solutions for the Inviscid SQG Equation, 2020

1291 **Vasilis Chousionis, Jeremy Tyson, and Mariusz Urbanski,** Conformal Graph Directed Markov Systems on Carnot Groups, 2020

1290 **Harold Rosenberg and Graham Smith,** Degree Theory of Immersed Hypersurfaces, 2020

1289 **Pavel M. Bleher and Guilherme L. F. Silva,** The Mother Body Phase Transition in the Normal Matrix Model, 2020

1288 **Alexander Blokh, Lex Oversteegen, Ross Ptacek, and Vladlen Timorin,** Laminational Models for Some Spaces of Polynomials of Any Degree, 2020

1287 **Mike Hochman,** On Self-Similar Sets with Overlaps and Inverse Theorems for Entropy in $\mathbb{R}^d$, 2020

1286 **Andrew J. Blumberg and Michael A. Mandell,** Localization for $THH(ku)$ and the Topological Hochschild and Cyclic Homology of Waldhausen Categories, 2020

1285 **Zhaobing Fan, Chun-Ju Lai, Yiqiang Li, Li Luo, and Weiqiang Wang,** Affine Flag Varieties and Quantum Symmetric Pairs, 2020

1284 **Rodney G. Downey, Keng Meng Ng, and Reed Solomon,** Minimal Weak Truth Table Degrees and Computably Enumerable Turing Degrees, 2020

1283 **Antonio Alarcón, Franc Forstnerič, and Francisco J. López,** New Complex Analytic Methods in the Study of Non-Orientable Minimal Surfaces in $\mathbb{R}^n$, 2020

1282 **David Carchedi,** Higher Orbifolds and Deligne-Mumford Stacks as Structured Infinity-Topoi, 2020

1281 **S. V. Ivanov,** The Bounded and Precise Word Problems for Presentations of Groups, 2020

1280 **Michael Handel and Lee Mosher,** Subgroup Decomposition in $\mathrm{Out}(F_n)$, 2020

1279 **Cristian Gavrus and Sung-Jin Oh,** Global Well-Posedness of High Dimensional Maxwell–Dirac for Small Critical Data, 2020

1278 **Peter Poláčik,** Propagating Terraces and the Dynamics of Front-Like Solutions of Reaction-Diffusion Equations on $\mathbb{R}$, 2020

1277 **Henri Lombardi, Daniel Perrucci, and Marie-Françoise Roy,** An Elementary Recursive Bound for Effective Positivstellensatz and Hilbert's 17th Problem, 2020

1276 **Victor Beresnevich, Alan Haynes, and Sanju Velani,** Sums of Reciprocals of Fractional Parts and Multiplicative Diophantine Approximation, 2020

1275 **Laurent Berger, Peter Schneider, and Bingyong Xie,** Rigid Character Groups, Lubin-Tate Theory, and (φ, Γ)-Modules, 2020

1274 **Gonzalo Fiz Pontiveros, Simon Griffiths, and Robert Morris,** The Triangle-Free Process and the Ramsey Number $R(3,k)$, 2020

1273 **Massimiliano Berti and Riccardo Montalto,** Quasi-periodic Standing Wave Solutions of Gravity-Capillary Water Waves, 2020

For a complete list of titles in this series, visit the
AMS Bookstore at **www.ams.org/bookstore/memoseries/**.